Microscopic Simulations of Complex Flows

NATO ASI Series

Advanced Science Institutes Series

A series presenting the results of activities sponsored by the NATO Science Committee, which aims at the dissemination of advanced scientific and technological knowledge, with a view to strengthening links between scientific communities.

The series is published by an international board of publishers in conjunction with the NATO Scientific Affairs Division

A	**Life Sciences**	Plenum Publishing Corporation
B	**Physics**	New York and London
C	**Mathematical**	Kluwer Academic Publishers
	and Physical Sciences	Dordrecht, Boston, and London
D	**Behavioral and Social Sciences**	
E	**Applied Sciences**	
F	**Computer and Systems Sciences**	Springer-Verlag
G	**Ecological Sciences**	Berlin, Heidelberg, New York, London,
H	**Cell Biology**	Paris, and Tokyo

Recent Volumes in this Series

Series B: Physics

Microscopic Simulations of Complex Flows

Edited by
Michel Mareschal
The Free University of Brussels
Brussels, Belgium

Plenum Press
New York and London
Published in cooperation with NATO Scientific Affairs Division

Proceedings of a NATO Advanced Research Workshop
on Microscopic Simulations of Complex Flows,
held August 23–25, 1989,
in Brussels, Belgium

Library of Congress Cataloging-in-Publication Data

NATO Advanced Research Workshop on Microscopic Simulations of Complex
  Flows (1989 : Brussels, Belgium)
   Microscopic simulations of complex flows / edited by Michel
Mareschal.
      p.   cm. -- (NATO ASI series. Series B, Physics ; v. 236)
   "Proceedings of a NATO Advanced Research Workshop on Microscopic
Simulations of Complex Flows, held August 23-25, 1989 in Brussels,
Belgium"--T.p. verso.
   "Published in cooperation with NATO Scientific Affairs Division."
   Includes bibliographical references and index.
   ISBN-13: 978-1-4684-1341-0   e-ISBN-13: 978-1-4684-1339-7
   DOI: 10.1007/978-1-4684-1339-7
   1. Fluid dynamics--Mathematical models--Congresses.
I. Mareschal, Michel. II. Title. III. Series.
QA911.N33   1989
532'.05'015118--dc20   90-46797
                                                                 CIP

© 1990 Plenum Press, New York
Softcover reprint of the hardcover 1st edition 1990
A Division of Plenum Publishing Corporation
233 Spring Street, New York, N.Y. 10013

SPECIAL PROGRAM ON CHAOS, ORDER, AND PATTERNS

This book contains the proceedings of a NATO Advanced Research Workshop held within the program of activities of the NATO Special Program on Chaos, Order, and Patterns.

PREFACE

This volume contains the proceedings of a workshop which was held in Brussels during the month of August 1989. A strong motivation for organizing this workshop was to bring together people who have been involved in the microscopic simulation of phenomena occuring on "large" space and time scales. Indeed, results obtained in the last years by different groups tend to support the idea that macroscopic behavior already appears in systems small enough so as to be modelled by a collection of interacting particles on a (super) computer. Such an approach is certainly desirable to study situations where no satisfactory phenomenological theory is known to hold, or where solutions of the equations are too hard to obtain numerically. It is also interesting from a more fundamental point of view, namely the investigation of the limits of validity of the macroscopic description itself.

The main technique used in bridging the gap between the macro and micro worlds has been the molecular dynamics simulations, that is the numerical solution of the equations of motion of the model particles which constitute the system under study, a gas, a liquid or even a solid. However, this technique is by no means the only one. The high computational cost of molecular dynamics has encouraged efforts made in order to simplify the models in order to have faster codes: one often does not need a high accuracy and resolution on microscopic scales (a few angstroems) and the motion which occurs on large scales (a few atomic distances) will not depend on the specific modelling of the interactions. This is the way chosen by the so-called lattice gas cellular automata to reproduce hydrodynamic behavior. Another attempt to simulate large scale phenomena at a lower cost than molecular dynamics has been the search for a stochastic rather than an exact treatment of the collisions. Although this technique seems to remain limited to dilute gases, its successes and possible extensions are to be seriously investigated.

We believe that the workshop has been very useful in this double task to confront the different techniques used and to show their present possible applications in various domains, with of course a special emphasis on flows. These proceedings therefore begin with a part dedicated to general presentations of the techniques used. The direct simulation of the Boltzmann equation is introduced by Graeme Bird (the method is also referred to as the Bird method). The molecular dynamics and cellular automata techniques are presented by experts in these fields: the validity of lattice gases simulations to reproduce flows is discussed, in particular the effect of the so-called spurious invariants. The nonequilibrium molecular dynamics is then introduced and evidence is given that the relation between thermodynamic flows and forces remain linear even with the very high constraints that are to be imposed on simulated systems. On a more technical level, the peculiarities of large systems programming for molecular dynamics is also presented. Finally a method which combines the Boltzmann simulation and the advantage of a lattice is presented.

The next part is dedicated to various illustrations where the microscopic modelling has been able to give further insight in the understanding of flows. The example of the two-phase flow along a solid surface is such a beautiful example of the ability of microscopic simulations that we have asked permission to reprint an article which originally appeared in Physics of Fluids. Other examples involve the computation of the viscosity of colloidal suspensions, the shock waves and hypervelocity impacts studies, the hydrodynamical and chemical instabilities, and also the possible non-Newtonian behavior in fluids under strong shear. We also added in this section the study of fluctuations in nonequilibrium systems and a proposition of a simple model to study a generic hydrodynamical instability. In all these examples, the microscopic simulation has been successfully compared with Navier-Stokes hydrodynamics in cases where both approaches are expected to agree and, besides, some additional information is obtained.

The following parts are more concerned with applications in other domains than flows. To start with, the problem of deriving irreversible behavior from reversible equations is once more examined. Ideas originally developed in the study of non-linear dynamical systems are applied to study the motion of open systems in phase space. The correlations built during the time of approach to equilibrium are computed in numerical experiments. The evolution in collapsing systems is reported, while the mathematical properties, that is the Lyapounov spectrum, of systems obeying Hamiltonian dynamics is examined. It looks as if the simulations of complex flows could lead to new ideas in the old problem of irreversibility.

Next are four papers concerned with transport properties. The mode coupling theories which compute transport coefficients in liquids can be developed in the case of lattice gases and, since lattice gases are cheaper to simulate, the predictions of these theories can also be checked against computer simulation with much more precision than in traditional molecular dynamics. The problem of a possible phase transition occuring in liquids under large stress is reviewed. The non equilibrium simulations of fluids under shear are used to compute the effect of vorticity on internal degrees of freedom.

The last two parts deal with applications in solid state and chemical systems. Molecular Dynamics has been used to look at the problems of amorphization and the motion of dislocation in simple solids. These nice applications might lead to solving very applied problems as is suggested by a study of mechanical deformation. It could (possibly) also clarify some aspects of cold fusion as suggested by another participant. Last, but certainly not least, the applicability of molecular dynamics to nonequilibrium chemical systems is critically reviewed. The method is also tested to study the possible effects of heat of reaction on the kinetics.

Obviously, the domains of the research reported have been scattered over many different areas of physics and chemistry, besides the original and unifying theme of hydrodynamics. However, the problems encountered in the modelling, and the solutions found to solve them are similar in many of these fields. This should give us confidence in the application of these techniques for future research.

Michel Mareschal
Brussels, February 1990

ACKNOWLEDGEMENTS

We are grateful to a number of Institutions that have made the workshop possible. First is the NATO Scientific Affairs Division which has provided the most important financial contribution for the workshop. In addition, we received money from the C.E.C.A.M. (Centre Européen de Calcul Atomique et Moléculaire, Orsay) which funded the meeting as a discussion meeting, from the Instituts Internationaux de Physique et de Chimie fondés par Ernest Solvay (Brussels) and from the Université Libre de Bruxelles.

The editor also wishes to thank the following individuals: Mr. Pierrot Kinet, for his invaluable technical assistance before, during and after the workshop, Mrs. Sonia Dereumaux-Wellens for her help in the organization of the meeting, Mrs. Nadia Sardo, Mrs. Irene Saverino, Mr. Maurice Adam and Jean-Claude Dereumaux, the technical staff. We also received assistance from the "Centre Audio-Visuel" of the Université Libre de Bruxelles which permitted the display of films and video tapes.

CONTENTS

FLOWS (I) : METHODOLOGY

FLOWS (II) : ILLUSTRATIONS

APPLICATIONS (I) : NONEQUILIBRIUM SYSTEMS

APPLICATIONS (II) : TRANSPORT AND MODE COUPLING THEORIES

APPLICATIONS (III) : SOLIDS

APPLICATIONS (IV) : CHEMISTRY

PART I

FLOWS (I): METHODOLOGY

THE DIRECT SIMULATION MONTE CARLO METHOD:

CURRENT STATUS AND PERSPECTIVES

G.A. Bird

Department of Aeronautical Engineering
The University of Sydney
N.S.W. 2006 Australia

INTRODUCTION

Most analytical and numerical studies of the flow of liquids and gases
have been based on models that regard them as structureless or continuous
fluids. The texts on fluid mechanics make, at most, a passing reference to
the underlying atomic or molecular structure of the fluid and then
concentrate exclusively on the continuum equations, generally the Navier-
Stokes equations, that provide an approximate mathematical model of the
fluid.

The Navier-Stokes equations adopt a fixed, laboratory, or Eulerian
frame of reference and express the conservation of mass, momentum, and
energy in the fluid. The approximations enter through the equation of state
which expresses the relationship between the fluid density, pressure, and
temperature and also through the terms for the transport properties. For
example, the incompressible fluid model involves the assumption of a
particularly simple limiting form of the equation of state. The viscosity
is the transport property that is most commonly included in the formulation,
followed by heat conduction and mass diffusion.

The advent of the digital computer has made possible the study of fluid
flows by following the motion of a large number of individual molecules.
The word "molecule" is used here as a generic term and includes both atoms
and molecules. These methods are essentially based on a physical model or a
"direct simulation" of the fluid and are independent of the traditional
mathematical models, although there may be common physical assumptions and
the two approaches can often be reconciled. One problem with the direct
simulations that has led to much confusion is that there are several
distinct classes of simulation and a very large number of variants of each
of these. The best known direct simulation methods are the molecular
dynamics (or MD) method (Alder and Wainwright, 1958) and the direct
simulation Monte Carlo (or DSMC) method (Bird, 1963).

Most applications of the MD method (Evans and Hoover, 1986) have been
to the study of the equation of state of and transport processes in dense
gases near the liquefaction point. On the other hand, the DSMC method has
been developed largely in the context of rarefied gas dynamics (Muntz,
1989). The essential distinction between them is that, with the exception
of the setting of the initial conditions, the MD method is deterministic,

while probabilistic procedures that involve random numbers are applied at all stages in the DSMC method (hence the "Monte Carlo" in the name). These probabilistic procedures assume that only binary collisions are important and take advantage of molecular chaos; i.e. the assumption that the two-particle velocity distribution function is the product of the two single particle distribution functions. This means that the DSMC method is restricted to dilute gases in which the mean free path is large compared with the mean molecular spacing and this spacing is, in turn, large compared with the effective molecular diameter.

The MD and DSMC methods are, therefore, largely complementary and they were developed by essentially separate "user communities". This changed when both methods were applied to a single problem in rarefied gas dynamics by Mieburg (1986). Although this paper was interesting and has proved to be a valuable contribution, the comparison that was made between the two methods was misleading in some respects (Bird, 1987a and 1989a). This paper is primarily concerned with the current capabilities of the DSMC method but, in view of the relationship with the MD method, comparative reference to the latter will be made whenever it is appropriate.

THE REQUIREMENT FOR A MOLECULAR MODEL

The degree of rarefaction of a gas flow is generally expressed through the Knudsen number which is the ratio of the mean free path to a typical dimension of the flowfield. The continuum flow model, expressed through the Navier-Stokes equations, may be assumed to be valid when the Knudsen number is very small in comparison with unity, and the limiting case as the Knudsen number tends to zero may be identified with the inviscid limit expressed through the Euler equations. The opposite limit as the Knudsen number tends to infinity corresponds to collisionless or free molecule flow in which intermolecular collisions may be neglected. The flow regime between free molecule flow and the limits of validity of the Navier-Stokes equations is generally referred to as the transitional flow regime. A Knudsen number of 0.1 has traditionally been quoted as the boundary between continuum and transition regime flows, but the characteristic dimension of flowfields may be specified in many different ways and the use of an "overall Knudsen number" can be misleading.

The criteria for the boundaries between the flow regimes may be spec- ified unambiguously if a "local Knudsen number" is based on the scale length of the appropriate macroscopic property. The conservation equations of fluid mechanics are valid for all flow regimes, but the Navier-Stokes equations depend on the Chapman-Enskog theory for the shear stresses, heat fluxes, and diffusion velocities as linear functions of the velocity, temperature, and concentration gradients, respectively. The Chapman-Enskog theory assumes that the velocity distribution function is a small perturb- ation of the equilibrium or Maxwellian function and the theory is based on the first term of an expansion in the local Knudsen numbers (e.g. Bird, 1988a). As shown in Fig. 1, the Navier- Stokes equations cease to be valid when the local Knudsen number exceeds about 0.2.

A great deal of work has gone into attempts to find alternatives to the Chapman-Enskog transport properties as closure relations for the conserv- ation equations, and thereby extend the range of validity of the continuum formulation. It now appears that this will not be possible and that there is no alternative to the recognition of a rarefied gas as a set of discrete particles, or molecules. The relevant equation for a dilute gas is then the Boltzmann equation. This is an integro-differential equation with the velocity distribution function as the only dependent variable. However, the number of independent variables is equal to the dimensions of phase

2

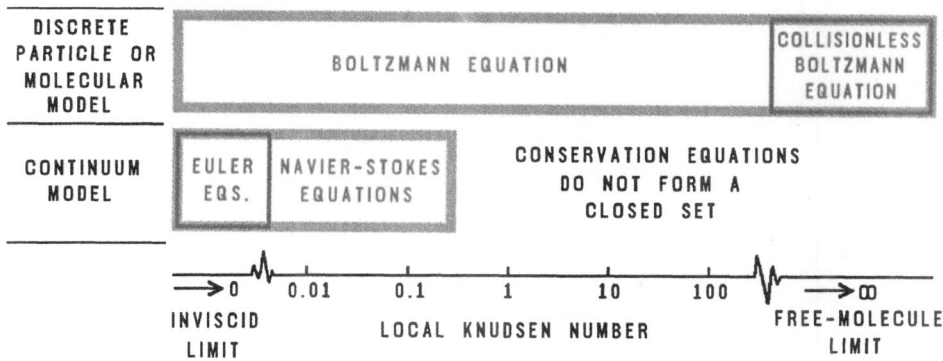

Fig. 1 Limits on the validity of the mathematical models.

space plus, in the case of an unsteady flow, time. A one-dimensional steady flow of a monatomic gas has an axially symmetric velocity distribution function so that the problem is three-dimensional in phase space. A five-dimensional mesh would be required for a two-dimensional steady flow. The specification of the computational mesh or grid in velocity space is particularly difficult because the unbounded nature of velocity space, and the one-dimensional steady flow has proved to be the practical limit for numerical solutions. In addition, the practical problems often involve physical effects, such as chemical reactions and thermal radiation, that have not yet been incorporated into the Boltzmann formulation.

Because of these difficulties, it has not been possible to find analytical solutions of the Boltzmann equation for complex nonlinear flows. Numerical solutions of the full nonlinear Boltzmann equation have been largely confined to one-dimensional steady flow studies using the Hicks-Yen-Nordsieck (HYN) method (Yen, 1984). This method employs a finite difference method for the "fluid-like" terms on the left hand side of the Boltzmann equation and a Monte Carlo method for the collision term on the right hand side. The restriction to relatively simple flows is primarily due to the computational requirements of any numerical method that has to work with a grid in phase space.

The alternative is to use a direct molecular simulation approach, with the MD and DSMC methods being the principal alternatives. The implications of the choice may be clarified through reference to the typical molecular magnitudes in Fig. 2. A typical problem would involve 10,000 to 100,000 model molecules in a volume of 1000 to 10,000 cubic mean free paths, or about 10 molecules to a cubic mean free path. It can be seen that these conditions occur in real gases at densities well above standard density and MD calculations for dense gases can employ realistic molecular diameters. However, if the MD method is be applied at low densities, the large number of small molecules are modelled by a small number of large molecules. For example, Meiburg's (1986) calculation was for the flow past an inclined flat plate with a chord length of 75 mean free paths. The molecular diameter was 0.75 of a mean free path or 1% of the chord of the plate. These conditions would apply at a density of 125 times the atmospheric density. The Boltzmann equation would not apply under these conditions, the equation of state would be different from that in a dilute gas, and the miss distance impact parameters for collisions in the critical region near the ends of the plate would be unrealistic (Bird, 1989a). The normalization procedures that were used in the early presentations (Bird, 1976a) of the DSMC method led to apparent molecular sizes similar to those involved in the MD method.

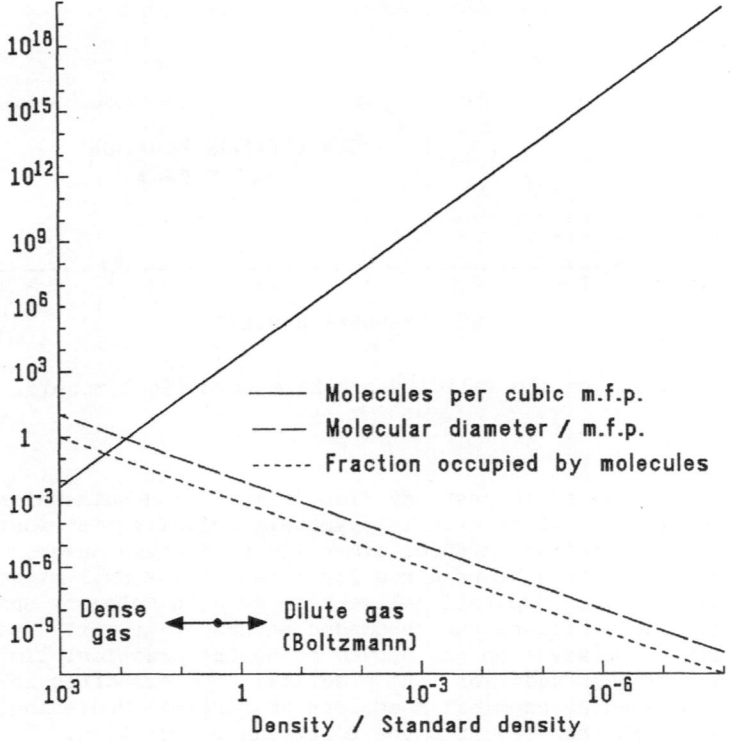

Fig. 2 Typical molecular magnitudes.

However, the molecules are selected for collisions on a probabilistic basis and, while the molecular size appears in the procedures for establishing the collision rate, it does not affect the collision parameters. This point is made clear in the more recent applications (Bird, 1981) of the DSMC method to engineering problems, in that the real molecular sizes are employed in conjunction with a parameter that explicitly states the number of real molecules that are represented by each simulated molecule. In contrast to the MD method which is most suited to dense gases, the DSMC method always simulates the limiting case of a very dilute gas and becomes more exact as the gas density decreases.

DESCRIPTION OF THE DSMC METHOD

As noted above, the DSMC method models the gas flow by a large number of simulated molecules in the computer and these are representative of the very much larger number of molecules in the real flow. Current implement-ations of the method require approximately 1MB of memory for each 20 to 30 thousand molecules, depending on the chemical complexity of the molecular model. The position coordinates, velocity components, internal state, and chemical identity of each simulated molecule are stored in the computer and these are modified with time as the molecule encounters a realistic set of intermolecular collisions and boundary interactions. The intermolecular collisions are always calculated as three-dimensional events, but advantage may be taken any symmetries in the flowfield to reduce the magnitude of the computation. The calculations are unsteady, but steady flow may be attained as the large time state. The flowfield is divided into computational cells with linear dimensions less than the local mean free path. The magnitude of the computational task can be estimated from the requirement that each of these cells should contain at least five to ten simulated molecules. Given

4

sufficient storage capacity, the task may be run with time or ensemble averaging until the sample size at each location is sufficient to give the required accuracy. There is a statistical scatter in the results with standard deviation given approximately by the inverse square root of the sample size, but numerical instabilities are completely absent. While high demands can be made on computer resources, particularly for three-dimensional flows, it is generally possible to take the solution to sufficiently high densities to overlap with the region of validity of the continuum equations.

While the simulation can be regarded essentially as a computer experiment with the procedures based on physical modelling, the basic assumptions of a dilute gas and molecular chaos are common to the Boltzmann equation. The consistency between the two was demonstrated at an early stage (Bird, 1970a) through the derivation of the Boltzmann equation from the DSMC method, and the relationship between the two has since been investigated in some detail by Nanbu (1986). This paper also analyzes the relationships between the many variants of the DSMC that have been proposed over the years. The major advantage of the 'conventional' procedures (Bird, 1976a,1989a) is that the computing time is linearly proportional to the number of simulated molecules. Nanbu (1983) modified the procedures to conform even more closely to the Boltzmann formulation. In particular, because the dependent variable in the Boltzmann equation is the distribution function for just one of the molecules in a collision, the velocity components of only one of the simulated molecules in a collision are changed as a result of the collision. Not only is this inefficient from the computational point of view, but a random walk is introduced and, as noted below, this can have a most deleterious effect on the quality of the simulation (Bird, 1989a). It is not rational to unnecessarily introduce undesirable effects into the simulation of a physical process merely to make the procedures more consistent with a particular mathematical model of that process. The mathematical model is itself an approximation and the objective should be to obtain the best possible physical model of the real gas flow. While the common approximations mean that the DSMC solutions can be considered to be equivalent to hypothetical solutions of the Boltzmann equation, the DSMC procedures often go beyond the limitations of the Boltzmann formulation. For example, simulations have included (Bird, 1977a) some dense gas effects such as ternary collisions. Moreover, the method is routinely applied (Bird, 1987b) to problems involving chemical reactions and thermal radiation and these effects are also beyond the current formulations of the Boltzmann equation.

A cell structure need be employed in the MD method only for the sampling of flow properties, but it is required in the DSMC method for the selection of collision pairs also. In addition, there is a discrete time step over which the molecular motion and the collision processes are uncoupled. The simulation becomes more exact as the time step and cell size tend to zero. It should be noted that disturbances can propagate at speeds faster than the ratio of the cell size to time step, and there is no stability criterion analogous to that associated with finite difference methods in continuum gas dynamics. During the early stages of the development of the method, some concern had been expressed (Bird, 1976a) about large values of this ratio. It was feared that coincidental collisions between molecules at opposite sides of the cells, coupled with the immediate migration of these molecules to the next cell, might cause false disturbances to propagate with speed equal to the ratio of cell size to time step. In practice, this ratio has been set to many times the local speed of sound and there has been no sign of false disturbances.

If the cell dimensions are small in comparison with the mean free path, they are also small compared with the scale length of the macroscopic flow gradients. The simulated molecules in the cell are then regarded as representative of the real molecules at the location of the cell, and the relative locations of the molecules within the cell are disregarded in the selection of collision partners. It follows that the flow gradients within a cell that are caused by the migration of molecules between cells will tend to be destroyed by the collisions. Meiburg (1986) drew attention to this process in a flow with vorticity and noted that the angular velocity of the collision partners about the center of the collision is not necessarily conserved. This is essentially a cell size effect and it is well established (Bird, 1977b; Moss and Bird 1985) that the cell size must be very small in comparison with the mean free path in regions with large gradients, such as Knudsen layers. A very small cell size is readily achieved in one-dimensional flows and calculations can be made at sufficiently low Knudsen numbers for the Navier-Stokes equations to be valid, so that comparisons can be made with exact analytical results. It has been found that, as long as the cell size is sufficiently small, good results are obtained for flows with high vorticity. The problem with Meiburg's DSMC calculation was that the cells had a linear dimension of three mean free paths which was excessively large by a factor of at least ten. On this point, it can be noted that conventional continuum computational fluid dynamics finite difference and finite element methods do not allow any vorticity at a grid point or within an element. The vorticity is modelled by the changes in velocity between the mesh points. To the extent the DSMC method preserves some vorticity in the cells (most of it when small sub-cells are used as noted below), it is superior in this respect to the continuum CFD methods.

As noted above, the DSMC method uses the cell system for the sampling of the macroscopic properties and for the selection of collision partners. The sampled density is used in the procedures for establishing the collision rate and it is desirable to have the number of molecules per cell as large as possible, generally around ten or twenty. On the other hand, in the selection of collision pairs, it is desirable to have this number as small as possible in order to reduce the mean separation of collision pairs and thereby minimize the smearing of gradients. These conflicting requirements can be reconciled by subdividing the sampling cell structure into a set of sub-cells for the selection of collision pairs. The sub-cells are chosen to contain an average of two or three molecules, so that all collisions approach the "nearest-neighbour" ideal. Should there be only one molecule in a sub-cell, the potential collision partner is selected from an adjacent sub-cell. The computing time penalty associated with these additional procedures is negligible. In fact all collisions could be between nearest neighbours, but this does not lead to the most efficient calculation from the computational point of view. The most efficient calculations for low Knudsen number flows employ cells that are narrow in directions with large gradients, but elongated in directions with small gradients. Very thin cells are often used in boundary layers. This is facilitated if the time step and number of real molecules represented by each simulated molecule are both reduced in the boundary region. If the reduction ratio is the same in each case, the flux of molecules between the boundary and outer regions is matched. This procedure effectively calculates the inner and outer regions with different time scales and is suitable only for steady flows.

The procedures for the probabilistic selection of a representative set of collisions at each discrete time step have been responsible for much of the comment on the DSMC. It is still useful to follow Derzko (1972) and define three distinct methods to serve as the basis of the discussion, which are:

1) The "collision rate" method in which the nonequilibrium collision rate is calculated from the standard kinetic theory result that

involves the average of all possible relative velocities. The required
number of collisions is then given by the product of this rate and the time
step. The collision pairs must be chosen with probability proportional to
the product of the relative velocity and collision cross section. The
disadvantage is that the computation time is proportional to the square of
the number of molecules.

2) The "time-counter" method in which the collision pairs are
accepted with the same probability as in method 1 and, at each collision, a
cell time is advanced by an appropriate amount (Bird, 1976a). Sufficient
collisions are calculated to keep the cell time concurrent with the flow
time. The advantage of this method is that the computation time can be
linearly proportional to the number of molecules.

3) The "direct" or Kac method in which all possible pairs in each
cell are considered and the probability of collision within the time step is
equal to the ratio of the volume swept out by the cross section (moving with
the relative velocity) to the cell volume. The disadvantage is again that
the computation time is very nearly proportional to the square of the number
of molecules.

Because of its efficiency, the time counter method has been the most
used, but there are problems in highly nonequilibrium regions such as the
front of a very strong shock. The acceptance of an unlikely collision can
advance the cell time by an interval that is much larger than the time step
and the overall collision rate can be distorted. The direct method can be
modified by reducing the number of sampled pairs by some factor and increasing the collision probabilities by the same factor. If the factor is such
that the maximum collision probability of any pair is unity the procedures
are similar to those in the time counter method, except that the time
counters are effectively replaced by the explicit relation for the number of
molecule pair choices. The computing time remains directly proportional to
the number of simulated molecules and the method is called (Bird, 1989a) the
"no time counter" or NTC method.

The operation of the collision procedures depends on the details of the
coding, and mathematical representations of the procedures are generally
incapable of providing an adequate model. Also, criticism may be based on
imagined rather than the actual procedures. For example, the null-collision
technique of Koura (1986) addresses itself to a hypothetical problem
associated with chemically reacting flows that had not, in fact, been
present in the previous applications (Bird, 1977a; Moss and Bird, 1985) of
the DSMC method to these flows. Similarly, the critical study of the now
obsolete time counter method by Belotserkovskii et al (1980) is based on an
oversimplified mathematical model of its operation. The cell time does not
have to be synchronised with the flow time at the end of each time step and
an averaged number density can be used for one of the components in the
product the number of molecules and the number density in each cell. The
claimed accumulation of error can be (and has been) readily avoided by
sensible coding.

As noted above, a major advantage of the DSMC method is that stability
problems are completely absent. On the other hand, the statistical consequences of the replacement of the extremely large number of real molecules by
a much smaller number of simulated molecules must always be kept in mind.
The statistical scatter generally decreases as the square root of the sample
size and, in order to attain a sufficiently small standard deviation, either
time averaging or ensemble averaging is employed for steady flows and
unsteady flows, respectively. An exception occurs when weighting factors
are applied to a disparate concentration gas mixture in order to increase
the representation of the trace species. Energy is then conserved on the

average in collisions, but not in each collision. There is therefore some energy transfer between the species at each collision and this has the characteristics of a random walk. The mean departure of the species temperature from equipartition and the time between reversals of the direction of the departure increases as the square root of the number of collisions.

The most serious statistical problem is when a significant effect in the real gas is a consequence of the few molecules towards the extremities of the distribution. The number of model molecules in the simulation might be so small in comparison with the number of real molecules that the important part of the distribution would not be adequately populated. This problem generally arises in connection with chemical reactions or, more particularly, thermal radiation. A recent (Bird, 1987b) solution to it was to assign a distribution of electronic states, rather than a single state, to each simulated molecule. Of all the flows that have been simulated, the only non-reacting flow that has proved (Bird, 1976b) to be sensitive to the number of simulated molecules is the hypersonic flow within a conical molecular beam skimmer. This is because a single molecule at a sufficiently large angle to the axis to strike the inside of the cone can produce such a large number of similar molecules, when reflected back into the beam, that the hypersonic beam breaks down to a subsonic flow. The time to breakdown is longer in the simulated flow than in the real flow.

PHYSICAL APPROXIMATIONS

The preceding discussion has been concerned with the computational procedures and approximations associated with the simulation. A typical application requires the calculation of millions of intermolecular collisions and gas-surface interactions. This requires molecular models that are computationally efficient and some degree of physical approximation is unavoidable.

Monatomic gas models

Early applications of the direct simulation methods employed the well established models of classical kinetic theory (e.g. Chapman and Cowling, 1952). The hard sphere model is the easiest to apply, and both it and the Maxwell model were widely used for early studies in which the primary objective was to make comparisons with other numerical or analytical methods employing the same model. These models are associated with unrealistic viscosity-temperature power laws and became inadequate when serious comparisons were to be made with experiment. The inverse power law of repulsion model was then implemented (Bird, 1970b), and some calculations were made (Sturtevant and Steinhilper, 1974) with the Exp-6 and LennardJones models that include the long range attractive force. The simpler model was found to be generally adequate for engineering studies. In all the successful correlations between experiment and simulation, the molecular impact parameters were chosen to match the coefficient of viscosity of the real gas.

Experience was gradually accumulated on the effects of molecular model on representative flows. It was found that the observed effects could be fully correlated with the variation, due to the relative velocity in the collision, of the differential cross-section of the molecules. The form of the deflection angle scattering law of the molecules appeared to be comparatively unimportant. These considerations led to the introduction (Bird, 1981) of the variable hard sphere, or VHS, molecular model. This is essentially a hard sphere with a diameter that varies as some inverse power of the relative velocity in the collision. The VHS model is the simplest one that is capable of modelling the viscosity coefficient, including its temperature dependence, of real molecules. It has been found (Bird, 1983 and 1986) that, as long as this is done, the results are almost independent of the molecular model. The VHS model is therefore recommended for engineering calculations, although it should be kept in mind that some of the more complex classical models provide a closer physical representation of real molecules.

Internal degrees of freedom

The lack of dependence of monatomic gas flows on the details of the scattering process means that phenomenological models should be satisfactory for molecules with internal degrees of freedom. The model should reproduce the degree of excitation of the internal modes and the relaxation time, while continuing to provide the correct values of the coefficients of viscosity and heat conduction. The model introduced by Borgnakke and Larsen (1975) meets these requirements and also satisfies the principle of detailed balancing. This is important because detailed balancing is a necessary as well as a sufficient condition for equilibrium. It should be noted that, while detailed balancing might appear to be an equilibrium thermodynamics concept that could be inappropriate to highly nonequilibrium rarefied flows, it is actually based on quantum mechanical considerations that are applicable to individual molecular interactions in the real gas.

The essential feature of the Larsen-Borgnakke model is that a fraction of the collisions are regarded as completely inelastic and, for these, new values of the translational and internal energies are sampled from the distributions of these quantities that are appropriate to an equilibrium gas. The remainder of the molecular collisions are regarded as elastic and are calculated as if the gas was monatomic. The fraction of inelastic collisions can be chosen to match the real gas rotational relaxation rate. Although the method has given satisfactory results in all applications, it is clearly physically unreal at the microscopic level. Efforts have been made (Davis et al, 1983) to develop more realistic models, but the approximations that are necessary for their practical implementation has meant that they do not satisfy detailed balancing sufficiently well to produce acceptable equipartition at equilibrium. Similar problems occur if the fraction of inelastic collisions is made energy dependent in order to model the temperature dependence of the excitation collision number.

The Larsen-Borgnakke model has been extended to partially excited vibrational modes in such a way that the vibrational energy is restricted to the values appropriate to the widely spaced vibrational levels. Alternatively, these levels may be regarded as separate molecular species in a chemically reacting gas.

Chemical reactions

Nonequilibrium chemical reactions are readily handled by procedures that are essentially extensions of the elementary collision theory of chemical physics. In this, the binary reaction rate is obtained as the product of the collision rate for collisions with energy in excess of the activation energy and the probability of reaction or steric factor. Since the formulation is in terms of kinetic theory, the model can be incorporated directly into the DSMC method.

The chemical data for gas phase reactions is almost invariably quoted in terms of the reaction rate constant. A form of the collision theory that is consistent with the VHS model can be used (Bird, 1979) to convert the temperature dependent rate constants of continuum theory into collision energy dependent steric factors or reaction cross-sections. Alternatively, the reactive cross-sections may be available directly from the thermophysical database. While this is more in keeping with the simulation approach, it leads to increased difficulties in ensuring that the data for the forward and reverse reactions is consistent with chemical equilibrium in situations where the reactions should proceed to equilibrium.

Recombination reactions are not important at very low densities where the probability of three-body collisions is extremely small. They become important in the continuum overlap region and these are dealt with by assigning a lifetime to each binary collision and regarding the ternary collision as a further binary collision between the pair of molecules in the original collision and a third molecule. These procedures can be incorporated into the DSMC method in a form (Bird, 1977) that guarantees the correct equilibrium state.

Electronic excitation and ionization

The introduction of electrons necessitates a number of modifications to the DSMC procedures. As noted earlier, each simulated molecule represents an extremely large number of real molecules and this means that the flow fluctuations in the simulation are orders of magnitude larger than those in the real gas. While this causes few problems in a neutral gas, independent fluctuations if ions and electrons in a partially ionized gas would lead to spurious electric fields that would be far stronger than the real electric fields. This problem may be avoided if the electrons are not allowed to move freely, but are kept in proximity to the ions in order to enforce charge neutrality. This is valid for flows in which the characteristic dimensions are large in comparison with the Debye distance. Alternatively, the electric field may be specified by considerations external to the direct simulation procedures, for example through a parallel solution of the Poisson equation.

If the enthalpy of the gas is sufficiently large to produce some ionization, the higher electronic states will also be significantly populated, and thermal radiation must also be taken into account. As with the vibrational states, the electronic states may be regarded either as separate molecular species with the transitions through chemical reactions or as internal states through phenomenological procedures.

For slightly ionized flows with a degree of ionization of the order of one part in a million, the sampling problem associated with a calculation with much less than a million simulated molecules can be overcome (Bird, 1989b) by associating an "ionized fraction" with each simulated molecule. This is physically reasonable because each simulated molecule generally represents many billions of real molecules.

Gas-surface interactions

The classical diffuse interaction model is appropriate to "engineering surfaces" that have not been exposed for a long period to an ultra-high vacuum. Experiments under ultra high vacuum conditions have raised the possibility that there could be a significant lack of thermal accommodation and retention of parallel momentum at the surface of space vehicles. A combination of diffuse reflection with the other classical model of specular reflection is generally used to provide qualitative information on these effects, but this does not provide an adequate detailed model. The reason for persisting with the inadequate classical models is that they satisfy the principle of detailed balancing, or reciprocity, while the more complex theories based on simplified physical models or complex sets of accommodation coefficients do not (Kuscer, 1974). There is a need for a more adequate simulation model that satisfies the reciprocity condition and is in qualitative agreement with the measurements on clean surfaces.

APPLICATIONS AND VALIDATION

Most applications of the DSMC method have been to problems in rarefied gas dynamics, but there have been a number of calculations for the internal structure of shock waves. Since the thickness of a shock wave is a function of the upstream mean free path, the DSMC solutions are feasible at all densities up to the dilute gas limit. The problem has proved to be a particularly valuable test case because of the availability of reliable experimental data and the fact that it does not involve the uncertainties associated with gas-surface interactions. Early studies (e.g. Sturtevant and Steinhilper, 1974) showed that there was good agreement as far as the thickness of strong shock waves is concerned. Recent comparisons (Erwin et al, 1989) have been for the highly nonequilibrium velocity distribution function within a Mach 25 shock wave in helium. The Maitland-Smith molecular model was used for the DSMC calculation and the agreement with experiment was excellent.

The accuracy of early calculations was limited by statistical scatter, but it is now feasible to calculate one-dimensional flows such as the shock wave structure with a sample of millions at each point and a cell size very much smaller than the mean free path. This permits accurate calculations of the structure of weak waves, although care must be taken with the boundary conditions (Bird, 1989a). Erwin et al (1989) have examined the ratio of the upstream to the downstream curvature of the density profiles in weak to medium strength waves. They have shown that these calculations indicate that the Burnett equations (which include the next term of the expansion in local Knudsen number for the transport properties) are superior to the Navier-Stokes equations. There is experimental evidence (Alsmeyer, 1976) to support this for medium strength waves. The degree of noneqilibrium increases with the shock Mach number and the limit of validity of the Navier-Stokes equations is at a shock Mach number less than two.

The experimental data base for shock waves in gas mixtures and gases with internal degrees of freedom is much less adequate than that for the simple monatomic gases, but is in good qualitative agreement with the calculations. The comparisons for these cases serve more as a test of the consistency of the physical data such as the relaxation and chemical rates, rather than as a validation test of the computational method. However, the agreement for the simple gas case is such that the DSMC method can be used with some degree of confidence for the deduction of physical and chemical data from measurements of the profiles of macroscopic flow properties in shock waves that involve gas mixtures, internal degrees of freedom, and chemical reactions.

The breakdown of equilibrium in the steady expansion of gas flows provides another test case that is unaffected by solid boundaries. An early calculation (Bird, 1970c) led to the proposal of a breakdown criterion that has been supported by subsequent experiments (Cattolica et al, 1974). Very low temperatures are encountered in this problem and Chatwani (1976) employed the quantum cross-section data for helium and neon which was stored in tabular form in the computer.

Matched sets of DSMC calculations and experiments in low density supersonic wind tunnels that have been made at Imperial College, London (Harvey, 1986). These have been the most careful and extensive tests for two-dimensional and axially-symmetric flows and good agreement was obtained. Many isolated comparisons have been made for with wind tunnel and flight data and also with solutions of the Navier-Stokes equations when the Knudsen

number is sufficiently low for the latter to be valid. One point that
should be taken into account in comparisons with continuum solutions for the
flow of gas mixtures is that the simulation includes (Bird, 1988b) thermal
and pressure diffusion that have long been understood (Chapman and Cowling,
1952) but are very rarely included in the Navier-Stokes equations. The
conventional continuum equations make no distinction between a simple gas
and a gas mixture with similar properties and their solutions can be
seriously deficient for gas mixtures.

Increases in computer speed and storage now allow some serious
calculations of three-dimensional flows and permit the two-dimensional
calculations to be pushed further into the continuum regime. There is a
strong incentive for this in that the continuum methods appear to be unable
to cope with contiuum flows with extremely large gradients, such as the
expansion around the lip of a rocket nozzle. The DSMC method is not suited
to vector computers because the outcome of molecular movement routines
depend strongly on the location of the molecule in the flow. Also, the
efficiency of routines such as the Larsen-Borgnakke model for internal
degrees of freedom comes from confining the additional computation to just a
few of the collisions. so that the conditional steps that upset vectorization are difficult to avoid in complex problems. The method is, however,
well suited to parallel computers. Much effort is now being devoted to
maximising the computational efficiency of the simulation of complex flows.

REFERENCES

Alder, B.J. and Wainwright, T.E., 1958, Molecular dynamics by electronic
computers, in "Transport Processes in Statistical Mechanics", I. Prigogine,
ed., Interscience, New York.
Alsmeyer, H., 1976, J. Fluid Mech., 74:497.
Belotserkovskii, O.M., Erofeev, A.I., and Yanitskii, V.E., 1980, A
non-stationary method of direct statistical modelling of rarefied gas flows,
USSR Comput. Math. and Math. Phys., 20:82.
Bird, G.A., 1963, Approach to translational equilibrium in a rigid sphere
gas, Phys. Fluids, 6:1518.
Bird, G.A., 1970a, Direct simulation and the Boltzmann equation, Phys.
Fluids., 13:2676.
Bird, G.A., 1970b, Aspects of the structure of strong shock waves, Phys.
Fluids, 13:1172.
Bird, G.A., 1970c, Breakdown of translational and rotational equilibrium in
gaseous expansions, Phys. Fluids, 13:1172.
Bird, G.A., 1976a, "Molecular Gas Dynamics", Oxford University Press,
London.
Bird, G.A., 1976b, Transition regime behavior of supersonic beam skimmers,
Phys Fluids, 19:1486.
Bird, G.A., 1977a, Direct molecular simulation of a dissociating diatomic
gas, J. Comput. Phys., 25:405.
Bird, G.A., 1977b, Direct simulation of the incompressible Kramers problem,
Progr. in Astro. Aero., 51:323.
Bird, G.A., 1979, Simulation of multi-dimensional and chemically reacting
gas flows, in "Rarefied Gas Dynamics", R. Campargue ed., CEA, Paris.
Bird, G.A., 1981, Monte Carlo simulation in an engineering context, Progr.
in Astro. and Aero., 74:239.
Bird, G.A., 1983, Definition of mean free path for real gases, Phys.
Fluids., 26:3222.
Bird, G.A., 1986, Low density aerothermodynamics, Progr. in Astro. and
Aero., 103:3.
Bird, G.A., 1987a, Direct simulation of high-vorticity gas flows, Phys.
Fluids, 30:364.
Bird, G.A., 1987b, Nonequilibrium radiation during re-entry at 10 km/s, Am.

Inst. Aero. Astro. Paper 87-1543.

Bird, G.A., 1988a, Direct simulation of gas flows at the molecular level, Comm. in App. Numerical Methods, 4:165.

Bird, G.A., 1988b, Thermal and pressure diffusion effects in high altitude flows, Am. Inst. Aero. Astro Paper 88-2732.

Bird, G.A., 1989a, The perception of numerical methods in rarefied dynamics, Progr. in Astro. and Aero., in press.

Bird, G.A., 1989b, Computation of electron density in high altitude re-entry flows, Am. Inst. Aero. Astro Paper 89:1882.

Borgnakke, C. and Larsen, P.S., 1975, Statistical collision model for Monte Carlo simulation of polyatomic gas mixture, J. Comput. Phys., 18:405.

Cattolica,R., Robben, F., Talbot, L., and Willis, D.R., 1974, Translational nonequilibrium in free jet expansions, Phys. Fluids, 17:1793.

Chapman, S. and Cowling, T.G., 1952, "The Mathematical Theory of Non-Uniform Gases", Cambridge University Press, London.

Davis, J., Dominy, R.G., Harvey, J.K., and Macrossan, M.N., 1983, An evaluation of some collision models used for Monte Carlo calculations of diatomic rarefied hypersonic flow, J. Fluid. Mech., 135:355.

Derzko,N.A., 1972, A review of Monte Carlo methods in kinetic theory, UTIAS Review 35, University of Toronto.

Erwin, D.A., Muntz E.P., and Pham-Van-Diep, G., 1989, A review of detailed comparisons between experiments and DSMC calculations in nonequilibrium flows, Am. Inst. Aero. Astro., Paper 89-1883.

Evans, D.J. and Hoover, W.G., 1986, Flows far from equilibrium via molecular dynamics, Ann. Rev. Fluid Mech., 18:243.

Harvey, J.K., 1986, Direct simulation Monte Carlo method and comparison with experiment, Progr. in Astro. and Aero., 103:25.

Koura, K., 1986, Null-collision technique in the direct simulation Monte Carlo method, Phys. Fluids, 29:3509.

Kuscer, I., 1974, Phenomenology of gas-surface accommodation, in "Rarefied Gas Dynamics", Becker and Fiebig eds., DFVLR Press, Porz-Wahn.

Meiburg, E., 1986, Comparison of the molecular dynamics method and the direct simulation technique for flows around simple geometries, Phys. Fluids, 29:3107.

Moss, J.N. and Bird, G.A., 1985, Direct simulation of transitional flow for hypersonic re- entry conditions, Progr. in. Astro. Aero., 96:113.

Muntz, E.P., 1989, Rarefied Gas Dynamics, Ann. Rev. Fluid Mech., 21:387.

Nanbu, K., 1983, J. Phys. Soc. Japan, 52:3382

Nanbu, K., 1986, Theoretical basis of the direct simulation Monte Carlo method, in "Rarefied Gas Dynamics", V. Boffi and C. Cercignani eds., B.G. Tuebner, Stuttgart.

Sturtevant, B. and Steinhilper, E.A., 1974, Intermolecular potentials from shock structure experiments, in "Rarefied Gas Dynamics", K. Karamcheti ed., Academic Press, New York.

Yen, S. M., 1984, Numerical Solution of the Nonlinear Boltzmann Equation for Nonequilibrium Gas Flow Problems, Ann. Rev. Fluid Mech., 16:67.

THE NONEXISTENCE OF NONLINEAR LAWS FOR SIMPLE FLUIDS BY
NONEQUILIBRIUM MOLECULAR DYNAMICS

Giovanni CICCOTTI

Dipartimento di Fisica
Università di Roma "La Sapienza"
Piazzale Aldo Moro 2, 00185 Roma (Italy)

ABSTRACT

After introducing and discussing the methods most commonly used to study transport coefficients we give a broad assessment of the main findings. In particular we present evidence for an exceptionally large region of validity of the linear laws. For simple fluids there is virtually no breakdown of their validity. Finally we provide some evidence for the legitimacy of NEMD simulations

1. INTRODUCTION

A statistical mechanical system (N interacting particles in a volume V where $N,V \to \infty$ with $N/V = \rho$) subject to a constant external field, $X \neq 0$, and thermalized (with thermalization rate $\nu_N \neq 0$), produces a stationary state, possibly a nonequilibrium one (depending on boundary conditions) with macroscopic flows $J \neq 0$.

The quantities

$$L = \lim_{\substack{X \to 0 \\ \nu_N \to 0 \\ N,V \to \infty, N/V = \rho}} \frac{J}{X} \qquad (1)$$

are the transport coefficients. They have attracted much attention in the molecular dynamics (MD) community since the famous paper by Alder, Gass and Wainwright[1], in 1970, followed shortly after by the second famous paper, by Levesque, Verlet and Kürkijari[2], in 1973, in which it was convincingly shown that transport coefficients can be actually computed by equilibrium MD with the help of a theoretical trick, Linear Response Theory[3]. Later on, due to technical (the slow convergence of Green-Kubo integrands[4]) and rheological (shear-rate dependence of the shear viscosity for complex, e.g. polymeric liquids) reasons, also the quantities:

$$L' = \frac{J}{X} = L' \,(X, \nu_N, N) \qquad (2)$$

began to be investigated. Moreover only the quantities L' can be computed by nonequilibrium MD (NEMD). The major aims of this paper are:

(i) to recall the various methods that can be used to compute transport coefficients (either L or L')

(ii) to discuss the behavior of $L'\,(X, \nu_N, N)$, found in many recent NEMD investigations and, in particular, the well-founded of the extrapolation

$$L' \to L \qquad (3)$$

(iii) to counter criticisms sometimes raised against NEMD methods.

2. GREEN-KUBO FORMULAS AND LINEAR RESPONSE THEORY

It is a very well-known result that in the linear regime of nonequilibrium thermodynamics the coefficients appearing in the constitutive relations (i.e. the transport coefficients) can be obtained by suitable integrals of time correlation functions wich have the general form

$$L = \frac{1}{k_B T} \int_0^\infty dt \; <\hat{J}(t)\, \hat{J}(o)>_{eq} \qquad (4)$$

where the $<...>_{eq}$ means an equilibrium average and $\hat{J}$ are microscopic phase functions whose average gives the macroscopic flow J. The $\hat{J}$ are generally time derivatives of suitable variables, $\hat{J} = \dot{O}$ where, e.g.,

$$O = \begin{cases} \Sigma_i q_i \mathbf{r}_i & \text{electrical conductivity} \\ \Sigma_i \mathbf{r}_i \mathbf{p}_i & \text{viscosity} \\ \Sigma_i \mathbf{r}_i h_i & \text{thermal conductivity} \end{cases} \qquad (5)$$

In eq. (5) $\mathbf{r}$ and $\mathbf{p}$ are the usual phase space variables, while q_i is the charge possibly associated to particle i and h_i is the microscopic energy associated to i.

Time correlation functions and their time integrals, although slowly converging, are computed in a straightforward way by equilibrium MD. Instead it is more interesting to comment on how Green-Kubo formulas are derived. Indeed to obtain eq.(4) one has to compute the macroscopic response of a system to a (weak) external perturbation. *Microscopically* this means that the evolution of the system is given by the following eqs. of motion

$$\dot{\mathbf{r}} = \frac{\mathbf{p}}{m} + \mathbf{C}\,(\{\mathbf{r}, \mathbf{p}\})\,\mathbf{X}(t)$$

$$\dot{\mathbf{p}} = \mathbf{f} + \mathbf{D}\,(\{\mathbf{r},\mathbf{p}\})\,\mathbf{X}(t) - \nu_N\zeta\dot{\mathbf{p}} \tag{6}$$

$$\dot{\mathbf{z}} = \nu_N\left(\frac{\Sigma\,\dfrac{p^2}{2m}}{3Nk_BT} - 1\right)$$

where one can easily identify the standard equilibrium part, while the $\mathbf{C}$ and $\mathbf{D}$ terms are the mechanical coupling with the external field $\mathbf{X}(t)$ and whatever term containing ν_N represents the contribution to the evolution eqs. coming from the presence of the Nosé-Hoover thermostat[4].

By writing the Liouville equation corresponding to the dynamics given by eqs. (6) and by applying by now standard techniques (see however ref. [5] for the nuisances introduced by the presence of the thermostat) one finds for the linear response of the system

$$<\hat{J}>_t^{neq} = \frac{1}{k_BT}\int_{-\infty}^t ds <\hat{J}(t)\,[\Sigma(\dot{\mathbf{p}}\mathbf{C}+\dot{\mathbf{r}}\mathbf{D})]\,(s)>_{can}\,\mathbf{X}(s) \tag{7}$$

where $<...>_t^{neq}$ means a nonequilibrium average at time t and $<...>_{can}$ an equilibrium canonical (constant temperature) average. It is now a standard matter to obtain from eq.(7) the Green-Kubo formula. It is sufficient to identify $\Sigma\,(\dot{\mathbf{p}}\mathbf{C}+\dot{\mathbf{r}}\mathbf{D})$ with $\hat{J}$, take for the time dependence of the external field the Heaviside form, $\mathbf{X}(t) = \mathbf{X}_0\,\theta(t)$ and extend the upper limit of the integral (to reach the stationary state) to infinity.

NEMD is the other way around: it simply performs numerically the conceptual experiment implicit in eqs. (6) without using the Liouville equation. This is feasible if we can

(i) identify the suitable $\mathbf{C}$ and $\mathbf{D}$;

(ii) find some kind of periodic boundary conditions compatible with $\mathbf{C}$ and $\mathbf{D}$.

Condition (ii) is more a practical than a conceptual question. However for the small systems normally used in simulation it is more or less compelling and I will not refer to inhomogenous NEMD methods in the sequel. Condition (i) is more subtle since it requires one to invent a mechanical perturbation even for coefficients like thermal conductivitly that appear in flows driven by thermodynamic, not mechanical forces. However, this too is an old story completely solved now thanks to the contributions by Luttinger[6], Mazur[7], Gillan[8], Hoover and coworkers[9], Evans[10]. These perturbations are generally called homogeneous since they are consistent with some variety of periodic boundary conditions and the ensuing techniques are called synthetic[11] to recall that the *mechanical* perturbations used are generally fictitious and bear no direct relationship with the physical gradient (i.e. the thermodynamical force) they simulate.

We will now turn to the discussion of how to implement NEMD.

3. APPROACHES TO NEMD

There are two ways to compute the response of a system subjected to a constant external field switched on at a certain time. The first can be called *stationary*, the second *dynamic*. In the stationary method one generates a very long trajectory, necessarily thermalized, for the perturbed system. Often, although not necessarily, the initial condition of this trajectory is an equilibrium configuration. After some time to be discarded (the time interval needed for the system to reach the stationary state) the macroscopic properties are obtained, as usual, by time averages. To apply this method it is compulsory

(i) to have a thermostat with a thermalizing rate large enough to extract the heat produced in the system by the dissipation of the driving force;

(ii) to apply perturbations strong enough to induce a response detectable with the statistics that a computer experiment can produce; this means perturbations enormous on the macroscopic scale.

The first problem can be easily handled, although the introduction of a thermalization mechanism in the microscopic dynamics has produced criticisms on the physical reliability of the approach. The second problem can be even more serious since it requires an extrapolation over an enormous interval to obtain the value of the zero-field transport coefficients. To give an idea of the orders of magnitudes involved, in the study of thermal conductivity the minimun synthetic "thermal gradient" one can use is of the order of 10^6 K/cm and one has to extrapolate to zero field thermal conductivities obtained for perturbations larger or at most equal to that. We will see that both these problems are less serious than they can appear at first.

In the dynamic method one uses the first part of a perturbed trajectory, the part just discarded in the previous approach, which begins from an equilibrium initial condition (generally called a segment). The averages are then performed over many segments. The justification of the method, simple for the case of Hamiltonian perturbations without thermostat, goes back at least to Kubo[12]. It has been recently extended to general perturbations in presence of a thermostat by Holian and Evans[13]. The time evolution of a non-equilibrium ensemble, $\rho^N(t)$, obtained by solving the full Liouville equation with an equilibrium ensemble, $\rho^{N,eq}$, as initial condition is given by

$$\rho^N(t) = S^+(t,0) \, \rho^{N,eq} \tag{8}$$

while the evolution of any observable O by

$$O(t) = S(t,0) \; O \tag{9}$$

where S^+ is the operator adjoint of S. Therefore one has

$$J(t) = \; <\hat{J}>_t^{\,neq}$$

$$= \int dr dp \; \hat{J} \; \rho^N(t)$$

$$= \int dr dp \; \hat{J} \; S^+(t,0) \rho^{N,eq} \tag{10}$$

$$= \int drdp \, [S(t,0) \, \hat{J}] \, \rho^{N,eq}$$

$$= <(S(t,0)\hat{J})>_{eq}$$

where the super-subscripts of the brackets, neq and eq, respectively refer to nonequilibrium and equilibrium averages and $\hat{S}(t,0)$ is the evolution operator associated with full dynamics of eq.(6). Eq.(10) provides also a microscopic derivation of Onsager's regression hypothesis, but that is only of marginal interest here.

The dynamic method has an advantage over the stationary one , since it gives not only the stationary response, but also its dynamical features. Moreover it leads naturally to an useful improvement which permits one to study, at least for short times, the response to very small perturbations.
I am referring to the subtraction technique[14-17], that I will now introduce.

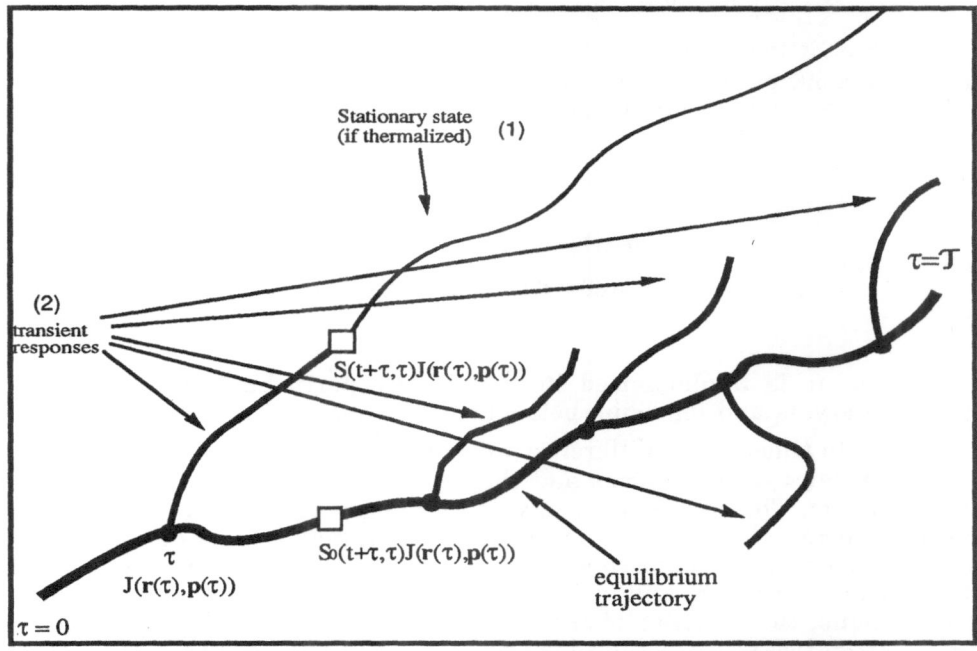

FIGURE 1. Sketch of the phase space trajectories needed to perform the various NEMD methods: (1) Long trajectory perturbed by a constant external field (and thermalized to reach a stationary state) over which one takes a time average. Notice that the first part of the trajectory is discarded. (2) Various short perturbed side-trajectories (segments), constructed by leaving the equilibrium trajectory at the running time t. The averages are performed i. directly over the segments computed up to time t (large perturbation), ii. over the difference between the perturbed and unperturbed segments, i.e. the squares of the figures (subtraction technique).

Since at equilibrium $J = < S_0(t,0)\hat{J} >_{eq}$, where $S_0(t,0)$ is the unperturbed time evolution operator, is zero, it is formally equivalent to calculate the macroscopic flux at time t by eq.(10) or by taking the equilibrium average of the variable.

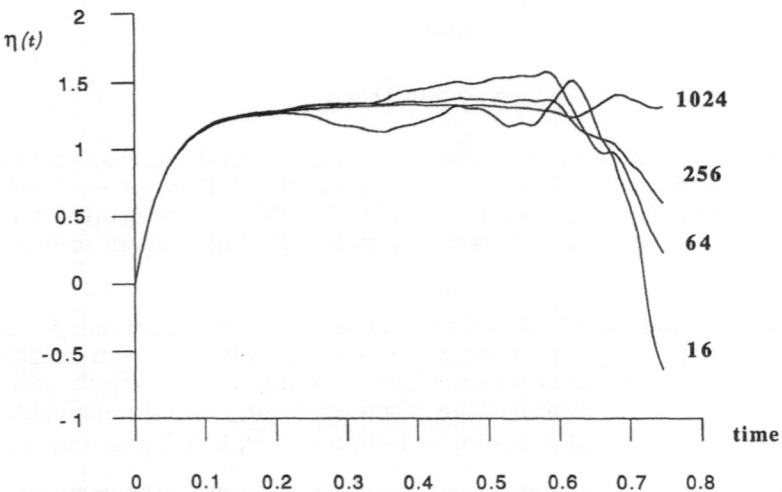

FIGURE 2. Statistical convergence of the mean response η(t) towards a plateau behavior for a 500 particles NEMD experiment.[17] The number of segments included in the average is indicated next to the corresponding curve.(units are usual reduced units).

$$\hat{\delta J}(t) = S(t,0)\hat{J} - S_0(t,0)\hat{J} \tag{11a}$$
$$J(t) = <\hat{\delta J}(t)>_{eq} \tag{11b}$$

The technique is implemented in practise by perturbing the system at regular intervals and following both perturbed and unperturbed trajectories in time up to time t. The difference variable $\hat{\delta J}$, eq.(11a), is computed for a number of pairs of trajectories and then averaged. See fig. 1 for a pictorial representation. The statistical equivalence of eqs. (10) and (11) should not hide their dramatic difference from the computational point of view. While the variance of the macroscopic current in eq. (10) is independent of the perturbation strength and generally larger unless one uses enormous perturbations, the variance of the difference, eq. (11), is of the order of the perturbation strength square[16-17] at least for times shorter than the correlation time between perturbed and unperturbed trajectories. The limitation just recalled could ruin any applicative interest. However many applications have shown that the time span within which the two trajectories remain correlated is most often sufficient to compute not only the first part of the dynamical response but all of it, so that finally one can use eq. (11) to compute transport coefficients in the linear domain by NEMD. The typical trend of the response to a constant external field switched on at time 0 is depicted in fig. 2 as a function of the accumulating statistics (number of segments over which the average is performed). In fig. 3 we summarize, instead, the various technique to perform nonequilibrium MD and the conditions required. In fig. 4 finally, we represent the range of validity of the various NEMD methods together with the regions in which the techniques superpose.

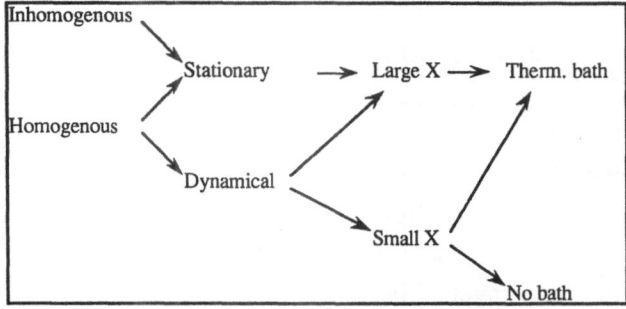

FIGURE 3. Conditions required to apply the various Homogenous and inhomogenous NEMD approaches. X indicates the strength of the perturbating field.

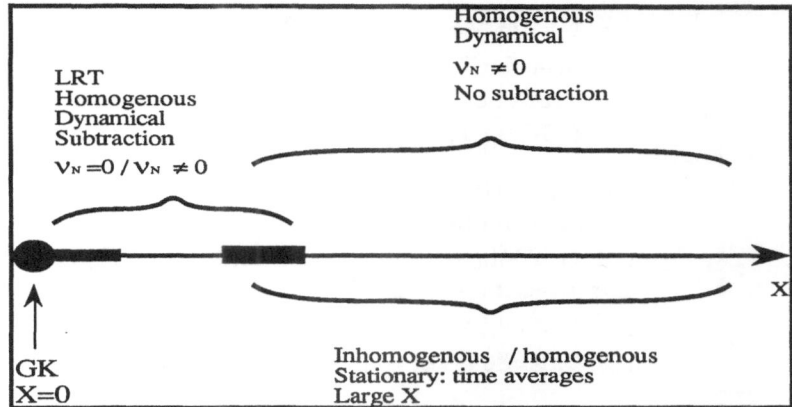

FIGURE 4. Ranges where the various NEMD method apply. The units are arbitrary and the regions of superpositions are intensified.

4. RESULTS

In the past ten years there has been much work, careful and less than careful, to estimate transport coefficients inside and outside the linear region. Much time has been also spent to discover all possible dependences on N, the number of particle of the system, and ν_N, the thermalizing rate. The last two tests are important to understand how much the results of the simulations represent reliably the thermodynamical limit and the usual experimental condition. In table 1 I summarize the common wisdom gained with this work. In particular, I want to stress the fact that from wathever point of view one looks to transport coefficients, one finds an unbelievably smooth dependence on the independent variables.

TABLE 1

TABLE 1
LENNARD-JONES RESULTS FOR λ' AND η'

A. Thermal conductivity, λ'.

- Triple point[16]:
       $\lambda'(X, N=108, \nu_N =0) = \lambda$ for $X \le 10^9$ K/cm
       moreover λ' is a regular function of X
       $\lambda'(X \approx 0, N, \nu_N =0) = \lambda$ $N \ge 108$

B. Shear viscosity, η'.

- Supercritical region, $\rho=0.7$, $T=2.75$ (reduced units)[17-18]:
       $\eta'(X, N=108, \nu_N =0) = \eta$ for $X \le X_0 = 2 \cdot 10^{12}$ sec^{-1}
       moreover for $X > X_0$ one finds $\eta'=1.292 - 0.13\, X^2$, i.e. an
          analytical dependence on the shear rate
       $\eta'(X \approx 0, N, \nu_N =0) = \eta$ $N \ge 108$
 	
       - Triple point[19]:
       no long time tails (the so-called molasses);
       $\eta'(X, N=2048, \nu_N = 10^{13}$ sec$^{-1}) = \eta$ $X \le X_0 = 2 \cdot 10^{10}$ sec^{-1}
       moreover the dependence on the shear rate is analytical
       $\eta'(X \approx 0, N, \nu_N =0) = \eta$ $N \ge 864$
       $\eta'(X \approx 0, N=864, \nu_N) = \eta$ $\nu_N \le 5 \cdot 10^{13}$ sec^{-1}.

5. CONCLUSIONS

NEMD has proved to be a powerful tool to compute transport coefficients of fluids and to simulate a large variety of nonequilibrium processes. The various methods introduced are independent and permit one to explore regions in the strength of the perturbation that have systematically non-zero superpositions, see fig. 4. Even the zero-perturbation limit can be studied by NEMD, using the dynamic approach implemented with the subtraction technique. The good consistence always found when using different methods, including Green-Kubo formulas, gives a very convincing and stringent argument in favour of the validity of NEMD methods.

ACKNOWLEDGEMENTS

Methods and results discussed in this paper have been obtained in collaboration with A. Bellemans, B. L. Holian, G. Jacucci, I. R. McDonald,

C. Massobrio, G. V. Paolini and more particularly M. Ferrario and J. P. Ryckaert. It is a pleasure to thank them all.

REFERENCES

[1] B. J. Alder, D. M. Gass and J. E. Wainwright, J. Chem. Phys. **53**, 3813 (1970)

[2] D. Levesque, L. Verlet and J. Kürkijarvi, Phys. Rev. **A7**, 1690 (1973)

[3] L. P. Kadanoff and P. C. Martin, Ann.Phys. (N.Y.) **24**, 419 (1963)

[4] S. Nosé, Mol. Phys., **57**, 187 (1986)

[5] B. L. Holian and D. J. Evans, J.Chem. Phys. **83**, 3560 (1985)

[6] J. M. Luttinger, Phys. Rev. **135**, A 1505 (1964)

[7] P. Mazur, in: *Cargèse Lectures in Theoretical Physics*, B.Jancovici, ed., Gordon and Breach, N.Y. (1966)

[8] M. J. Gillan, AERE, Harwell Rep. R9332 (1978)

[9] W. G. Hoover, D. J. Evans, R. B. Hickman, A. J. C. Ladd, W. T. Ashurst and B. Moran, Phys. Rev. **A22**, 1690 (1980)

[10] D. J. Evans, Phys. Lett. **A91**, 457 (1982)

[11] D. J. Evans and G. P. Morris, Comp. Phys. Rep. , 297 (1984)

[12] R. Kubo, J. Phys. Soc. Jpn. **12**, 570 (1957)

[13] B. L. Holian and D. J. Evans, J. Chem. Phys. **83**, 3560 (1985)

[14] G. Ciccotti and G. Jacucci, Phys. Rev. Lett. , 789 (1975)

[15] G. Ciccotti, G. Jacucci and I. R. McDonald, J. Stat. Phys. **21**, 1 (1979)

[16] G. V. Paolini, G. Ciccotti and C. Massobrio, Phys Rev. **A34**, 1355 (1986)

[17] J. P. Ryckaert, A. Bellemans, G. Ciccotti and G. V. Paolini, Phys. Rev. **A39**, 259 (1989)

[18] J. P. Ryckaert, A. Bellemans, G. Ciccotti and G. V. Paolini, Phys. Rev. Lett. **60**, 128 (1988)

[19] M. Ferrario, G. Ciccotti, J. P. Ryckaert and B. L. Holian, to be published.

LATTICE GAS AUTOMATA:

A NEW APPROACH TO THE SIMULATION OF COMPLEX FLOWS

Jean Pierre BOON

Faculté des Sciences, CP 231
Université Libre de Bruxelles
B-1050 Bruxelles, Belgium

Turbulence is a complicated phenomenon which has remained one of the last mysteries of classical physics. This quotation from a 1969 paper by Elliott Montroll [1] has kept a substantial part of its truth: twenty years later —and one century after Osborne Reynolds established his famous *dynamical similarity law* (1883)— the mystery of turbulence is still far from being fully unveiled. Yet, over the past two decades, theoreticians and experimentalists, physicists and mathematicians, have accomplished considerable progress towards the understanding of complex flow phenomena. This achievement relies to a large extent on the formidable development of modern computational tools [2]. In particular very interesting perspectives have appeared since 1985 for highly parallel computation in fluid dynamics, based on a thoroughly new method: *lattice gas hydrodynamics*. Two laboratories in France (l'Observatoire de l'Université de Nice, and l'Ecole Normale supérieure de Paris) and one in the United States (the Center for Nonlinear Studies at Los Alamos) have pioneered this research, which has attracted the interest and the collaborative efforts of several groups in the U.S. and in Europe (Shell Research Laboratorium Amsterdam, le Groupe de Physique Non–linéaire de Bruxelles, Politecnica Madrid, le Centre de Recherches en Combustion de Marseille, IBM Research Rome, etc...). The development of lattice gas hydrodynamics has been technically connected to cellular automata but, as it has been pointed out [3], lattice gases should not be referred to as cellular automata. In many respects, lattice gases have become a field *per se* with strong interfaces with kinetic theory, algorithmics, computer and supercomputer simulations, and dedicated machine development. Many of these aspects are presented and discussed in two recent volumes devoted to the subject [4,5]. In the present paper we shall restrict ourselves to a review of lattice gas automata and illustrative examples of their application to fluid dynamics.

The microscopic description of a fluid system involves the virtually infinite complexity of the many-body problem, which can be bypassed by statistical mechanical methods. At large scale —that is for wavelengths large compared to the molecular size— the fluid can be treated as a continuous medium and is therefore adequately described by classical hydrodynamics. Now, complexity is also reflected at large scale by the nonlinearities in the hydrodynamical equations, which, except for particular (usually oversimplified) cases, cannot be solved explicitly. The connection between microscopic level —the domain of *molecular dynamics*— and macroscopic level —the domain of *hydrodynamics*— is established by Liouville-Boltzmann *kinetic theory* [6]. Correspondingly, three computational approaches have been developped for the numerical study of fluid dynamics.

(i) The approach via continuous medium description is to solve numerically the Navier-Stokes equations, which raises the usual difficulties associated with the numerical treatment of partial differential equations. In practice, feasibility is achieved by finite elements methods and finite difference equations; these methods, which use quite involved numerical techniques, have produced spectacular results. However, they require considerable computational power and so turn out to be rather prohibitive.

(ii) The molecular dynamics approach starts from a microscopic modeling of the fluid, simulating a real system of interacting particles. This method has been used extensively for studying thermodynamic and transport properties as well as small scale dynamical behavior of fluid systems[6,7]. Recently it has been extended to investigate systems subject to external constraints [8,9]. The major difficulty here arises from the ratio of time scales and spatial scales, i.e. the ratio of the characteristic hydrodynamic time versus the molecular interaction time, and the ratio of hydrodynamic wavelength versus intermolecular potential range. Both quantities assume large values; as a result molecular dynamics simulations require long computation times and large systems (i.e. large number of particles), and consequently costly computational means.

(iii) Quite recently, the development of an "oversimplified version" of the molecular dynamics approach has been stimulated by progress and perspectives in parallel computers.Similarly as for molecular dynamics simulations, the prediction of flows in fluids will follow from a microscopic description of interacting particles, but here the particles are confined to points moving along the links of a regular lattice, and interactions reduce to simple mathematical rules. The motivation for using a *lattice gas* (in fact, a well known model system in Statistical Physics) to simulate hydrodynamics stems from the idea that the details of the microscopic properties should be unimportant to the macroscopic behavior of the fluid. So whether the fictitious microworld one uses is a caricature of a real fluid does not matter as long as it produces correct hydrodynamics.

In a sense, the lattice gas approach to hydrodynamics simulation appears as intermediate between the two other numerical methods ((i) and (ii) above) in the way kinetic theory establishes the connection between molecular dynamics and hydrodynamics: the hydrodynamic equations can be obtained from kinetic theory by multiscale expansion. However, the question arises as to how a kinetic model can be constructed to simulate hydrodynamics? Such a program will require to define (i) proper mathematical objects, and (ii) appropriate rules governing them.

THE LATTICE GAS

In constructing the lattice gas model, one introduces a major simplification (of considerable computational convenience) by discretizing space (point particles on a lattice), time, and velocity. Each node on the lattice will behave as a Boolean processor updated at each time step according to the rules "connecting" neighboring nodes (via the lattice links), which rules must satisfy *conservation laws* (mass, i.e. particle number; momentum; and energy). Such a system bears resemblance to a *Cellular Automaton* [10] with interactions restricted to first neighbors according to a set of *collision rules* to be specified.

A 2-D square lattice model was first proposed by Hardy, de Pazzis, and Pomeau (HPP) [11] in the mid-seventies to investigate the ergodic problem and

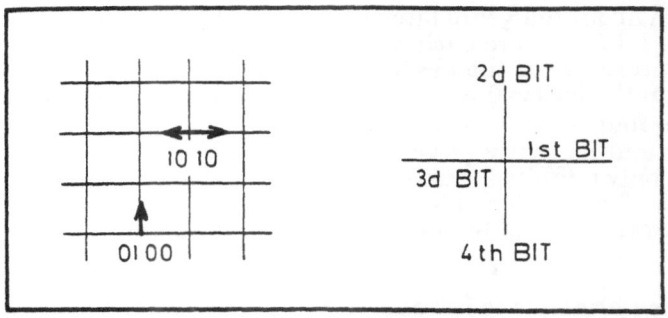

Fig.1. b-bit word representation of node state (here $b = 4$)

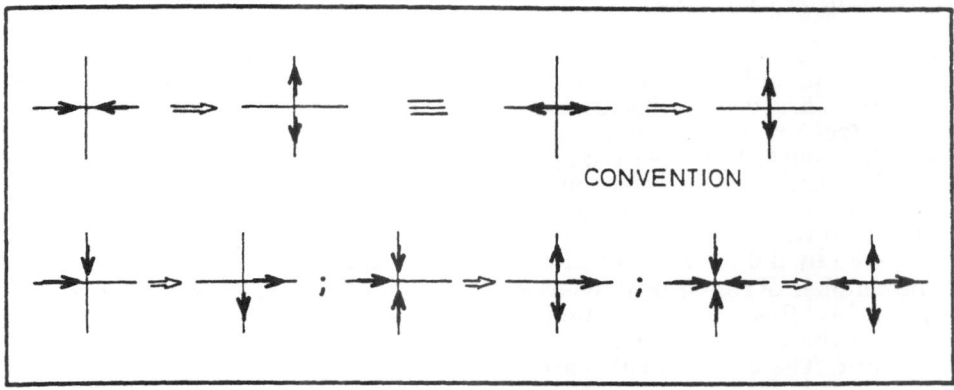

Fig.2. Collisions for HPP model. Note that only collisions of the first type are efficient (i.e. produce momentum transfer).

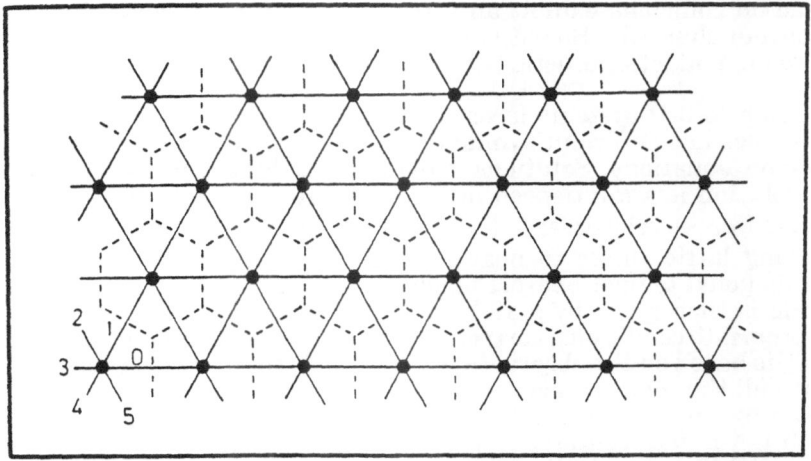

Fig.3. Triangular lattice with hexagonal symetry for FHP model. The unit area around each node is $\sqrt{3}/2$ (for link length = 1).

was reactivated about ten years later to simulate hydrodynamics [12]. Consider a plane square lattice where each node has its state defined by a 4-bit word to represent the presence (or the absence) of particles with discrete velocities (1 = particle with unit velocity; 0 = no particle) on each of the four links connecting the node to its four neighbors (see Fig.1). So each node has 2^4 possible input configurations and as many possible output configurations, which yields 16^{16} possible rules, only a limited number of them being acceptable according to conservation laws. The collision rules are shown in Fig.2; note that the convention of outgoing arrows is usually adopted to indicate to which node particles are associated.

An *exclusion principle* is introduced in that no two particles with same velocity $\bar{C}_i$ can occupy simultaneously the same link (or site i = 0,1,2,3). Obviously, collision rules must be constructed so as to satisfy conservation laws, i.e. number of particles and momentum. Note that energy conservation is degenerate here ($C^2 = 1$), which is unimportant for incompressible isothermal flows. (Such a lattice gas is a model fluid with equal specific heats, $C_P = C_V$, and equal compressibilities, $X_S = X_T$).

For symmetry reasons that will become clear below, the square lattice is not appropriate for hydrodynamic simulations. The two other basic regular two–dimensional tiling geometries are the triangular lattice and the hexagonal lattice (which in fact are reciprocal to each other). Frisch, Hasslacher, and Pomeau (FHP, [13]) proposed to use a triangular lattice with hexagonal symmetry which –as we shall see– has sufficient symmetry to ensure the required isotropy in the macroscopic equations of fluid mechanics. Here each node has regular hexagonal neighborhood (i.e. 6 links and 6 first neighbors, Fig.3). So the state of a node will be given by a 6 (or 7, if one allows for rest particles at the nodes)–bit word, and the number of configurations associated to one node is 2^6, which yields 64^{64} possible rules. Restriction to a limited number follows from conservation laws; in addition the exclusion principle and efficiency of collisions are to be taken into account. The collision rules are illustrated in Fig.4.

Although the probability of actual triple collisions in a real gas is quite small compared to the probability of binary collisions, triple collisions are very important here. Indeed head-on collisional processes conserve particle number and momentum, but also difference in particle number in opposite directions, which yields a total of 4 conservation laws in a 2-D system! Therefore triple collisions are crucial in that they remove the spurious invariant. On the other hand, head-on collisions exhibit an interesting feature because they have two possible output channels. So with a random choice of output configurations, the model is made nondeterministic (while preserving mirror-symmetry).

Extension to 3-D systems faces the problem that none of the fourteen 3-D Bravais lattices has sufficient symmetry to produce the required isotropy of the hydrodynamic equations. Solutions to bypass this difficulty have been proposed. The first solution is a multispeed model on a cubic lattice, where particles can have three different velocities: $0, 1, \sqrt{2}$, zero for particles at rest, one for particles moving along lattice links to nearest neighbor, and $\sqrt{2}$ for particles moving along the diagonal to next-nearest neighbor. As represented in Fig.5a, the state at one node is then given by a 19-bit word. This model is shown [14] to yield, under appropriate conditions, correct inviscid isotropic hydrodynamics. A second model [14], is based on the observation that in 4-D, there exists a regular Bravais lattice with all the required symmetries. Indeed the 24-hedron, with 24 vertices represented by the Schläfli symbol $\{3, 4, 3\}$ can be used to tile regularly the 4-D space with a 4-D face-centered-hypercubic lattice (FCHC). A 3-D projection of the 4-D FCHC (i.e. one lattice site wide in the 4th dimension) produces a 3-D lattice with the required symmetries. A representation of the 3-D projected FCHC is given in Fig.5b. This is a single-speed model, with all lattice nodes

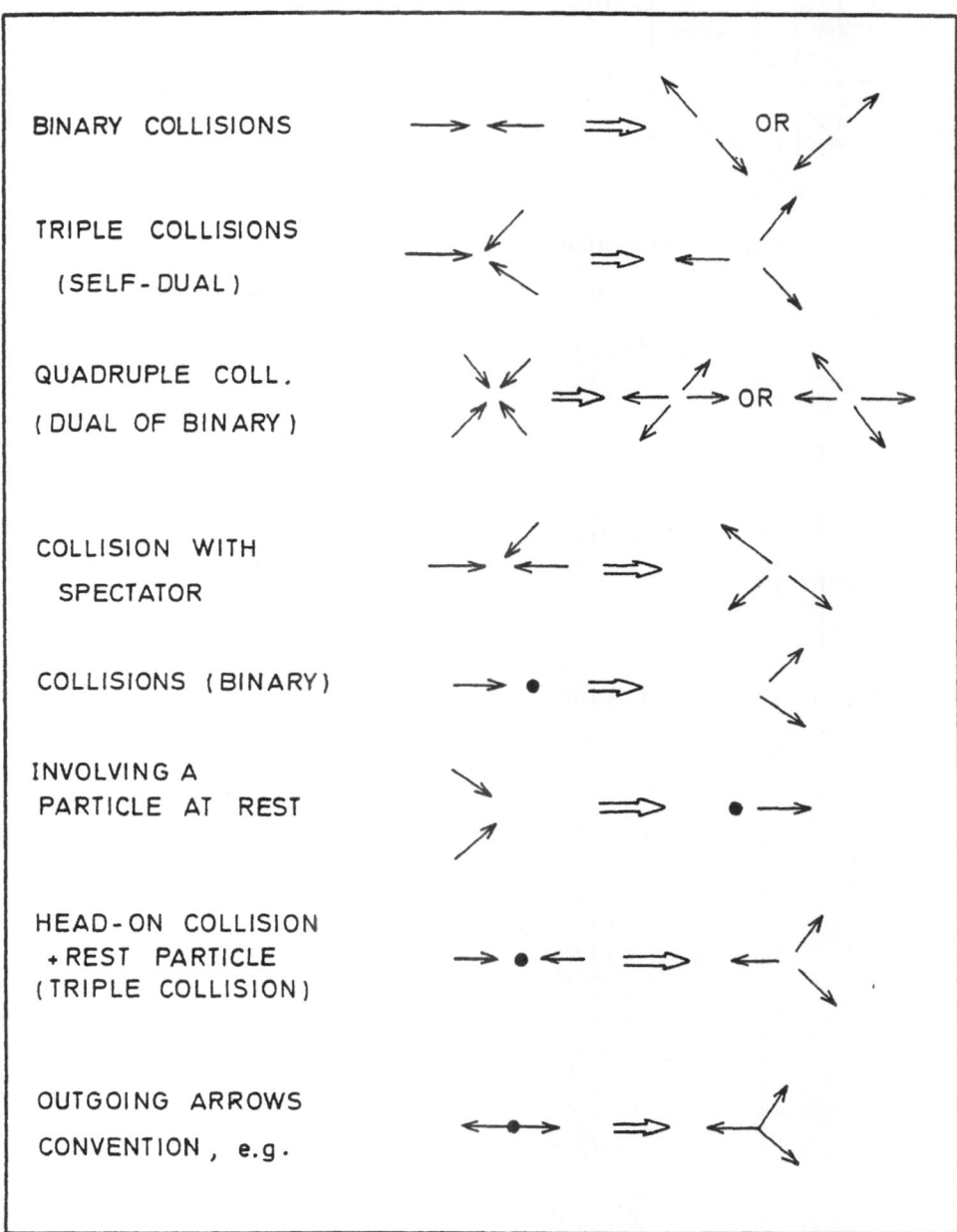

BINARY COLLISIONS

TRIPLE COLLISIONS
 (SELF-DUAL)

QUADRUPLE COLL.
(DUAL OF BINARY)

COLLISION WITH
 SPECTATOR

COLLISIONS (BINARY)

INVOLVING A
PARTICLE AT REST

HEAD-ON COLLISION
+REST PARTICLE
(TRIPLE COLLISION)

OUTGOING ARROWS
CONVENTION, e.g.

Fig.4. Collision rules for FHP model

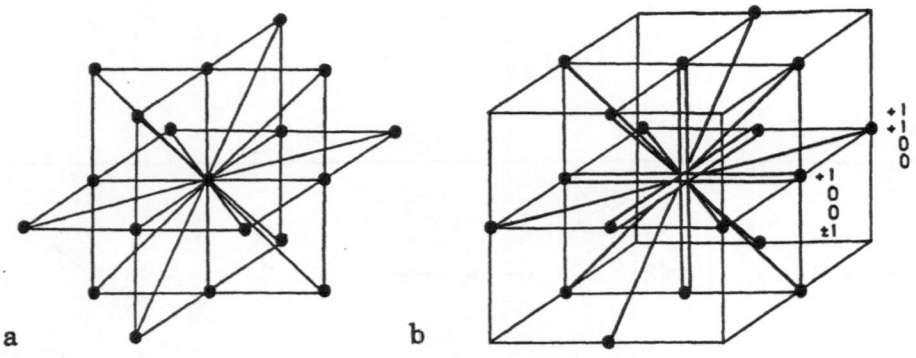

a   b

Fig. 5. 3-D models for lattice gas hydrodynamics.

(a) multispeed model on cubic lattice;

(b) 3-D projection of 4-D FCHC.

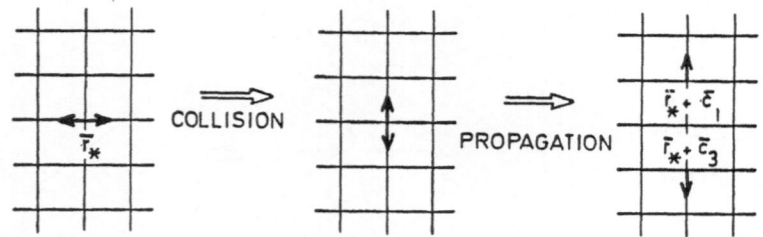

Fig. 6. Example of uptdating rule for the HPP model (Eqs.1-3) :

$$n_i(t* + 1, \bar{r}* + \bar{C}_i; i = 1, 3) = 0 + 1 - 0 = 1$$

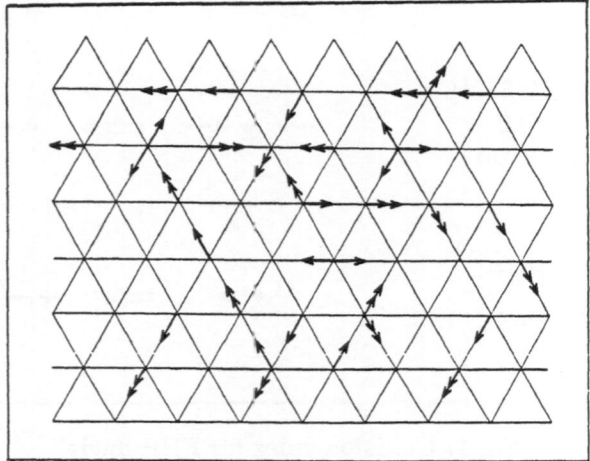

Fig.7. Example of evolution from time $t*$ (single arrow) to time $t*+1$ (double arrow) on the hexagonal lattice gas, according to the rules of Fig.4. (Note the convention of outgoing arrows in the collision representation).

connected via links with unit length $C = \sqrt{2}$. The state at each node is given by a 24 (or 25 if particles at rest are included)-bit word.

FROM MICRODYNAMICS TO MACRODYNAMICS

One defines the state of a node at time t_* and position $\bar{r}_*$ on the lattice $\mathcal{L}$ by the Boolean field:

$$n.(t_*) = \{n_i(t_*, \bar{r}_*); \bar{r}_* \in \mathcal{L}\}$$

where i denotes the site bit (direction) and $*$ indicates discrete variables. The updating rule for the cellular automaton follows from a 2-step process: collision, followed by propagation.

(1) Collision: the state of site i after collision is given by its state before collision minus the depopulating contribution plus the populating contribution;

$$n_i' = n_i + \Delta_i \tag{1}$$

As a simple example, Δ_i , for the HPP square lattice model, reads

$$\Delta_i = -n_i n_{i+2} \bar{n}_{i+1} \bar{n}_{i+3} + n_{i+1} n_{i+3} \bar{n}_i \bar{n}_{i+2}$$

where $\bar{n}_i = 1 - n_i$.

(2) Propagation: after collision particles are shifted one unit lattice length over one time step, so that the complete evolution is given by

$$n_i(t_* + 1, \bar{r}_* + \bar{C}_i) = n_i(t_*, \bar{r}_*) + \Delta_i(t_*, \bar{r}_*) \tag{2}$$

These are the *microdynamical equations* governing the lattice dynamics. Operationally the system is handled as a cellular automaton: at each time step the output state is computed from the input state according to the specific rules. Fig.6 illustrates a simple case for the HPP model. An example of global evolution after one time step for the FHP model is shown in Fig.7. These considerations are easily generalized to a b-nearest-neighbor D-dimensional lattice, using a set of $\bar{C}_i (i = 0, \ldots, b-1)$ vectors with components $C_{i\alpha}(\alpha = 1, \ldots, D)$ and modulus C. (For the HPP model: $b = 4, D = 2, C = 1$).

As usual in Statistical Mechanics, we go to a probabilistic description [15] by defining the phase space Γ as the set of all possible assignments (or configurations) $s(.) = \{s_i(\bar{r}_*)\}$ $(i = 0, \ldots, b-1)$ of the Boolean field $n_i(r_*)$. $P(t_*, s(.))$ is the probability, at time t_*, for assignement $s(.)$ with

$$\sum P(t_*, s(.)) = 1, \ (s(.) \in \Gamma)$$

Starting from an ensemble of initial conditions, each configuration evolves according to the updating rules of the automaton. This is formally expressed by the "*Liouville equation*"

$$P(t_* + 1, \mathcal{E}s(.)) = P(t_*, s(.)) \tag{3}$$

with $\mathcal{E}$, the evolution operator, which can be written as $\mathcal{E} = \mathcal{S} \circ \mathcal{C}$, where $\mathcal{S}$ is the streaming operator and $\mathcal{C}$, the collision operator. In order to indicate explicitly the two-step sequence of the automaton evolution, one can rewrite Eq.(3) as

$$P(t* + 1, \mathcal{S}s(.)) = P(t*, \mathcal{C}^{-1}s(.)) \tag{4}$$

For non-deterministic systems, the description must be generalized so as to include all possible choices of the Boolean variables $\xi_{ss'}$, giving the transition selection from state s to state s'. Each transition being assigned a probability $A(s \to s')$, one has

$$\langle \xi_{ss'} \rangle = A(s \to s'), \forall s, s'$$

in accordance with the "semi-detailed balance" assumption

$$\sum_s A(s \to s') = 1, \ \forall s' \tag{5}$$

which expresses that, if all states have equal probabilities before collision, they also have equal probabilities after collision. Given that the ξ 's take values independent of each other at each time, defining in this way a Markov process, and given that the ξ values may be assumed to be independently chosen at each node, Eq.(4) becomes

$$P(t* + 1, \mathcal{S}s(.)) = \sum_{s(.)} \prod_{\bar{r}*} A(s(\bar{r}*) \to s'(\bar{r}*)) \, P(t*, s(.)) \tag{6}$$

This *Master Equation* describes the evolution of a probabilistic cellular automaton by expressing the probability for a (propagated) configuration s'(.) in terms of all possible initial configurations s(.) weighted by the transition probabilities. Note that in the deterministic case, $A(s \to s')$ reduces to $\mathcal{C}^{-1}s'(.)$, and Eq. (6) becomes simply the Liouville equation (4).

With the transition variable $\xi_{ss'}$, we can rewrite the microdynamical equation as [15]

$$n_i(t* + 1, \bar{r}* + \bar{C}_i) = n_i(t*, \bar{r}*) + \Delta_i$$

$$\Delta_i = \sum_{s,s'} (s_i' - s_i)\xi_{s,s'} \prod_j n_j^{s_j} (1 - n_j)^{1-s_j} \tag{7}$$

Having introduced a probabilistic description, we define mean quantities by ensemble averaging over P; in particular the average population is defined as the occupancy probability of site i at node $\bar{r}*$ at time $t*$ i.e.

$$N_i(t*, \bar{r}*) = \langle n_i(t*, \bar{r}*) \rangle \tag{8}$$

It follows from conservation of mass and momentum that

$$\sum_i [N_i(t* + 1, \bar{r}* + \bar{C}_i) - N_i(t*, \bar{r}*)] = 0$$

$$\sum_i [\bar{C}_i N_i(t* + 1, \bar{r}* + \bar{C}_i) - \bar{C}_i N_i(t*, \bar{r}*)] = 0 \tag{9}$$

Then, with the Boltzmann approximation that particles involved in a collision have no prior correlations, by taking the ensemble average of Eq.(7), one obtains the *lattice Boltzmann equation*

$$N_i(t* + 1, \bar{r}* + \bar{C}_i) = N_i((t*, \bar{r}*) + \Delta_{i\,Boltz}$$

$$\Delta_{i\,Boltz} = \sum_{s,s'} (s'_i - s_i) A(s \to s') \prod_j N_j^{s_j} (1 - N_j)^{1 - s_j} \qquad (10)$$

The more familiar differential version of (10) is obtained by Taylor expansion of the finite differences to first order to yield

$$\partial_t N_i + \bar{C}_i . \nabla N_i = \Delta_{i\,Boltz} \qquad (11)$$

Next one defines the density and the mass current respectively by

$$\varrho(\bar{r}*, t*) = \sum_i N_i(t*, \bar{r}*) \; ; \qquad \underline{j}(\bar{r}*, t*) = \sum_i \bar{C}_i N_i(t*, \bar{r}*) \qquad (12)$$

or equivalently the density per site $d = \varrho/b$, and the mean velocity $\underline{u} = \underline{j}/\varrho$.

We now sketch how the macroscopic equations are obtained for these averaged quantities (For details, see the articles by Frisch et al [15] and by Wolfram [16]).

Most important is that ϱ and $\underline{j}$ (or d and $\underline{u}$) are slowly varying quantities, that is they vary over a spatial range and on a time scale which are large compared to the microscopic space and time scales. Now the dynamical behavior of the system involves characteristic times related to (i) relaxation to local equilibrium (τ_o); (ii) sound propagation (τ_s), and (iii) dissipation (τ_D), with $\tau_o << \tau_s << \tau_D$. Considering the spatial scale expansion parameter $\mathcal{E}$, τ_s and τ_D will scale as $\mathcal{E}^{-1}$ and $\mathcal{E}^{-2}$ respectively, whereas local equilibrium relaxation is independent of scaling (i.e. $\tau_o \sim \mathcal{E}^o$). So a multiscaling follows with time variables $t*, t_1 = \mathcal{E}t*, t_2 = \mathcal{E}^2 t*$ and space variables $\bar{r}*, \bar{r}_1 = \mathcal{E}\bar{r}*$. Consequently, not too far from equilibrium the population distribution function N_i may be expanded as

$$N_i = N_i^{(o)}(t, \bar{r}) + \mathcal{E} N_i^{(1)}(t, \bar{r}) + \mathcal{O}(\mathcal{E}^2) \qquad (13)$$

with $N_i^{(o)}$, the equilibrium distribution function [15]

$$N_i^{(o)} = [1 + \exp(h + \underline{q}.\bar{C}_i)]^{-1} \qquad (14)$$

where the Lagrange multipliers h and $\underline{q}$ can be expressed in terms of ϱ and $\underline{j}$. Considering the physical nature of the lattice gas model, it is quite logical that a system with built-in exclusion principle has a Fermi-Dirac equilibrium distribution function (14).

Starting from the conservation equations (9), one can perform a multiscale expansion à la Chapman-Enskog. To first order, O($\mathcal{E}$), one obtains the Euler equations. The next order calculation yields the macrodynamical equations in the form

$$\partial_t \varrho + \nabla(\varrho \underline{u}) = 0 \qquad \qquad \textit{continuity equation} \quad (15)$$

$$\partial_t(\varrho \underline{u}) + \nabla . \mathbf{P} = \nabla . \mathbf{S} + h.o.t. \qquad \textit{momentum equation} \quad (16)$$

where $\mathbf{P}$ is the momentum flux tensor and $\mathbf{S}$ the viscous stress tensor. This result is important because Eq.(16) bears striking resemblance to the Navier-Stokes equation. The question now arises as to whether, the lattice gas constitutes a model fluid that produces correct hydrodynamic behavior; in other words is the macrodynamical equation (16) really Navier-Stokes?

A satisfactory answer to this question would require a technical discussion of the explicit form of the tensorial quantities in (16). This is where the symmetry of the lattice comes into play. In fact for the square lattice, the tensors are not isotropic and so produce anisotropic viscous dissipation. On the other hand the 2-D lattice with hexagonal symmetry (FHP model) and the 3-D projection of the FCHC lattice have sufficient symmetry to ensure the required macroscopic isotropy. Then –when centers (i.e. particles at rest) are included– Eq.(16) reads [15]

$$\partial_t(\varrho u_\alpha) + \partial_\beta(G(\varrho)\varrho u_\alpha u_\beta) = -\partial_\alpha P(\varrho) +$$

$$+\partial_\beta(\nu_s\partial_\beta(\varrho u_\alpha)) + \partial_\alpha\left(\left(\frac{D-2}{D}\nu_s + \nu_B\right)\nabla.(\varrho\underline{u})\right) \tag{17}$$

with

$$P(\varrho) = c_s^2\varrho + \mathcal{O}(G(\varrho)u^2/C^2)$$

$$G(\varrho) = \frac{D}{D+2}\quad\frac{\varrho}{\varrho_m}\quad\frac{1-2d}{1-d} \tag{18}$$

where ν_s and ν_B are the kinematic shear and bulk viscosities respectively, c_s is the sound velocity, and ϱ and ϱ_m the total density and the density of mobile particles.

An important difference between Eq.(17) and the usual Navier-Stokes equation is the presence of the factor $G(\varrho)$. This factor arises as a consequence of space disretization in the lattice gas description. Note that at low speeds it should be unimportant. But it raises a more serious problem as it breaks Galilean invariance. For pure Navier-Stokes flows, Galilean invariance is recovered at the macroscopic level by appropriate scaling:

$$\tau = G(\varrho)t, \qquad \nu = \nu_s\,G^{-1}(\varrho), \qquad p = c_s^2\,\varrho\,G^{-1}(\varrho)$$

However for compressible fluids, and for multicomponent systems, time scaling is different for the momentum equation and for the mass and concentration equations. As a result momentum and mass do not propagate on the same time scale. Solutions to this problem have been proposed [18]; in particular by increasing the number of rest particles, $G(\varrho)$ can be set equal to unity [19], restoring in this way a "pseudo" Galilean invariance.

Among the most interesting applications of the method are the simulations at high Reynolds number, $Re = l_o u\nu^{-1}$, with l_o, the characteristic hydrodynamic length, and ν, the rescaled kinematic viscosity. Obviously high Re simulations are more easily performed with systems with low viscosity. So the optimization of the Reynolds number should be viewed not only in terms of large systems and high speeds, but also in terms of those factors that minimize the kinematic viscosity, that is via an optimization of the collision rules. It has been shown recently by Hénon [20] that, within the Boltzmann approximation, and without semi-detailed balance (see Eq.(5)), "exotic" collision rules can be set up for the FCHC model with additional rest particles, such that the kinematic viscosity becomes extremely small (and even negative). There is a critical density, $d_c \simeq 1/2$, such that when d approaches d_c (by lower values), the viscosity tends to zero (by upper values). Thus by fine-tuning of the viscosity, it should be possible to work at very high Reynolds numbers. To date however, lattice gas simulations of Navier-Stokes flows have not exceeded Reynolds numbers of a few thousands.

Simulations and computations have been performed to check the basic properties of lattice gases, the linear response to small perturbations and the behavior of the transport coefficients. Good agreement is obtained between theory and experiments for sound propagation and for the density dependence of the viscosity coefficients (see references [21] to [24]). A most celebrated problem, known in Statistical Mechanical jargon as "long time tails", concerns the long time persistence of correlations, in particular, in two–dimensional systems, where a logarithmic divergence is predicted for the Green–Kubo integrands, questioning in this way the validity of 2-D hydrodynamics. A direct measurement of such effect is most easily performed for the velocity autocorrelation function and FHP gases are particularly well suited for its computation. Recent careful simulations [25] have indeed shown that the long time tail effect could be detected by direct time measurements and by spectral density analysis of the computation data.

Before presenting illustrative examples of lattice gas simulations of hydrodynamic flows, a few words are of order to briefly describe their implementation. We shall consider the two–dimensional triangular FHP model, with collision rules including particles at rest, a system that has been widely used for 2-D hydrodynamic simulations. The state of the system at time $t*$ is given by a L_1 x L_2 matrix (size of the cellular automaton universe) of 6 (or 7)-bit words assigned to each node. The bit value 1 or 0 reflects the presence or the absence of a particle at site i(= 0, ..., 5 or 6) with velocity C_i. Updating the universe is performed by "solving" the microdynamical equations (collisions + propagation) by a sequence of logical operations. Boundary conditions and initial conditions are set according to the problem studied. For instance, in a sound propagation experiment, a uniformly random distribution of particles and velocities is realized as initial condition, and periodic boundary conditions are imposed, which confine the system on a torus (particles escaping the universe at one boundary are reinjected symetrically at the opposite boundary). On the other hand, a directed flow simulation experiment requires an initially biased velocity distribution along a given direction, with boundary conditions ensuring steady incoming flow of particles at the input side and constant density condition at the output boundary. In experiments such as channel flow and flow behind obtacles, their shape is designed according to the lattice geometry by specific collision rules, with reflection conditions corresponding to free-slip boundaries (specular reflection; Fig.8a), no-slip boundaries (bounce–back reflection, Fig.8b), or rough surfaces (combination of specular and bounce-back reflections with equal probabilities, Fig. 8c). The obstacle size l_o must be small compared to the size L of the CA Universe in order to avoid artefacts. In turn large L implies large numbers of particles.

Presently typical 2-D lattices are of the order of 3×10^6 nodes (e.g. 1024 x 3072) populated with 6×10^5 particles, i.e. with a density $d \sim 0.2(\varrho \sim 1.4)$. Streamline maps are obtained by representing the velocity field vectors associated to the fluid elements, i.e. by averaging the particle velocities over a number of nodes (e.g. 8 x 8, 32 x 32,...depending on the problem under investigation). Technical restrictions as to the universe size, the minimal kinematic viscosity, and the velocity u (which must be small compared to the upper limit C) are determinant; within these limits, presently achievable flows are for Reynolds numbers not exceeding a few thousands. As discussed in the previous section, it is to be expected that fastly progressing developments will overcome these limitations in the near future.

Poiseuille flow in a channel. Flow at the inlet of a 2-D duct was simulated by d'Humières and Lallemand [26] on a 512 x 3072 FHP lattice gas with $d = .22$ and average velocity $u = .30$. Velocity profiles so obtained are presented in Fig.9 for the region close to the input boundary (Fig.9a) and for a region located

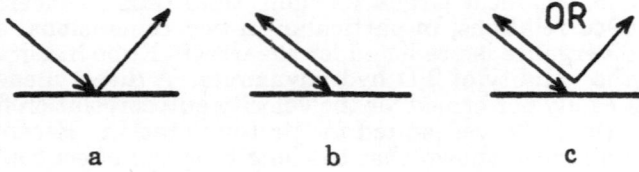

Fig.8. Boundary reflections : (a) free-slip; (b) no-slip; (c) combination of (a) and (b).

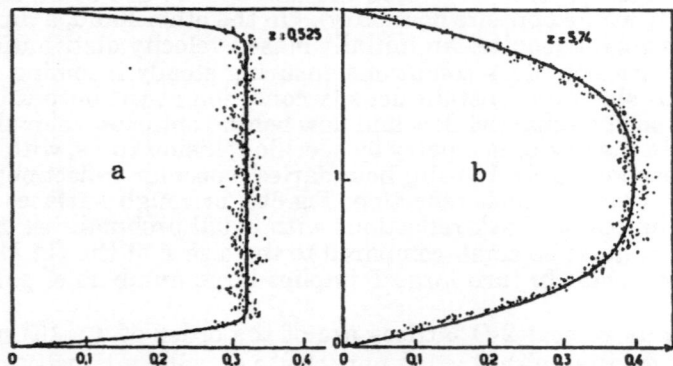

Fig.9. Velocity component profile in a channel (a) close to the inlet, and (b) further downstream (relative distances from inlet are $\sim$ .5(a) and $\sim$ 6. (b)).

(d'Humières and Lallemand, 1986).

about ten times further downstream (Fig.9b) where a characteristic Poiseuille profile has developped. Fig.9 shows good agreement between the simulation and the profiles computed by the Slichting method. This was the first example of quantative comparison between lattice gas flow and classical fluid mechanics for a hydrodynamic system involving both viscous dissipation and non–linear behavior.

von Karman streets. Among the first hydrodynamic flow "experiments" on a lattice gas were those performed in 1985 by d'Humières, Lallemand, Shimomura, and Pomeau [21,27], then by Salem and Wolfram [28], to simulate wakes behind a plate. An example of von Karman street developping behind a flat plate at Re = 300 is given in Fig.10, where two successive velocity maps are shown, indicating the unsteady nature of the flow. Similar experiments have also been performed to simulate flow around a stationary cylinder.

Flow around a wing profile. The example in Fig.11 showing streamlines around an airplane wing profile at various inclinations with respect to the mean flow direction illustrates the ability of the method to conveniently realize more complex obstacle shapes and modify their orientation without computational difficulty.

Channel flow in expanded geometry. This phenomenon was studied by simulating flow in a channel with sudden expansion (Fig.12) where recirculation (back flow) takes place behind the step profile. Isomach curves map the velocity field at Reynolds numbers Re=50 (Fig.12a) and Re=150 (Fig.12b); the latter is also shown on an expanded scale (Fig.12c) along with the corresponding isodensity curves (Fig.12d). The arrow in Fig.12b indicates the location of the reattachment point as evaluated from Navier-Stokes finite elements finite differences computations; as seen, good agreement is obtained.

Boundary layer. The evaluation of the velocity profile in a fluid flow near solid boundaries is an important problem e.g. for moving obstacles and for Brownian motion. Two lattice gas approaches have been developed to investigate boundary interactions: simulation via microdynamics and computation from the lattice Boltzmann equation[30]. Both methods yield time evolution of the velocity profile in agreement with phenomenological hydrodynamic theory.

RECENT DEVELOPMENTS

Important progress has been realized recently in lattice gas simulations of hydrodynamic flows to study problems like jets in periodic channels, flame fronts, and the Kelvin–Helmoltz and Rayleigh–Taylor instabilities [29]. Such problems involve the introduction of two–species particles. In this respect, current research activity on lattice gas hydrodynamics has led to interesting variations of the FHP model. One of the most promising versions is a two–species model where a "color" bit is added to the particles. The automaton is then a 14-bit model which uses the FHP collision rules between identical particles, but additional rules for color exchange between colliding particles of different species [17]. If "color" is to be conserved during collisions, the model is for mutual diffusion of two equivalent non reacting gases. As the two species are then perfectly miscible, the system eventually reaches homogeneous state. On the other hand, interfaces can form between different species if reactive collisions change the relative number of particles of each type. A simple chemical reaction is described by the majority rule for autocatalytic transformation:

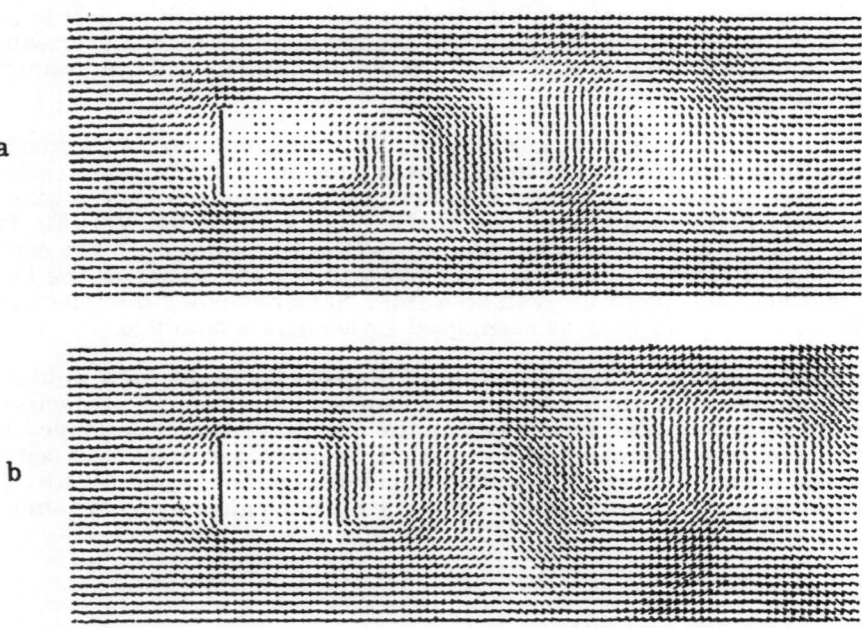

Fig.10. von Karman street formation behind a "bounce-back" (see fig.8) flat plate in 512 x 1024 CA wind tunnel experiment at $Re = 70$; time (b) = time (a) + 1500 time steps. (d'Humières and Lallemand, 1986)

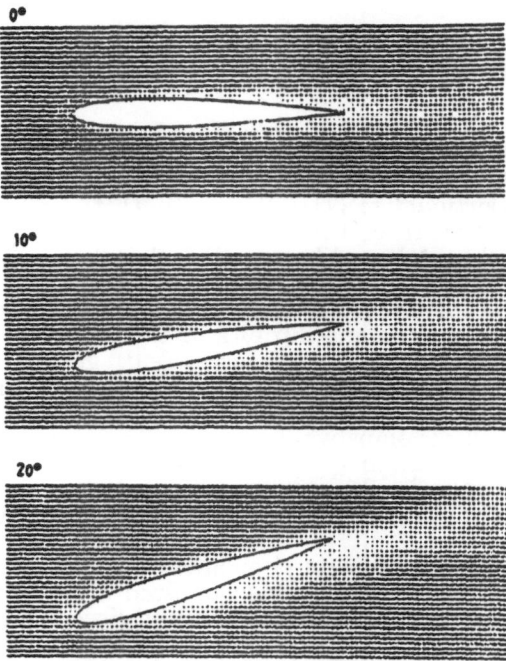

Fig.11. Flow around an airplane wing at various inclinations with respect to the direction of average input flow. (Lallemand and d'Humières, 1986).

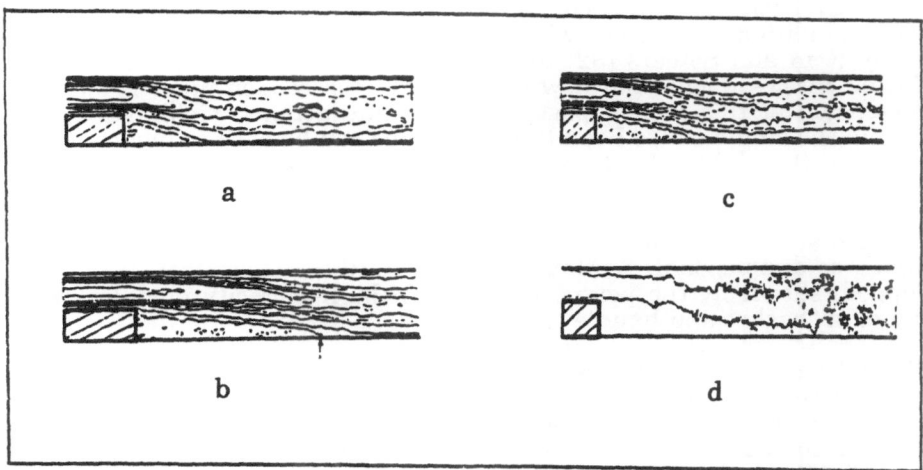

Fig.12. Channel flow with sudden expansion. Isomach curves for velocity component along the average input flow direction at $Re = 50$(a) and $Re = 150$ (b,c); $\triangle u = 10\%$. Isodensity curves (at $\varrho = .95\ \varrho_{max}$) at $Re = 150$(d). (Note 2x expanded spatial scale in c and d). Arrow (in b) indicates reattachment point. (Noullez, Lallemand, and d'Humières, 1986)

$$2A + B \xrightarrow{k_1} 3A$$

$$A + 2B \xrightarrow{k_2} 3B$$

Such a transformation rule induces a phase separation between A-dominant and B-dominant regions. These regions are separated by interfaces whose stationary length is proportional to $(D/k_i)^{1/2}$, with D, the mutual diffusion coefficient, and k_i the reaction rate. Different collision rules can be used to model other types of chemical reactions, like e.g. diffusion flames [19].

Body-forces can be introduced by including collision rules that do not conserve momentum. These collisions flip bits in the required direction with the correct probability to simulate external forces (e.g. gravity effects). With different propabilities for different particle species, gravitational instabilities can be simulated. As an example, Fig.13 shows the 2-D simulation of the Rayleigh-Taylor instability, which develops when a heavy fluid penetrates a lighter fluid layer [29]. Another illustration is the Kelvin-Helmoltz instability where two fluid layers moving in opposite directions with respect to each other, develop, by shear constraint, a roll up at the interface, as shown in Fig.14.

Viscous fingering is a complicate phenomenon which arises as the result of instabilities that occur when a dense viscous fluid filling a porous material is forced to be displaced by the local injection of a less viscous fluid. This instability is analogous to the Saffman-Taylor instability observed in Hele-Shaw cells; yet in porous media the underlying mechanism is basically two dimensional and so the phenomenon is well suited for lattice gas investigation [31]. Here by selecting appropriate collision rules for the two fluids, their respective viscosities can be set to a ratio fo ~ 33. The porous medium can be simulated by a random spatial distribution of static scattering nodes. The low viscosity fluid is then pushed through the medium initially filled with the highly viscous fluid. Viscous penetration develops as shown in Fig.15 with a characteristic wavelength of the order of a fraction of the transverse size of the lattice.

Most promising are the recent developments of 3-D lattice gases. As mentioned earlier, work is still very much in progress, but spectacular realizations have already been accomplished. A variation of the FCHC model, including boundary conditions consistent with the Navier-Stokes equation, has been designed by Rem and Somers [32] to run on a transputer network machine for simulating three–dimensional flow in a square crosssection pipe. A two-phase flow extension of this model, based on local interactions, yields surface tension, while the single-phase boundary conditions can be used to model adhesive surfaces and thereby to study wetting processes. Among the most recent realizations are the simulations of fully three-dimensional flows. Using the FCHC model with improved collision rules, Rivet et al [33] studied external flow past a circular plate at Re 190. The flow is found to evolve from axi-symmetrical to fully three-dimensional. As illustrated in Fig.16 a 3-D vortical structure has developped, shown by the perspective view of high–vorticity modulus regions displaying a "basket and handle" structure.

Transition to three–dimensionality in external flows is considerably more complicated than transitions in convective internal flows, and in fact, is very poorly understood. Lattice gas methods should in this respect offer attractive possibilities for quantitative studies in 3-D "numerical wind tunnels". Many other applications, that we have just barely mentioned (e.g. lattice gas reactive systems)[36] are now being developped and investigated in detail. Parallel to these efforts are those devoted to the development of dedicated machines. The first realization was the Cellular Automata Machine (CAM) built at MIT [34], a general purpose machine not perfectly suited for hydrodynamic simulations. A more recent realization, the "Réseau d'Automates Programmables" (RAP, Ecole

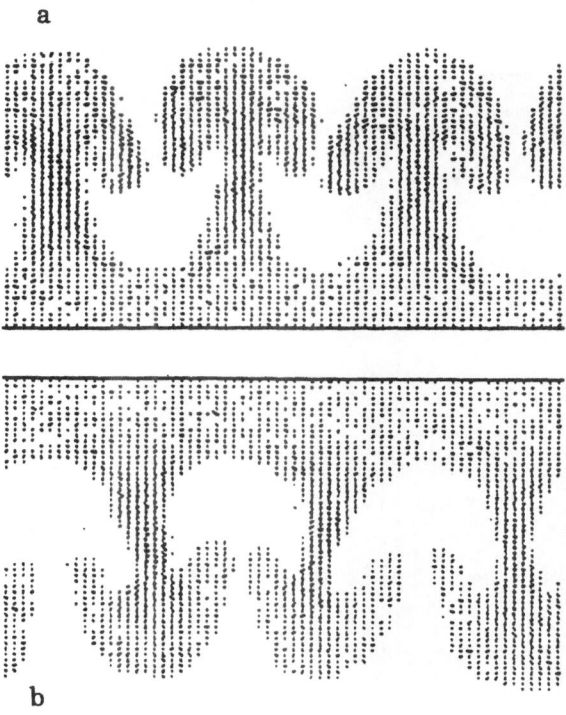

a

b

Fig.13. 2-D lattice gas simulation of the Rayleigh-Taylor instability. Maps of A-particle flux (a) and of B-particle flux (b) after $t = 1600$.

(Clavin, d'Humières, Lallemand, and Pomeau, 1986)

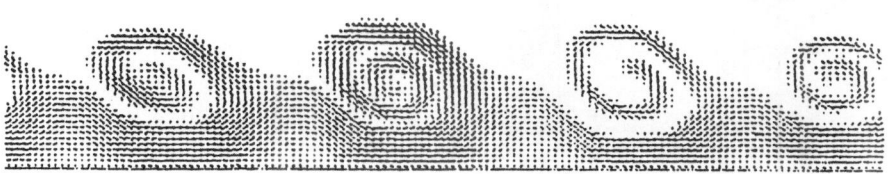

Fig.14. 2-D lattice gas simulation of the Kelvin-Helmoltz instability. Average velocities are $+u$ (left to right) and $-u$(right to left) in lower half and in upper half of channel respectively. Map shows flux of particles of one species (d'Humières, Lallemand, and Searby, 1987)

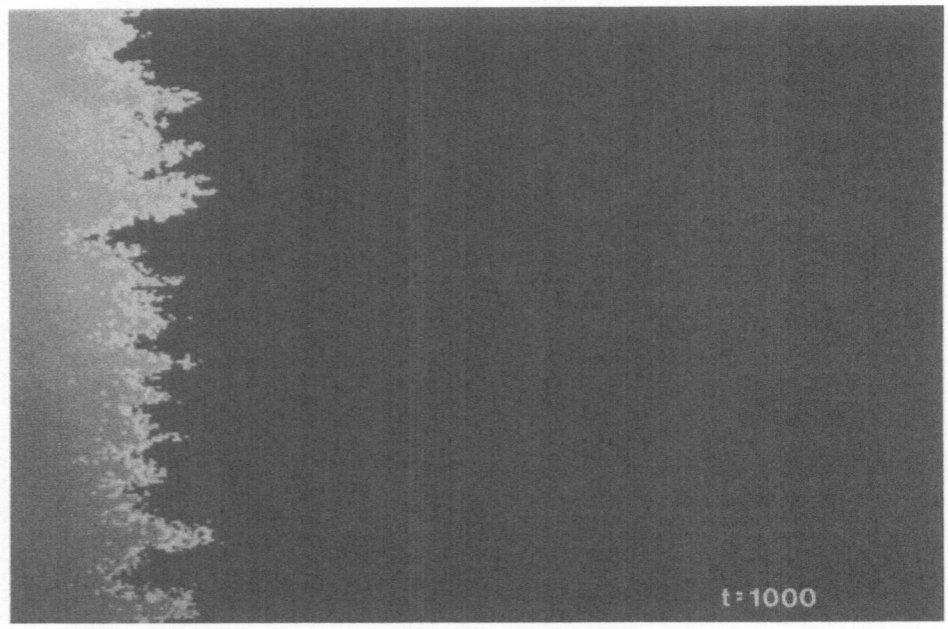

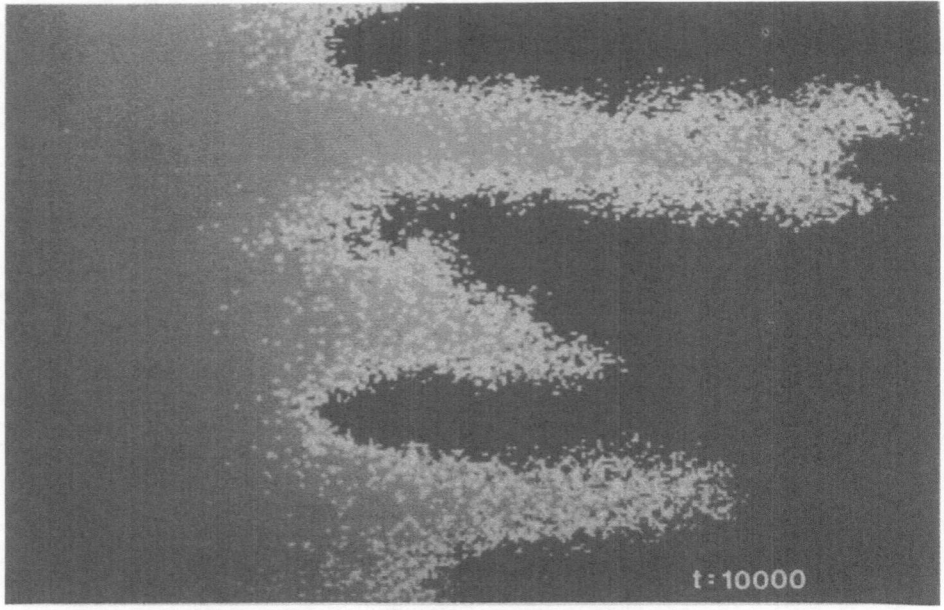

Fig.15. Viscous fingering in 2-D porous lattice. Dark region indicates the low viscosity fluid. Time is given in lattice gas
time step units. Notice tip-splitting at $t = 10000$. (Bonetti,
Noullez and Boon, 1988).

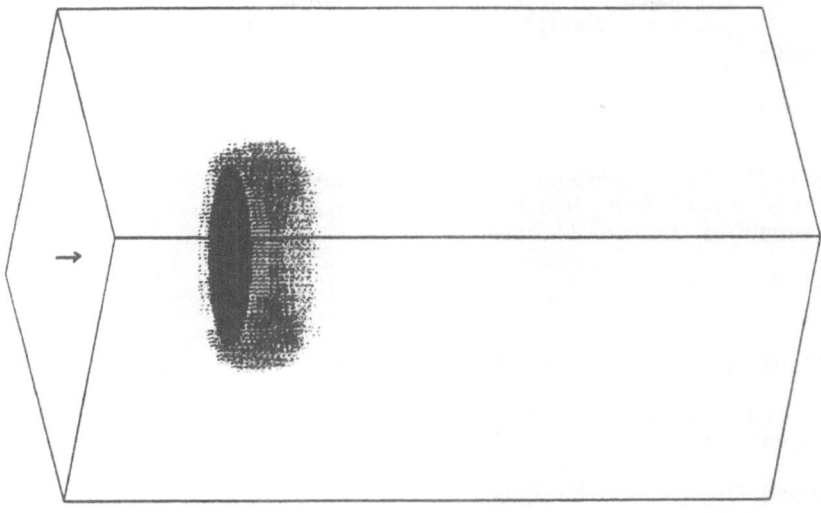

$$t = 400 = 0.62 \ \tau_R$$

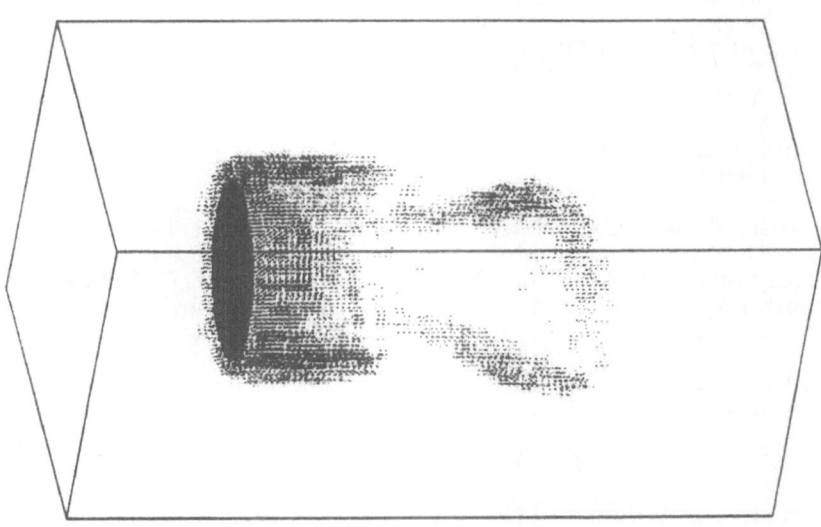

$$t = 4000 = 6.2 \ \tau_R$$

Fig.16. Perspective view of the high vorticity regions developed at $Re = 190$ in a 3-D flow passed a circular plate (shown in black) (Rivet, Hénon, Frisch, and d'Humières, 1988).

Normale Supérieure, Paris) [35] has a slightly different architecture, specially designed for fluid dynamics applications. Alternative techniques based on transputer networks [32] have also proved efficient. Further progress is to be expected in the near future concerning not only these computational techniques, but also the theoretical aspects and the experimental methods developped for lattice gas hydrodynamics.

Acknowledgements We acknowledge support from the "Fonds National de la Recherche Scientifique" (FNRS, Belgium). Part of the work reported here was supported by European Community Grant ST2J–0190.

REFERENCES

1. E.W. Montroll, *Contemporary Physics*, **1**, 273 (1969).

2. See e.g. the special issue of *La Recherche* : "Les nouveaux ordinateurs" (November, 1988).

3. M. Hénon, "On the Relation between Lattice Gases and Cellular Automata", in Reference [5].

4. *Complex Systems* (August, 1987); for a review see also J.P. Boon and A. Noullez "Lattice Gas Hydrodynamics" in *Modern Approaches to Turbulence Modeling*, edited by D. Olivari and P. Bergé, (AGARD, Paris, 1987).

5. *Kinetic Theory, Lattice Gases, and Foundation of Hydrodynamics* edited by R. Monaco, (World Scientific, 1989).

6. J.P. Boon and S. Yip, *Molecular Hydrodynamics* (Mc Graw Hill, New York, 1980); P. Résibois and M. DeLeener, *Classical Kinetic Theory of Fluids* (Wiley, New York, 1977).

7. J.P. Hansen and I.R. Mc Donald, *Theory of Simple Liquids* (Academic Press, London, 1976).

8. M. Mareschal and J.P. Ryckaert, *Physicalia*, **10**, 184 (1988).

9. C. Trozzi and G. Cicotti, *Phys. Rev.* **A29**, 916 (1984); D.C. Rapaport and E. Clementi, *Phys. Rev. Lett.* **57**, 695 (1986); M. Mareschal and E. Kestemont, *Nature* **329**, 427 (1987).

10. S. Wolfram, *Theory and Applications of Cellular Automata* (World Scientific, Singapore, 1986).

11. J. Hardy, O. de Pazzis, and Y. Pomeau, *Phys. Rev.* **A13**, 1949 (1976).

12. N. Margolus, T. Toffoli, and G. Vichniac, *Phys. Rev. Lett.* **56**, 1694 (1986).

13. U. Frisch, B. Hasslacher, and Y. Pomeau, *Phys. Rev. Lett.* **56**, 1505 (1986).

14. D. d'Humières, P. Lallemand, and U. Frisch, *Europhys. Lett.* **2**, 291 (1986).

15. U. Frisch, D. d'Humières, B. Hasslacher, P. Lallemand, Y. Pomeau, and J.P. Rivet, *Complex Systems*, **1**, 648 (1987).

16. S. Wolfram, *J. Stat. Phys.* **45**, 471 (1986).

17. see e.g. D. d'Humières, P. Lallemand, J.P. Boon, D. Dab and A. Noullez, "Fluid Dynamics with Lattice Gases", in *Chaos and Complexity* (World Scientific, 1988).

18. D. d'Humières, P. Lallemand, and J. Searby, *Complex Systems*, **1**, 632 (1987).

19. V. Zehnlé and J. Searby, "Lattice Gas Experiments on a Non–exothermic Diffusion Flame in a Vortex Field", to appear in *J. Physique*.

20. M. Hénon, "Optimization of Collision Rules in the FCHC Lattice Gas", in Reference [5].

21. d'Humières, P. Lallemand, and T. Shimomura, Los Alamos Report LAUR 854051 (1985).

22. J.P. Rivet, and U. Frisch, *C.R. Acad. Sci. Paris*, **302**, 11 (1986).

23. U. Frisch and J.P. Rivet, *C.R. Acad. Sci. Paris*, **303**, 1065 (1986); J.P. Rivet, *Complex Systems*, **1**, 838 (1987).

24. D. d'Humières and P. Lallemand, *Physica* **140A**, 337 (1986).

25. J.P. Boon and A. Noullez, "Lattice Gas Diffusion and Long Time Correlations", "Long Time Correlations in Lattice Gases", preprint 1989; D. Frenkel and M.H. Ernst, "Simulation of Diffusion in a 2-D Lattice Gas Cellular Automaton : a Test of Mode-Coupling Theory", preprint 1989

26. D. d'Humières and P. Lallemand, *C.R. Acad. Sci. Paris*, **302**, 983(1986).

27. D. d'Humières, Y. Pomeau, and P. Lallemand, *C.R. Acad. Sci. Paris*, **301**, 1391 (1985).

28. J.B. Salem and S. Wolfram, in reference [10], p.362.

29. P. Clavin, D. d'Humières, P. Lallemand and Y. Pomeau, *C.R. Acad. Sci. Paris*, **303**, 1169 (1986); d'Humières, P. Lallemand, and G. Searby, *Complex Systems*, **1**, 632 (1987); P. Clavin, P. Lallemand, Y. Pomeau, and G. Searby, *J. Fluid Mech.* (1987).

30. P. Lavallée, J.P. Boon, and A. Noullez, "Boundary Interactions in a Lattice Gas", in Reference [5].

31. M. Bonetti, A. Noullez, and J.P. Boon, "Lattice Gas Simulation of 2D Viscous Fingering", in Reference [5].

32. P.C. Rem and J. Somers, "Cellular Automata on a Transputer Network", in Reference [5].

33. J.P. Rivet, M. Hénon, U. Frisch, and D. d'Humières, *Europhys. Lett.* **7**, 231 (1988).

34. T. Toffoli, *Physica* **100**, 195 (1984).

35. A. Clouqueur and D. d'Humières, *Complex Systems*, **1**, 584 (1987).

36. D. Dab and J.P. Boon, "Cellular Automat Approach to Reaction Diffusion Systems", in "*Cellular Automata and the Physics of Complex Systems*", ed. P. Manneville (Springer Verlag, Berlin), **1**, (1989).

LATTICE GAS AUTOMATA:
COMPARISON OF SIMULATION AND THEORY

Gianluigi Zanetti

Program in Applied & Computational Mathematics
Princeton University
Princeton NJ 08544
USA

PROLOGUE

In this chapter I will review some of the work that Leo Kadanoff, Guy McNamara and I have done on the study of the hydrodynamic behavior of Lattice Gas Automata, a technique that was recently proposed by Uriel Frisch, Brosl Hasslacher, and Yves Pomeau[1,2] for the numerical solution of the incompressible Navier Stokes equation.   I will give more details later, but, for the time being, it is enough to describe the LGA as a collection of particles that inhabit the sites of a regular lattice, are allowed to hop from a lattice site to its nearest neighbors, and can experience a primitive form of collision at the lattice sites. The "collisions" between particles are implemented as a set of deterministic collision rules that redistribute the particles at each site between the possible hopping directions. The set of rules actually used is a technical detail; what is important is that the rules are constructed to conserve the total number of particles and the total linear momentum present at each site. These conservation laws reappear in the macroscopic dynamics* as differential conservation laws. It is usually argued that, when the underlying regular lattice has been properly chosen (*i.e.* an hexagonal lattice for the two dimensional LGA) the form of the hydrodynamic equations appropriate for the LGA is very similar to that found for simple fluids. In fact, immediately following Ref. 1 many authors (see the large bibliography in Ref. 3) applied this new method to various test cases and obtained rather promising qualitative, and sometimes semiquantitative, results.

In our work we have been mainly concerned with testing the theoretical understanding of the LGA model. The strategy we decided to follow was to probe the hydrodynamic behavior of the LGA first with a gross check, based on a two dimensional channel flow, later with a rather delicate check based on a peculiarity of two dimensional fluids: the infrared divergence of transport coefficients. The result of our tests are mixed. The LGA is in fact able to reproduce quite accurately the parabolic momentum density profile expected in the simulation of the channel flow. However, in the second check there is a discrepancy of about 30% between the simulation results and theory.

We have been able to explain this discrepancy, but in doing so we have discovered that the hydrodynamic behavior of this simple lattice gas automata it is actually more complicated that the one of simple fluids. Some of the implications of this will be treated in the concluding section.

---

* Described by coarse grained conserved densities, e.g., momentum density, obtained by averaging their microscopic equivalents over subregions of the lattice.

Microscopic Simulations of Complex Flows
Edited by M. Mareschal, Plenum Press, New York, 1990

INTRODUCTION

The motivations for the study of fluid flows range from engineering problems to deep theoretical questions about the character of turbulent flows. The Navier Stokes equation is used to describe the flow of simple fluids. It is the continuum limit version of the mass and linear momentum conservation laws microscopically obeyed by the fluid,

$$\partial_t \rho = -\partial_i g_i$$
$$\partial_t g_i = -\partial_j \Pi_{ij}$$
(1)

together with the assumption that the momentum density current, Π_{ij}, has the structure[*]

$$\Pi_{ij} = \frac{1}{\rho} g_i g_j + p \delta_{ij} - \nu \left[\partial_i g_i + \partial_j g_i \right] - (\zeta - \nu) \partial_l g_l \delta_{ij}.$$
(2)

In the equation above, ρ, and g_i are, respectively, the mass and momentum density, p is the pressure, while $i, j, k, \cdots$ run over $\{x, y\}$. The nonlinear term in Eq. 2 corresponds to the convection of the momentum density by itself and the last two terms describe its diffusion. The transport coefficients, ν, and ζ are, respectively, the kinematic viscosity and the second viscosity.[4] Note that the coefficient of the convective term is $1/\rho$, and thus the Navier Stokes equations are Galilean invariant, i.e., the form of the equations does not change when we do a Galilean transformation. This is, of course, a trivial observation, but it will be relevant in the following.

Unfortunately, the presence of the non-linear convective term in Eq. 2 makes Eq. 1 very difficult to solve analytically, and the only way to tackle the problem is via numerical techniques.

The numerical integration of Eq. 1 has become a rather sophisticated industry. The ideas behind many of the schemes used are straightforward but the actual implementations are, due to considerations of efficiency and numerical stability, very complicated, and sometimes mysterious: the numerical integration of nonlinear partial differential equations is a difficult art that requires years to master. Note that many of the problems one encounters in the numerical integration of Eq. 1 have little or nothing to do with the actual physical flow that one wants to simulate, rather they are byproducts of the specific numerical scheme used.

The scheme suggested in the previous paragraph directly translates the physical problem we would like to solve, say swirls in the tub drain, into the numerical solution of the Navier Stokes equation with appropriate boundary conditions. As an alternative, one could take a bathtub full of water and directly do the experiment. The computer analog of the latter technique is to simulate the fluid at the microscopic (i.e. molecular) level rather than the continuum limit description given by Eqs. 1,2. This can be done (as one can see from the appropriate chapters of this volume). However, the amount of computation required is, for the available computers, prohibitive.

Microscopic Definition

Frisch, Hasslacher, and Pomeau, designed a "molecular" model that can be efficiently simulated. They consider particles living on the sites of an hexagonal lattice (why hexagonal will be explained later) and endow each particle with a "velocity" chosen from the finite set

$$\vec{C}_a = (\cos(\frac{2\pi}{6}a), \sin(\frac{2\pi}{6}a)); \quad \vec{C}_6 = (0,0)$$
(3)

where $a = 0, \ldots, 5$. They then introduce the particle occupation number $f_a(\vec{r}, t)$, i.e., the number of particles, at lattice site $\vec{r}$ and at time step t, ready to hop in direction $\vec{C}_a$, and they define the microscopic densities, $n, \vec{g}$,

$$n(\vec{r}, t) = \sum_{a=0,6} f_a(\vec{r}, t)$$
$$\vec{g}(\vec{r}, t) = \sum_{a=0,6} \vec{C}_a f_a(\vec{r}, t).$$
(4)

---

[*] Here, and in the following sections, all the formulae are specialized to the two dimensional case.

We will call them the number and momentum density,[†] respectively. As a further simpification, the model does not allow more than one particle with a given velocity at a given site, and thus $f_a(\vec{r}, t)$ is either 0, or 1.

The time evolution of such a model can then be easily constructed as the composition of two operations. In the first, the "streaming", particles hop from the site where they are to the nearest neighbor in the direction of their velocities. In the second, the "collision", the particles are redistribuited between the possible velocity directions in such a way that both the microscopic number and momentum density are, site by site, conserved. In formulas,

$$n(\vec{r}, t+1) = \sum_{a=0,6} f_a(\vec{r} - \vec{C}_a, t)$$

$$\vec{g}(\vec{r}, t+1) = \sum_{a=0,6} \vec{C}_a f_a(\vec{r} - \vec{C}_a, t).$$

(5)

For what is discussed in this chapter, the actual implementation of the collisions, barring patological choices, is almost irrelevant, as far as they conserve the number and momenta of the particles involved.

Hydrodynamic behavior

The model we have defined above has two natural length scales, one microscopical, i.e., the particles mean free path ℓ, the other macroscopical, i.e., the size of the lattice L. Between these two extremes we should find an intermediate scale $\ell << \Lambda << L$, large enough that averaging on a region of linear size Λ reduces appreciably the statistical noise, small enough that there is a resonable resolution in space. We can now subdivide the lattice in subregions of linear size Λ and label each one of them with the average momentum and mass per site in the given subregion. Of course this procedure makes sense only if we assume that these averaged quantities are slowly varying on the length scale of the subregions, and that the local equilibrium description holds (that is that the probability distribution functions are well aproximated, after a short time needed for the thermalization, by the equilibrium probability distribution consistent with the values of the coarse grained densities). The time evolution of these slowly varying fields should then be well represented by the differential conservation laws of Eq. 1. To close this set of equations we will assume that it is possible to give constitutive relations for the stress tensor Π, that is that we can express it in terms of the conserved densities and their gradients. Said so, we can try to establish the structure of the stress tensor by writing it as an expansion in the (local equilibium) momentum density and in the gradients of the (local equilibrium) conserved densities, around its thermodynamic equilibrium value for $\rho = \rho_0$, $g_m = 0$,

$$\Pi_{ij}^{LGA} = p\delta_{ij} + A_{ijkl}g_k g_l + B_{ijkl}\partial_k g_l + \text{Higher Order Terms}$$

(6)

where I have kept all the relevant terms, i.e. the terms that are not zero by obvious symmetry considerations, up to the order of the standard analysis that goes into Eq. 2. The "Higher Order Terms" are proportional to $g^4, \partial_l\partial_k\rho, \cdots$.

The two tensors A_{ijkl}, and B_{ijkl} reflect the symmetry of the underlying lattice. It turns out[5] that the six-fold symmetry of the hexagonal lattice implies full rotational symmetry for the two fourth order tensor of Eq. 6. Thus

$$\Pi_{ij}^{LGA} = p\delta_{ij} + \lambda g_k g_l - \nu\left[\partial_i g_j + \partial_j g_i\right] - (\zeta - \nu)\partial_l g_l \delta_{ij}.$$

(7)

At this stage there are two essential differences between Eq. 2 and Eq. 7. The first is that the factor λ is not $1/\rho$, cf. Eq. 2, rather it has a non-trivial dependence[1] on ρ_0.

The other interesting fact is that, in Eq. 7, there are higher order corrections that depend only on powers of g_m.[*] Both differences have the same physical origin: the LGA does not

---

[†]   the "mass" of a particle is defined to be one.

[*]   The pressure, p, of Eq. 7 is also qualitatively different from the one of Eq. 2. It has a dependency in $g_i g_i$.

have Galilean invariance. On the other hand if the flow is weakly compressible this difficulty can be easily surmounted by an opportune rescaling[2] of g_i and time.[*] In fact, in the limit of small Mach numbers,[⋆] the density can be considered constant. We are then allowed to consider λ together with the transport coefficients, to be constant too. Thus we can rescale $\vec{g} \to G\vec{g}\,', \vec{r} \to L\vec{r}\,', t \to Tt', T = L/(\lambda G)$, and obtain

$$\partial_k g_k = 0$$
$$\partial_t g_i + g_j \partial_j g_i = -\partial_i p + \frac{\nu}{\lambda G L} \partial_k \partial_k g_i. \tag{8}$$

Where I systematically dropped the primes. We have thus been able to recast Eq. 7 as the incompressible Navier Stokes equation, by concealing the non-Galilean invariance in a redefinition of the Reynolds number of the simulation.

It is clear that the model I just described is rather artificial, mainly because the dynamics is intrinsically discrete. On the other hand, this model is very actractive from the computational point of view. In fact, since we do not have to track the single particles and the specified interaction is only local, the algorithm is completely parallel. Moreover, there are a finite number of velocities at each site, and thus the collisions can be given as a set of deterministic rules that transform a given incoming configuration in an outgoing one. (Of course both configurations should have the same number of particles and momentum)

We can thus interpret the LGA as an analog computer where we setup simulations and then obtain solution of the Navier Stokes equations by rescaling the resulting flows.

Just after Ref. 1 was published there was a large number of simulation who seemed to confirm, at least qualitatively (sometimes semiquantitatively), Eq. 7. (see Ref. 3 and in other chapters of this book.) Even though the results of these simulations were not particularly impressive for the modern standards of computational fluid mechanics they were still extremely promising, mainly because the amount of human work needed to setup the simulation was really minuscule compared to what is needed with conventional molecular dynamics simulations.

<u>Playing Cassandra</u>

Before I go on and start describing our quantitative tests, I think I should mention why things could go wrong in such a simple model.

Let us consider a simple case first. Let's assume that we allow only the following collision rule: when two particle come into the site along opposite directions they will leave the site along a line rotated 60° counter clockwise with respect to the incoming direction. In other words, we have collisions of fixed "chirality". The particles will then follow triangular paths, and thus restore the hexagonal simmetry (apparently lost by our choice of collision rules) necessary to obtain Eq. 7. On the other hand, if the isotropy of the underlying lattice is lost, for instance by setting up a "wall", this averaging on triangular paths will be affected with dire consequences on Eq. 7. E.g., in a channel flow simulations the expected parabolic profile for the momentum density will be skewed left or right depending on the chirality of the collisions.[22]

I will now describe something slightly more subtle.[1] Let us again consider only the two body collision described above. However, this time the actual chirality of the collision is irrelevant. We select then a line of lattice sites and construct the following quantity,

$$Q = \sum_{\{\text{sites on the line}\}} (\text{particles going} \to) - (\text{particles going} \leftarrow) \tag{9}$$

where $\to$, and $\leftarrow$ are alligned along the lattice line we have chosen. It is easy to see that Q is conserved by the dynamics. There is nothing special in this particular line of lattice sites

---

[*]   What I described so far is the two dimensional variant of the LGA. The model can be extended to three dimensional simulations[6,7] but at the price of some technical complications that are basically irrelevant to the current discussion.

[⋆]   that is when the typical fluid velocities are much smaller than the speed of sound for the gas

and thus to every one of them there will be an associated conserved quantity. This of course completely destroys our assumptions on what are the conserved quantities of the system and thus makes Eqs. 1,7 meaningless. The conserved quantity, Q, can be trivially removed by just enlarging the set of collision rules, e.g., by allowing three body collisions, on the other hand there could be other, less obvious, conserved quantities and thus it is worthwhile checking.

QUANTITATIVE TESTS

We have addressed the problem of testing the LGA as a two stages process. In the first we check the gross behavior of the lga fluid by simulating a two dimensional channel flow. In the second stage we do a rather delicate test based on the famous infrared divergence of the transport coefficients of two dimensional fluids. I.e., we measure, using the channel flow of the first test as viscometer, the dependence of the LGA kinematic viscosity in the size of the system, and then we compare it with the theoretical prediction.

We consider a very simple flow: two dimensional pipe flow, and we implement it in the LGA as an uniformly body forced channel flow.[o] We are using a channel flow because it is one of the few cases where Eqs. 1,2 can be solved analytically and, since the flow is (for small Reynolds numbers) time independent, we can obtain precise results by time averaging the macroscopic densities.

We want to simulate a two dimensional channel containing an uniformly forced fluid. This can be easily implemented in the LGA by bounding a rectangular region of the lattice between parallel walls, the no-slip boundary conditions at the walls can be very easily arranged, e.g., by back reflecting the particles hitting the wall. It is also very easy to impose an uniform body force on the lattice gas. A simple scheme is to randomly select a lattice site and, if possible, add one unit of momentum in the direction of the applied force.

The expected momentum density profile, assuming that Eq. 7 is correct, will then be parabolic.

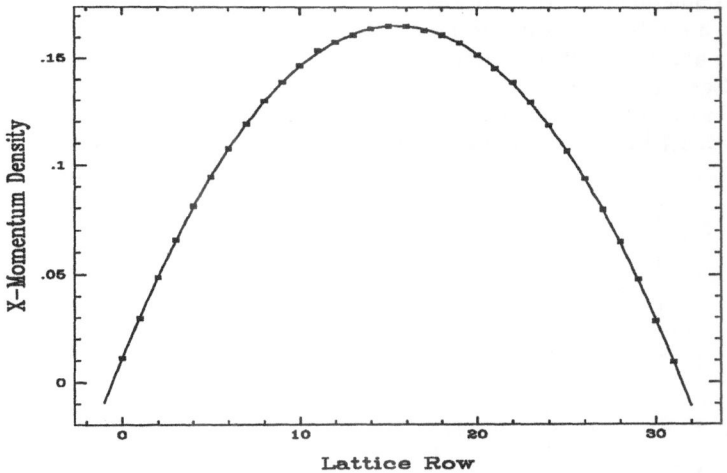

Figure 1. Parabolic profile obtained by the channel flow simulation. The symbols are the simulation data, the solid line is a parabolic fit.

Unfortunately the scheme I just described is effected by a technical difficulty. In fact, close to the walls, the hydrodynamic description breaks down and this introduce modifications

---

[o]   The standard approach, i.e., pressure driven flow, has some major technical disadvantages[8].

to the parabolic profile of the order of ℓ/W, where ℓ is the LGA mean free path, while W is the channel width. (This is an example of "Knudsen Layer", see Ref. 9)

To avoid this problem we used the following simulation setup. We considered again a rectangular region of the lattice, bounded with periodic boundary conditions on the opposite sites. We now divide it in two parts with an horizontal line and uniformly force toward right in the top part and in the opposite direction in the lower half.

The resulting flow, assuming we keep the forcing low enough[10], is then predicted to be two opposite parabolas. Indeed, this is the result of the simulations.[8,11] In Fig. 1 there is a plot of the longitudinal (i.e., along the channel) component of the momentum density, g_x, as a function of y. The agreement between the measured momentum profile and the predicted parabola is very good. The statistical error on the measurement of g_x can be controlled: is proportional to $\frac{1}{\sqrt{T}}$ where T is the time ran by the simulation. For instance, Fig. 1 was obtained by averaging over 2 million microscopic time steps, and the discrepancy between the data and the parabolic fit is smaller than 0.2%. From the parabolic fit and the knowledge of the forcing level we can measure ν, the kinematic viscosity of the LGA fluid, with very high precision (cf. Eqs. 1,7).

A delicate test

The effective kinematic viscosity, ν, measured in the simulations is the sum of two contributions. One is the the bare (or molecular) viscosity, the other is a renormalization of ν due to hydrodynamic effects. The characteristic time and length scales of the physical processes behind these two contributions are widely separated. Fast (time scale of the order of a mean time between collisions) and short distance (of the order of a mean free path) processes, in the realm of the molecular description of the gas, control the bare viscosity. The renormalization correction measures instead the interaction between macroscopic flow eddies at different length scales. The hydrodynamic correction is a nonlinear effect due to the presence of the convective term in Eqs. 2,7. It can be estimated[12,11,13] and the result, to first order in perturbation theory, is a small but finite correction in three dimensions, and an infrared divergence in two dimension. In the range of system sizes accessible to our simulation, the functional behavior of the hydrodynamic correction is well approximated by $\Lambda(\rho_0)\log(L)$ where L is the linear dimension of the system (i.e. the channel width), and ρ_0 is the equilibrium average number of particles per site. The amplitude $\Lambda(\rho_0)$ can be estimated, starting from Eqs. 1,7, in terms of the thermodynamic derivatives, e.g., $\lambda(\rho_0)$, and transport coefficients. Thus we can do a rather delicate test (with no free parameters in the evaluation of $\Lambda(\rho_0)$) of Eqs. 1,7 by measuring ν as a function of $\log(L)$ for various values of the average density ρ_0.

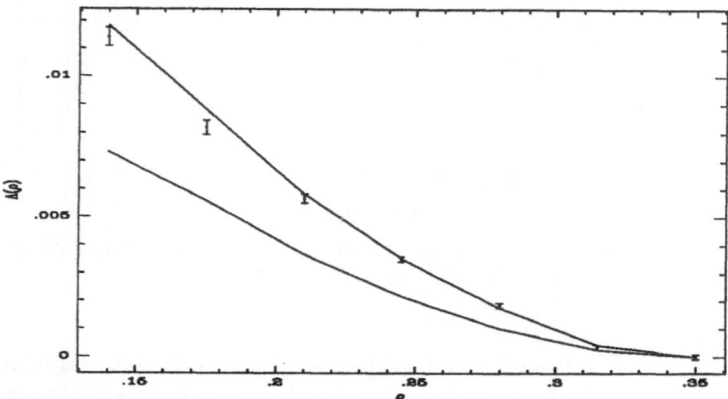

Figure 2. $\Lambda(\rho)$ *vs* ρ. The symbols are simulations data, the solid lines come from theory.

In Fig. 2 the data points are the measured value of $\Lambda(\rho)$ for different values of ρ, while the solid lines come from theory. For the time being please forget about the higher continuous curve and concentrate on the data points and the lower solid line. The solid line is what the perturbative calculation predicts, the data points are instead given by the simulations. Clearly they do not match. What is wrong?

Back to the LGA hydrodynamics

There is clearly something missing in Fig. 2 and possibly in Eq. 7. Eq. 7 is wrong. We have foundE[14,15] that the physics of the LGA is actually richer than Eq. 7 would indicate. In fact its dynamics allow, together with the conservation of number of particles and momentum, three more conserved quantities. The existence of these quantities is intimately related to the fact that the lattice gas is defined on a regular lattice and that the particles can only hop from one site to its nearest neighbors.

These conserved modes can be understood via a simplified example. Consider a one-dimensional lattice gas. We assign to each particle a "momentum" that can be either ± 1 depending on which direction the particle will jump in the next time step. We now divide the sites of the lattice into two subsets by labeling them either "even" or "odd", depending on their distance from the origin and, finally, we specify the time evolution of the system by letting the particles hop in the direction of their momenta. It is clear that the total number of particles, N, is conserved and, moreover, that the total momentum on the even sites, G_e, and the total momentum on the odd sites, G_o, are exchanged at each time step. Thus the dynamics allow three extensive conserved quantities: $N, G = G_e + G_o$, and $H = (-1)^t(G_e - G_o)$, where t labels the time step. The first two are the usual total number and momentum conservation, the third is a peculiarity of our extremely simplified dynamics.

The generalization of H to the two dimensional example described above is immediate. Let us now choose a direction, say $\vec{C}_1$, an arbitrary reference point, say $\vec{r} = \vec{0}$, and then divide the lattice in "even" and "odd" lines parallel to the chosen direction. It should be evident by now that the component perpendicular to $\vec{C}_1$ of the total momentum on the "even" lines is exchanged at each time step with the analogous quantity computed on the "odd" lines. There are three new conserved quantities that can be constructed this way. This implies a nontrivial modification to Eq. 7. In fact the new three H_a modes (the generalization of the H in the example) are nonlinearly coupled to the momentum density g_m.[†] The presence of a nonlinear coupling between h and g explains the difference between the simulation data and theory cited before. (The higher solid line in Fig. 2 is the new, improved, theoretical prediction.) However, it has a less pleasing effect on the application of the LGA to the integration of the Navier Stokes equation. In fact the new version of the stress tensor,[15] which replaces Eq. 7, includes a term quadratic in h but with no direct dependence on the momentum density g. This implies that h can act as a source for g. Hence, if the h_a are not strictly zero, the LGA flow is not related to a solution of the Navier Stokes equation.

CONCLUSIONS

When the LGA algorithm was first proposed there was hope that it could represent a break trought in computational fluid mechanics. The hope was based on the observation that the algorithm is intrinsically parallel and very well suited to be implemented on dedicated computer architectures. There was also the feeling that, since the model could be interpreted as a gas of particles (albeit with a drastically simplified time evolution), the hydrodynamic behavior of the LGA was somehow "deeply" connected to the hydrodynamic behavior of simple fluids.

It turned out that the intrinsic limitations of the model, (basically, the only free parameter that can be used to adjust the Reynolds number of the simulation is the system size)

[†] This is vaguely reminiscent of the coupling between the staggered magnetization density ($\leftrightarrow h$) and the magnetization density ($\leftrightarrow \vec{g}$).[16,17] The staggered magnetization and the h modes, however, are fundamentally different. h is a true conserved density while the staggered magnetization is not.Ref. 17

do not make it computationally competitive[18] when compared to the more traditional approaches, e.g. pseudospectral methods, see however Ref. 19. Of course, this does not apply to situations were LGA related models are the only technique available, e.g., certain immiscible fluid flows simulations and flows in porous material.

The main result of this chapter is that the simple LGA model described in the introduction, and all related models, are affected by artifacts. These artifact are due to the extreme discretization of the model and make the hydrodynamic behavior of the LGA fluid qualitatively different from the hydrodynamic behavior of simple fluids. Fortunately, a part from pathological situations, these artifacts do not effect in a discernable way the macroscopic flow of the LGA fluid. Moreover, it appears that these artifacts can be removed at the price of a slight increase in the complexity of the model. In fact, it is easy to write a twelve velocities model (six of speed one and six of speed two) that will not conserve the staggered momentum densities described in the text. The surprising thing, at least to me, is that this new model does not seem to have other spurious conserved quantities.[23] Even thought it is debatable if the LGA is a convenient technique to do fluid mechanics simulations, the LGA is clearly a very efficient tool in the study of problems in which the relation of micro and macro scales is of interest. On the one end, this includes attempts to test constitutive laws,[24] on the other end, this also includes the simulative verification of the techniques used in non equilibrium statistical mechanics. The theoretical curve of fig., for instance, was predicted by using a very general technique (called mode-mode coupling) which is based on a reasonable, but nonetheless uncontrolled, perturbative scheme. Thus, the agreement shown in fig.2, between the theory and the simulation result can be interpreted both as a check of eq. and as a verification of the mode-mode coupling pertubation technique. (see also the work of...). Why is the LGA such a good tool? First of all, it is very "cheap" to simulate, expecially compared to standard molecular dynamics simulations. Secondly the lattice gas automaton is very noisy. This is a source of problems when it is applied to the solution of the incompressible Navier Stokes equation[20,21] but it is an advantage in our case since it enhances the thermodynamic derivative which tipically appear in the mode-mode calculations.

REFERENCES

1. U. Frisch, B. Hasslacher, and Y. Pomeau, Physical Review Letters **56**, 1505 (1986).

2. U. Frisch et al., Complex Systems **1**, 649 (1987).

3. G. Doolen, editor, *Lattice Gas methods for partial differential equations*, Addison-Wesley Publishing Co., 1989.

4. L. Landau and E. Lifshitz, *Fluid Mechanics*, Pergamon Press, 1986.

5. L. Landau and E. Lifshitz, *Theory of Elasticity*, Pergamon Press, 1986.

6. J. Rivet, M. Henon, U. Frisch, and D. dHumieres, Europhys. Letters **7**, 231 (1988).

7. D. dHumieres, P. Lallemand, and U. Frisch, Europhys. Letters **2**, 291 (1986).

8. L. Kadanoff, G. McNamara, and G. Zanetti, Complex Systems **1**, 791 (1987).

9. C. Cercignani, *Theory and Application of the Boltzmann Equation*, Elsevier, 1975.

10. Z. S. She, Large scale dynamics and transition to turbulence in the two-dimensional kolmogorov flow, in *Proceed. Fifth Intern, Beer-Sheva Seminar on MHD Flows and Turbulence, Israel, March 2-6, 1987.*, 1987.

11. L. Kadanoff, G. McNamara, and G. Zanetti, Physical Review B **40**, 4527 (1989).

12. D. Forster, D. Nelson, and M. Stephen, Physical Review Letters **36**, 867 (1976).

13. Y. Pomeau and P. Résibois, Physical Reports **19**, 63 (1975).

14. G. McNamara and G. Zanetti, Physical Review Letters **61**, 2332 (1988).

15. G. Zanetti, Physical Review B **40**, 1539 (1989).

16. R. Freedman and G. Mazenko, Physical Review B **13**, 4967 (1976).

17. B. Halperin and P. Hohenberg, Physical Review **188**, 898 (1969).

18. S. Orszag and V. Yakhot, Physical Review Letters **56**, 1691 (1986).

19. S. Zaleski, Weakly compressible fluid simulations at high reynolds numbers., in *Discrete kinematic theory, lattice gas dynamics, and foundations of hydrodynamics*, edited by R. Monaco, pages 102–113, World Scientific, 1989.

20. J. Dahlburg, D. Montgomery, and G. Doolen, Physical Review B **36**, 2471 (1987).

21. S. Orszag and V. Yakhot, Physical Review Letters **56**, 1691 (1986).

22. G. McNamara and G. Zanetti, unpublished.

23. G. Zanetti in preparation.

24. D. Rothman, to be published in the Journal of Geo. Res..

VECTORIZED MOLECULAR DYNAMICS FOR SYSTEMS WITH SHORT-RANGE INTERACTIONS

D.C. Rapaport

Physics Department
Bar-Ilan University
Ramat-Gan, Israel

ABSTRACT

Methods for carrying out molecular dynamics simulation on vector supercomputers are discussed. The techniques normally used in these computations are not suited to the pipelined processing employed by this class of processor and alternative approaches are needed. The enhancements to the basic computational algorithm address both the issues of processing speed and memory utilization, and have proved capable of dealing with systems containing over two million particles.

1. INTRODUCTION

The successful application of molecular dynamics simulation methods to a broad range of problems over the past thirty years has shown that in a great many contexts [1,2] it is possible to obtain meaningful results using systems that are relatively small, typically a few hundred to a few thousand particles at most. On the other hand, there also exist applications where it is obvious that the longest length scales that can be reached in such small simulations are not adequate for faithfully representing the phenomena in question, and that systems of larger size -- tens of thousands of particles or more -- will be necessary. A representative example of a problem of this kind is a fluid that exhibits hydrodynamic instability [3]. The increasing availability of the latest generation of supercomputers [4] -- machines whose peak performance now reaches into the gigaflop range -- means that problems of this magnitude can now be realistically attempted.

It is a fact of life that algorithms suitable for small problems do not always prove suitable when the same problem is scaled up by one, or -- worse still -- a few orders of magnitude. This is especially true in the case of molecular dynamics. The algorithm used for simple systems of atoms (typically interacting via a Lennard-Jones or similar potential) is almost trivially stated. While such an algorithm is adequate, and perhaps even efficient for systems of up to a hundred or so particles, some form of improvement is required to deal with larger systems. It has long been recognized that the introduction of certain bookkeeping techniques [2] -- namely cell and/or neighbor lists -- can greatly enhance performance, even by orders of magnitude. Until recently, however, it was though that such techniques were only effective on processors whose performance was not achieved by means of vectorization (in other words, all machines except supercomputers), but it has since been demonstrated [5,6] that both bookkeeping schemes can indeed be used effectively with vector computers.

Use of an enhanced algorithm has made possible extensive simulations of systems containing as many as 200,000 particles [3] taking full advantage of the benefits of vector processing. The bookkeeping schemes exact a penalty however, namely one of storage, and the substantial price paid in terms of computer memory requirements can make it impossible to consider even larger systems unless an inordinate amount of storage is utilized. A reexamination of the algorithm reveals that this demand for memory is one that can be overcome at the expense of further algorithmic complication, but, more importantly, with only minor impact on the efficiency of the computation. In the limit of very large system size (a desired but unreachable goal) the fraction of storage required for the bookkeeping functions tends to zero. In this paper we survey the evolution of the molecular dynamics algorithm for simple particles. The presentation avoids mathematical technicalities; some of these details have already appeared in print [5], while those related to the more recent developments will appear shortly.

2. CONVENTIONAL METHODS

2.1 Molecular dynamics - a brief overview

The basic molecular dynamics approach [2] involves the solution of the coupled differential equations of motion of the individual 'particles' that constitute the system. In the simplest case, the so-called particles are pointlike entities that interact with one another by forces directed along the directions of mutual separation. In more complex cases the particles become rigid molecules that undergo rotational motion in addition to translation, and which interact, for example, by means of forces originating at several distinct fixed locations within the molecule (corresponding typically to the positions of atoms or charge centers). Additional realism -- as well as complexity -- appears through the introduction of flexibility into the molecular structure, such as might be required to model a chain polymer. Yet further realism can be included if features such as polarizability or three-body forces are added to the model.

For a broad range of problems it is sufficient to consider the simplest of particles, typically a particle whose only structural feature is its excluded volume. Such a particle is exemplified by the hard sphere (or hard disk in two-dimensional situations). While a truly hard sphere cannot be described by a differentiable potential function, the hard-sphere step potential can be replaced by a suitably shaped function that acts over a very limited range and diverges rapidly at small interparticle separations at a rate (typically an inverse-twelfth power law) that still allows the equations of motion of this system of nearly-hard spheres to be integrated in the normal way. A commonly encountered example of such a potential is one derived from the Lennard-Jones functional form by an appropriate shift and truncation; such a potential is very short-ranged and purely repulsive. While not suitable for studying the liquid-gas phase transition, it has been used extensively for studying freezing and melting, fluid structure, transport coefficients, fluid flow, and so on -- in fact any situation where there is reason to believe that the behavior is primarily attributable to the excluded volume of the particles and not to other details of the potential.

2.2 Simple approach for continuous potentials

The basic molecular dynamics algorithm involves the computation of the force acting on a given particle by considering all other particles in the system; once the total force acting on each particle has been computed the equations of motion are integrated by standard means. Starting from some initial state, the evolution of the system is followed over a sequence of time steps, typically of the order of ten thousand, although this can vary by an order of magnitude in either direction depending on the problem. The work is approximately halved if Newton's third law is applied, but this still leaves a problem requiring an amount of computation that varies quadratically with the number of particles. For systems with long-range interactions this is to be expected and cannot be reduced except by introducing approximations into the computation, but in the case of short-range interactions only the immediate environment of each particle really ought to be explored in order to establish the forces that act upon it; examining all particles in the system to compute these forces is obviously a gross waste of computational effort. Clearly some scheme for representing the system state in a manner that allows the determination of neighborhood relationships in an efficient manner is essential. Fortunately such a scheme is not difficult to come by.

2.3 Cell partitioning

The most obvious way to reduce computational effort is to subdivide the system into a set of relatively small boxes, or cells. If the edge length of each cell exceeds the interaction range then it is apparent that each particle can interact only with those particles that are (at that instant of time) in either the same cell or in one of the immediately adjacent cells (of which there are 26 in three dimensions, and 8 in two). Since determining the cell membership for each particle requires a fixed computational effort, and the interaction computations that rely on the cell partitioning involve a neighborhood whose size depends only on interaction range (but not on the total number of particles in the system), the entire computation grows linearly with system size -- the slowest growth rate possible.

Beyond the requirement that the minimum cell edge exceed the interaction range there is no precise recipe for determining optimal cell size. The preferred size is one in which the mean cell occupancy is close to unity; in a high density system this will actually be the smallest size allowed, but at lower densities the criterion of unit cell occupation rather than minimum size might prove more effective, in order to avoid excessive processing of empty cells. The best approach is the empirical one: vary the cell size for the actual system at the state point to be studied and determine the value at which the simulation runs fastest.

There is also the question of how to represent the information describing cell membership. Given the allowed range of cell occupancies (from zero to a value that is the maximum number of particles that can be squeezed into a cell under extreme conditions) the most effective approach from the storage point of view is to use a linked list of cell occupants, with a separate list being associated with each cell. The storage needed for the entire set of lists can be taken from a common pool, whose overall size is equal to the number of particles. To complete this particular data organization a set of pointers -- one pointer per cell -- is constructed, each indicating the first particle in the corresponding cell; starting from the first particle the linked list provides access to the remaining particles in the cell. Assuming the total number of cells to be similar in magnitude to the total number of particles, the storage required to implement this scheme is proportional to the number of particles.

While the linked list approach is normally very effective on scalar processors, the fact that it requires accessing memory in an essentially random fashion means that it cannot be employed effectively in a situation where optimal performance demands that data be read from and written to memory sequentially, a feature of all modern vector supercomputers. An alternative approach, that extends the cell technique in a way that allows vectorization, will be discussed later.

2.4 Neighbor lists

The observation that in a fluid of moderate to high density the environment of each particle changes only gradually (where the time scale is based on the mean time between collisions) suggests that information on neighborhood relationships continues to be valid, at least approximately, for a certain period of time subsequent to its original generation. The neighborhood is defined to be a spherical (circular in two dimensions) region with radius somewhat larger than the interaction range. If a list of all the particles present in the neighborhood of a given particle is prepared, then it is clear that this information will remain relevant -- in the sense that it still contains all the interaction partners of that particle -- for a period spanning several time steps; the actual duration of this period depends on the maximum velocity of the particles involved (as well as on the neighborhood radius itself). Continual monitoring of the particle displacements can be used as a means of determining when regeneration of the list of neighbors is required -- namely the earliest instant at which a particle not in the neighbor list could possibly become an interaction partner. Insofar as the representation of the lists of neighbors is concerned, the information could be stored in straightforward tabular form (possibly with supplemental data to permit condensing the tables by grouping according to one of the neighbors); the fact that the neighbor relationship is commutative halves the total storage requirement.

The amount of storage required for the neighbor data depends directly on the radius of the neighborhood region; this radius is set equal to the interaction range plus a value representing the thickness of the bordering shell (or annulus in two dimensions). The larger this surrounding border the less frequent the time-consuming neighbor data regeneration; however, once the proportion of particle pairs that are classified as neighbors but which actually lie outside the interaction range becomes substantial the performance will start to degrade. The optimal size must once again be determined by experimentation. The process of generating the neighbor lists can utilize the cell approach as a preliminary step, with the cell size of course being based on the chosen neighborhood radius rather than on the interaction range itself.

The storage penalty can, however, be substantial, and here there is a blatant tradeoff of memory to gain speed; for very large systems it may very well be the case that the neighbor list cannot be used because the computer simply does not have the memory required. The neighbor list approach is suitable for a vector computer without further modification provided each particle has a moderately large number of neighbors (this excludes the case of very short range forces), and further improvements in performance are possible using a variant of the approach that will be described later for the cell method. (The extension of the vectorized cell method described towards the end of the paper that is intended for dealing with extremely large systems will not be applicable to neighbor lists. The reason for this is that neighbor data is generated once for the entire system and then used over the course of several time steps, whereas the approach introduced for dealing with very large systems is based on subdivision, with only a minimal amount of data being retained for those parts of the system not under immediate consideration.)

3. THE NATURE OF VECTOR COMPUTING

3.1 Architecture of vector processors

The vector supercomputer represents a compromise between the computer designer's desire to achieve maximal computation rate for specific kinds of operations and the user's need for the fastest possible computations over a broad range of problems. The dominance of the former consideration is reflected in the fact that performance figures quoted by manufacturers are almost always beyond the reach of the user, often by orders of magnitude. While this design compromise has nevertheless proved to have considerable benefit to both parties in a great many kinds of computation in science and engineering, the situations where the performance potential of the supercomputer is far from realized are all too frequent.

What distinguishes algorithms that map effectively onto a vector computer is the manner in which data is accessed and the nature of the processing involved. Peak performance is achieved when the data is retrieved sequentially from storage, when certain combinations of the basic arithmetic operations (avoiding division if possible) are carried out, and when the results are returned sequentially to storage. The reason for this preferred mode of operation is that the processor is capable of handling both memory access and arithmetic in a pipelined fashion, with a resulting throughput substantially greater than what would be possible if each operation were to be carried out individually. The pipelining is only possible if an operation is performed repeatedly on a set of consecutive data items. Any deviation from a general operational pattern of this kind results in reduced performance.

However, with the exception of limited kinds of matrix and vector computation which follow this prescription precisely, this state of perfection is rarely (if ever) attained. The issue then is how to achieve the best performance given the preferred manner of operation of the hardware. It goes without saying that a great deal of effort has been spent in addressing this issue [4], since a vector supercomputer used inefficiently may not even be cost effective when compared with a more conventional machine.

In addition to the requirement that the data be sequentially ordered for vector processing, the design of the vector processor imposes a fixed startup period for each vectorized operation; this overhead is independent of the number of data items processed in the course of the operation, and may possibly be performed concurrently (fully or partly) with a previous vector operation. A paradoxical consequence of this overhead is that if the vectors are too short, vector processing is actually slower than the corresponding scalar operations (on the same computer). The minimal vector length requirements vary and depend on both the type of operation and the machine itself. Thus in addition to rearranging the data into the sequential form favored by the processor, it is also necessary to ensure that the resulting vectors are adequately long so that the fixed startup times do not outweigh expected performance gains. There is no guarantee that this can be achieved in all cases; there are indeed calculations for which effective vectorization is not possible.

3.2 Data organization for vector processing

To help the user implement an algorithm whose intrinsic structure bears little resemblance to the organization demanded by the vector processor, the instruction sets of most vector computers include the capability for reorganizing data at a relatively high rate; this rate tends to be intermediate between the vector and scalar processing speeds. There are different approaches to dealing with data reorganization, and not all approaches are to be found on all machines. Furthermore, even when a particular scheme for rearranging data is implemented at the hardware level, the questions of how fast such operations are carried out in comparison with the peak (vector) computation speed and whether the compiler is even capable of utilizing the hardware feature must be taken into account (at the time of writing this is not always the case).

The two principal schemes for reordering data are known as gather-scatter and compress-expand. The act of gathering data involves the use of a vector of indices to access -- in no particular order as far as the computer is concerned -- some or all of the elements of a set of data items (more than one access of a data item is allowed) and then store them consecutively in another vector. The scatter operation is just the opposite, in that an index vector is used to store a consecutive set of data items in some alternative order in another (possibly longer) vector; not all elements of the destination vector need be modified, and elements may actually be stored into several times (assuming this is meaningful for the particular context). Compression involves selecting an ordered subset of data items from a vector and storing them consecutively in another vector; expansion is just the reverse of this operation. Because data order is preserved under the compress and expand operations, information concerning which of the elements are to be extracted by the compression, or where the elements resulting from an expansion are to be placed, can be represented by means of a bit vector in which the set bits denote the elements involved; this serves as an extremely compact alternative to the more general index vector needed for gather and scatter. In the case that no reordering is required, the proportion of elements participating in an operation on a subset of a vector can determine whether gathering or compression is the faster operation -- provided that the computer (and the compiler) provides the choice; not all machines do.

3.3 Efficiency of automatic vectorization

Ideally, one of the tasks of the compiler should be to produce a machine translation of the source program that delivers close to optimal performance on the designated hardware. Unfortunately such an idealized situation is rare indeed. Judging by the achievements to date, compiler efficiency is an even more complex issue than hardware efficiency, and the performance of different compilers -- even from the same manufacturer -- exhibit considerable variation in this respect.

There are several factors contributing to this state of affairs. At one extreme is the possibility that the logical structure of the algorithm is simply too complex to be analyzed by any reasonable automated procedure. At the other extreme lies the compiler that has not been taught to recognize certain basic computational patterns (or templates). In the former instance it is the task of the programmer to rearrange matters to make the logic of the computation more obvious, while in the latter the sentiments of the user should be transmitted directly to the manufacturer in parallel with the inconvenience of either rewriting the program to satisfy the requirements of the compiler or perhaps using special subroutine calls (see further) that result in an efficient, although less intelligible and portable program.

Even in the fortunate situation where the compiler is very perceptive and competent at efficiently mapping the program onto the hardware, there are frequent situations in which certain relatively simple language constructs may, in principle, prevent the compiler from dealing effectively with parts of the program. One typical example involves the case of an algebraic statement whose general form implies a possible dependence on something that has only just been computed; such a case is not generally vectorizable because of the manner in which the pipelined vector processing restricts dependencies between the individual data items involved. In those instances where it is known to the software developer that such dependence either does not exist or will not produce erroneous results when vectorized, the capability of conveying such information to the compiler by means of directives (that are not actually part of the language) ought to help the compiler perform its task. The capacity for aiding the compiler in this way varies from one compiler to another.

The alternative to total reliance on the compiler is to directly invoke the machine instructions. This can always be done by programming in assembly language, but is best avoided in favor of a sometimes available alternative which allows access to machine instructions by means of subroutine calls from the normal programming language. The advantage of the latter approach is that it leaves the program in more intelligible form; for most of the program the compiler will prove adequate, and only certain critical sections might need to be handled in this way. The availability of this feature and the degree to which it is capable of enhancing performance varies between computers.

Even when it appears that the portion of the program where most of the execution time will be spent has been fully vectorized there is usually no indication given as to whether the machine code produced is the most efficient possible. The machine may, for example, be capable of achieving a given result in more than one way, and the judicious employment of temporary storage registers rather than main memory, or the simultaneous use of multiple functional units in the processor, can result in substantial performance gains. Only by reading the assembly language listing produced by the compiler and using one's knowledge of the processor architecture and performance (a rather arduous part of the programmer's general education) is it possible to determine whether the performance level reflects what the machine is really capable of achieving; it is doubtful whether such a thorough analysis is carried out very often (even if the compiler output includes information on the megaflop rate attained this may have little relevance to the actual efficiency of the computation).

3.4 Software portability

The issue of portability is especially acute when it comes to supercomputers. The emergence of comparatively standardized languages has facilitated transport of programs between different computer systems, and provided that the computer hardware is not overly sensitive to, for example, the order in which instructions are scheduled for execution, a program that performs reasonably efficiently on one kind of machine should run similarly well on another. The issue of optimization on scalar computers (presuming a competent compiler) is therefore principally one of algorithm efficiency.

Vector supercomputers, on the other hand, generally tend to be hypersensitive to program structure -- for reasons such as those stated earlier -- and the issue of lack of portability becomes a serious one: a program that runs well on one brand of machine fails to perform as expected on another. The problem is in fact more serious in that performance can change substantially between different models of a particular machine, depending on the kinds of instructions implemented in hardware, memory organization, and other more subtle factors -- such as timing of individual instructions -- that might be entirely unknown to the user. Performance can also change between different versions (or releases) of a compiler, and there is no guarantee that the code generation and optimization capabilities improve monotonically with time. These considerations apply especially to vectorized implementations of molecular dynamics algorithms which, as pointed out in the course of this article, tend to require machine-dependent adaptations in order to run efficiently.

4. A VECTORIZED ALGORITHM USING LAYERS

4.1 The problem with cells

The usual implementation of the molecular dynamics algorithm based on cells involves a set of linked lists, one list per cell, in which the list elements are the identities of the particles belonging to the cell at a given instant. As mentioned earlier, the reason for using a linked list, and not merely storing the data sequentially, is that the number of particles per cell can vary considerably, and sequential storage would require that an amount of storage equal to the maximum possible cell occupancy be reserved for each cell. Such excesses are to be avoided if memory is scarce, as is usually the case with processors not in the supercomputer class.

Linked lists are handled very inefficiently on a vector processor since the use of pointers to connect related data items forces all the affected data accesses to be carried out in scalar mode. The implication is that the cell technique, which has proved very effective on scalar computers, must -- at the very least -- be modified in a way that renders it vectorizable. Obviously it would not be desirable to abandon the cell approach entirely and return to the original method which considers all pairs of particles: though fully vectorizable, and even efficient for smaller systems, there comes a point at which the gain due to vectorization can no longer compensate for the additional work involved. The layer method of reorganizing cell data which will now be described provides a solution that retains the benefits of the cell framework.

4.2 Layers

In the cell version of the algorithm, the computations involve a series of nested loops. The outermost loop of the interaction computation is one that scans all the cells into which the particles have been placed. The next-to-outermost loop is over the possible offsets between pairs of neighboring cells (one particular instance being a cell paired with itself -- in other words an offset of zero); the order of these two loops is unimportant. The two innermost loops are over the pairs of occupants of each pair of cells specified by the outer loops, with the case of a cell paired with itself being treated in a slightly different manner to ensure that particle pairs are considered once only. Note that it is the innermost loops that have the fewest numbers of iterations since the mean cell occupancy was chosen to be low (typically of order unity). This fact completely rules out any possibility of vectorization in the case of very short range interactions.

In order to make effective use of a vector processor the well-known requirement of sufficiently long data vectors must be satisfied [4]. While the set of cell processing loops just described fails to obey this criterion, the reversal of the loop order, with the loop that scans all cells being innermost, would provide a satisfactory solution. In order to reverse the order in which the loops are nested, it is necessary to place the two loops over cell occupants on the outside, followed by the loop over cell offsets, and finally the long loop over cells. How is this realized in practice?

The scheme is actually quite simple both in principle and practice [5]. Instead of representing cell occupancy using linked lists, the identities of particles in the cells are placed in a set of arrays; each array contains one element per cell, and the total number of arrays is equal to the maximum cell occupancy. These arrays will be referred to as layers. While scanning the particles as part of the cell assignment process, the first particle encountered in a given cell is assigned to the corresponding element in the first layer, the second particle in the cell (if any) to the second layer, and so on. Unfilled layer elements are flagged with a value not corresponding to a valid particle identity number (such as zero). The interaction computation then consists of pairing layers using all relevant offsets (one at a time), and an innermost vectorizable loop that processes pairs of particles specified by the layer elements. Only in the case that both layer elements specify valid particles is the calculation actually carried out, but it is generally true that vector processors have some means of bypassing particular vector components in the course of a calculation -- in this case pairs of layer elements which do not both correspond to occupying particles. There may indeed be more than one way to formulate such a computation, but this is machine and compiler dependent in the extreme. Two versions of the computation have been constructed, one which represents those pairs of cells in the layers that require attention by means of a bit vector which is then used to organize the required particle coordinates by expansion of compressed data, the other which builds an explicit set of particle indices that are subsequently used to gather coordinates (in both cases the inverse operations are used to add the newly computed interaction terms to previously accumulated values).

The innermost loop that is processed in vector fashion involves occupants of cells that come from two layers (the two layers may actually be the same). A suitable choice of the range of offsets ensures that atoms are never paired with themselves. When both layers are fairly densely occupied the vector processing is efficient, but the later layers will of course be less densely populated -- with the final layer showing extreme sparsity -- and thus some layer pairings will produce very few (if any) interaction pairs. This is a shortcoming of the method; it might be worth investigating alternative schemes for handling the more sparse layers, but any conclusions would be machine dependent and more complicated than the present uniform treatment of all layers. Additional processing is required to handle the boundaries of the region; this is discussed in the original paper [5].

The pipelined nature of vector processing imposes certain restrictions on the data contained in the vectors. The most significant of these is that the processing of each item in the vector can be carried out independently of other data items. The implication of this statement for the interaction computations is that a particular particle can appear no more than once in a set of particles submitted for vector processing. The layer method guarantees this to be the case since a particle can only appear once in a layer (even when a layer is paired with itself, the two appearances of each particle are of a nature that no conflicts arise). The compiler is not aware of this fact however, since it is not apparent from the source program, and it is necessary to tell the compiler that certain loops involved in the layer processing can in fact be safely vectorized without encountering data dependence; this is done by means of the special compiler directives mentioned earlier. Timing measurements reveal the layer method to be several times faster than the non-vectorizable basic cell method. This gain more than compensates for the added complexity of the new algorithm.

4.3 Storage overheads for layers

There are significant storage overheads associated with the layer approach because each cell must have enough storage available to accommodate the maximum possible occupancy, which is generally several times the mean. This is precisely the reason that linked lists were introduced originally for the cell approach. There is a way to substantially compress this data (see below), but this requires additional processing, and may not be feasible on all kinds of hardware. The reason for using a subdivided system -- discussed later -- is to eliminate the bulk of this overhead.

In the case of a fluid of spheres or disks interacting with a very short range repulsive potential at moderate to high density, if the cell size is chosen to give unit average cell occupancy the first layer will be almost completely populated, and the occupancy of subsequent layers will drop sharply to zero after three to five layers. The final layers will be extremely sparsely populated, so that on some kinds of computers it may prove effective to compress this information. This is done using a bit vector which indicates the cells in a particular layer that are actually occupied; the list of particles in the layer then only has to hold information related to occupied cells (in that layer), and the correspondence between particle and cell is made using the bit vector. But, as already pointed out, this scheme requires that the hardware (and the compiler) support manipulations of this kind.

4.4 Layers and neighbor lists

The layer approach can also be applied to neighbor lists. One possibility would be to enlarge the cell size to equal the neighborhood range; the advantage is that the layer information can then be used over several time steps, but the storage requirements and sparse layers might not recommend this approach. An alternative technique is to associate the layers with the cells in the original way, but to use the layers to generate neighbor tables that are divided into segments such that no particle can appear more than once in any segment [6,7]. The processing of such sets of particles can then be fully vectorized. This scheme is in fact almost identical to the implementation of the layer approach in which particle indices are generated; the only difference is that in the neighbor list method the indices are stored for use over several time steps.

4.5 Performance

The layer-based method has proved effective on Cyber 205 and related ETA 10-Q computers, as well as on single processors of Cray XMP-48 and YMP systems. The interaction computations use table lookup to reduce the work required. Different approaches to layer storage were adopted in the two implementations. In the Cyber/ETA formulation considerable use was made of bit vectors in order to reduce storage requirements; extensive production using two-dimensional systems with as many as 200,000 particles has been carried out in exploring the applicability of molecular dynamics simulation to the modeling of fluid flow instability [3]. Test runs of up to 500,000 particles were also conducted. The time required per atom-step on the Cyber was of the order of 4-5 microseconds; the actual values were 4.8 if the potential energy was included in the computation, and 4.1 if omitted, irrespective of system size (beyond a minimum of several thousand particles), with the calculations using 32-bit floating point arithmetic. (On the ETA, which should have given similar performance figures, a substantial and as yet unexplained size dependence was noted.) The Cray tests used the alternative version of the method in which pairs of particle indices are generated as an intermediate step in computing the interactions; both the index generation and the subsequent interaction calculation vectorize completely. Only systems of a few thousand particles were tested using this method and no effort was made to compress the layer data (the subdivision scheme to be introduced below provides a more flexible alternative to layer compression). The computations used 64-bit arithmetic and resulted in similar timing estimates for the XMP - - actually 5.2 microseconds -- and a smaller value (3.4 or 3.9 -- depending on compiler) for the YMP. The time required when layers are not used is typically an order of magnitude larger.

5. SUBDIVISION FOR STORAGE ECONOMY

5.1 Dealing with a subdivided system

The next step in the development of a fast but storagewise economical algorithm involves subdividing the region of space occupied by the system into a set of subregions and processing each such subregion in a manner that is practically independent of the others. Indeed the approach has a great deal in common with the computational scheme which would have to be used in the event that the calculations were distributed over a set of coupled processors where each is responsible for a part of the overall effort. The savings achieved by this approach are in the temporary storage required for the layer representation of the cell data; if the system is subdivided then this is proportional only to the number of particles being considered at once rather than to the entire system size. The approach discussed in this section differs substantially from an alternative form of subdivision described previously [5] where the layer assignment is first carried out for the entire system and only then is the region subdivided; despite the fact that the scheme to be described here is slightly more complicated, it is less demanding in terms of the processor hardware required in its implementation.

The spatial subdivision scheme requires keeping track of which particles belong to which subregion, so that when particles move between subregions this must be reflected in the data organization. This represents one of the needs for information transfer between subregions; a second need for communicating between subregions is in order to allow each subregion to be provided with information about particles in neighboring subregions that are potentially within interaction range of its own particles. If this information is made available then the interaction processing and subsequent integration of the equations of motion can be carried out completely independently in each subregion. Since both kinds of data transfer involve comparatively small amounts of data (in the case of the information needed for interactions the assumption is that the forces are of short range compared with the smallest subregion edge), the additional computations needed to support the subdivided approach will result in only a minor performance degradation. The additional bookkeeping required to support this approach is tedious rather than profound.

5.2 Alternative spatial subdivisions

There are various ways in which the system can be subdivided, the simplest of which is just a series of slabs obtained when the system is sliced along planes parallel to one of the end faces (or edges in two dimensions). Of course any form of subdivision is allowed, but the slabs are the simplest to handle since each slab has only two neighboring slabs. The optimal subdivision is one which minimizes the surface to volume (or area) ratio, but the gain can be outweighed by the additional computations to handle particles near subdivision boundaries; this would have to be explored in greater detail. In the case of the slab subdivision, the slab thickness, and hence the number of slabs used, must be determined empirically. If the slab is too thick there will be a substantial storage penalty because of the dependence on the number of particles per slab, but if too thin excessive processing will be spent on dealing with slab boundaries and the interactions which cross them.

5.3 Performance

Timing measurements were carried out on the Cray YMP for a series of two-dimensional systems ranging in size up to 2.56 million particles. The additional computation time required to deal with subdivision was minimal, amounting to no more than five percent, and attributable primarily to the additional computations involved in processing those interactions that crossed slab boundaries which must be processed twice (once for each slab).

Less than 16 megawords of storage were required for the largest system (which was subdivided into a total of 80 slabs); if a simple leapfrog method is used for integrating the equations of motion only four words of storage must be reserved for particles not in the subregion being processed (or five words if the identities of the particles are also required, as is the case when the simulations are used in modeling certain kinds of flow problem).

5.4 Extension to shared-memory multiprocessors

While the approach based on spatial subdivision shares a lot in common with the implementation using a set of connected processors each with its own private storage, it was pointed out above that it can also form the basis for a version of the program designed to run on a multiprocessor system using shared memory (such as a multiprocessor Cray). One of the problems encountered when several processors attempt to use memory that is common to them all is access conflict. A spatial subdivision of the computation ensures that each processor will spend most of its time working on its own private data, and only when data must be transferred between subregions does the opportunity for conflict arise; the spatial subdivision approach should therefore prove effective in such an environment. In fact two levels of subdivision might be introduced in a multiprocessor implementation of this kind, one that splits the system among the processors, the other in order to economize on storage needed for the layer approach within each processor. This is a matter for future exploration.

6. CONCLUSION

Supercomputers, with their uncompromising demand for careful management of data, present a challenge when it comes to implementing algorithms whose data is not structured in the required way. Molecular dynamics simulation, especially in cases where the interactions are limited to very small interparticle separation, is an example of such a problem. However, by careful reformulation of the problem, it is possible to arrive at a computational scheme whose data is organized in a manner that can be processed efficiently by a vector computer. While the performance achieved in this way is still far from the theoretical maxima claimed for the machines owing to the considerable amount of data rearrangement that goes on throughout the computation, the performance is substantially better than what would otherwise be achieved, and the memory requirements can be kept close to minimal.

ACKNOWLEDGMENTS

Parts of the work described here were carried out during visits to the Universities of Georgia and Mainz, and to the KFA at Julich.

REFERENCES

1. 'Molecular Dynamics Simulation of Statistical Mechanical Systems',Proceedings of the Enrico Fermi International School of

Physics,Course XCVII, Varenna, 1985, eds. G. Ciccotti and W.G. Hoover (North-Holland, 1986).

2. M.P. Allen and D.J. Tildesley, 'Computer Simulation of Liquids' (Oxford, 1987).

3. D.C. Rapaport, Phys. Rev. A 36:3288 (1987), and to be published.

4. R.W. Hockney and C.R. Jesshope, 'Parallel Computers', 2nd edn., (Hilger, 1988).

5. D.C. Rapaport, Computer Phys. Repts. 9:1 (1988).

6. G.S. Grest, B. Dunweg and K. Kremer, preprint (1989).

7. D.C. Rapaport, in 'Computer Simulation Studies in Condensed Matter Physics - Recent Developments II', eds. D.P. Landau et al (Springer, 1989, in press).

CFD WITH THE LATTICE BOLTZMANN EQUATION

F.J. Higuera[a], S. Succi[b], R. Benzi[c]

[a]Dept. of Fluid Mechanics. Univ. Politécnica de Madrid
Pza. Cardenal Cisneros 3, 28040 Madrid, Spain
[b]IBM European Centre for Scientific and Engineering
Computing. Via Giorgione 159, 00147 Roma, Italy
[c]Universita' di Roma "Tor Vergata". Via Orazio Raimondo
00173 Roma, Italy

ABSTRACT

We review recent developments of the Lattice Boltzmann Equation
method (LBE) for incompressible hydrodynamics. Modifications of the
discrete Boltzmann equation for a lattice gas to improve its efficiency as
a numerical method are discussed. Estimations of the minimum lattice size
necessary for high Reynolds number simulations are given, showing that
this minimum arises from physical properties of the flows (range of scales
to resolve) and not from any intrinsic limitation of the lattice gas
method. Applications of the LBE to several problems are presented.

GENERALITIES AND THE LATTICE BOLTZMANN EQUATION

Lattice gases (LG) are completely discrete dynamical systems evolving
according to a set of cellular automata rules which are both simple and
local, in order to make them suitable for massively parallel computation.
They can be figured out as made of a large number of particles traveling
in a sincronised way along the branches of a regular lattice and colliding
at the sites of the lattice, when several particles meet there. The
possibility of using lattice gases to simulate hydrodynamical phenomena is
based on the conservation, during the microscopic evolution of the gas, of
quantities which can be interpreted as the mass and momentum densities of
a fluid. The large scale equations for these densities coincide with the
incompressible Navier Stokes equations when the lattice is sufficiently
symmetric and the macroscopic velocity of the flow is much smaller than
the velocity of the gas particles. See references[2,3] for further details.

The discrete analog of the Liouville equation describing the
evolution of a LG deals with a field of Boolean variables that define the

state of the system by indicating the presence or absence of a particle in each lattice site and direction. In direct LG simulations, this Boolean equation is *exactly* solved on a computer and then a space or space-time averaging over a sufficiently large number of lattice sites is performed to get the macroscopic flow variables. One of the main difficulties of these direct lattice gas simulations is the high level of fluctuations associated with the necessarily limited number of lattice sites that can be used with present computational resources. Estimations of the size of the lattice and of the amount of work necessaries to keep the fluctuations within tolerable bounds[3,4], show that a critical Reynolds number exists, depending on the type of flow, below which the smoothing of the fluctuations consumes most of the computational effort. Critical values of the Reynolds number typically lie beyond the range of present simulations, especially in 3D. An additional shortcoming of 3D simulations is the need to deal with a huge look-up table to perform the collision phase of the lattice evolution[5].

One possible solution is to use the Boltzmann equation for the lattice gas, instead of following the microscopic evolution of its individual molecules. In this way one deals with mean populations, which represent ensemble averages and which are, therefore, free from fluctuations. Even though many interesting features of the lattice gas microdynamics are not captured by the Boltzmann approach, it provides a convenient starting point for the numerical solution of the Navier Stokes equations. The reason is that these equations come out formally from the LBE in the double asymptotic limit of small Knudsen and Mach numbers. The solution of the Boltzmann equation suffices therefore to describe the macroscopic behaviour of realistic fluids. The Boltzmann equation for a lattice gas is[1,2],

$$N_i(\mathbf{x}+\mathbf{c}_i, t+1) = N_i(\mathbf{x}, t) + \Omega_i(N) \qquad (1)$$

where $0 \le N_i \le 1$, $i = 1 \ldots b$, are the mean particle populations in the b possible states per lattice site, and $\mathbf{c}_i$ are the corresponding particle velocities. The variable $\mathbf{x}$ runs over the discrete lattice sites. $\Omega_i(N)$ is the collision operator, depending on the mean populations and on the transition probabilities between couples of site configurations, that define the collision rules of the automaton, see[1,2]. Eq. (1) is a finite differences equation, with floating point variables, that can be solved within the computer precision by the usual two-phase procedure of lattice gas evolution, i.e. evaluating the changes of the populations at each site under the action of the local and instantaneous collisions and then shifting the new populations to the neighbouring sites pointed by the corresponding velocities. A global H-theorem can usually be proved for Eq. (1), guarantying that the method based on it is intrinsically unconditionally stable. (Ref. 1, Appendix F).

The restriction to no more than one particle per lattice site and direction, which is essential to describe the evolution of the gas in terms of Boolean variables, becomes unnecessary when dealing with mean populations in the Boltzmann approximation. When this exclusion principle is suppressed, the mean populations become unrestricted non-negative quantities and the Bose-Einstein statistic replaces the Fermi-Dirac one. An extension of the formalism to this case was carried out[6], including the introduction of an appropriate entropy function and the evaluation of the viscosity coefficient through the Chapman-Enskog expansion.

THE COLLISION OPERATOR AND SIMPLIFIED COLLISIONS

The probability of having a particular configuration at a lattice site is given, in the Boltzmann approximation, by the product of the probabilities of having or not having a particle in each of the b directions at that site. These probabilities coincide with the b mean populations, or their complements to the unity, at the site considered. The collision term $\Omega_i(N)$ in Eq. (1) is the sum over all the possible couples of *in* and *out* configurations of the *in* probability multiplied by the corresponding *in* → *out* transition probability. and by the population change (± 1, or 0) that that particular transition represents for the direction i. $\Omega_i(N)$ is therefore a high degree polynomial function of the mean populations, generally too complex for extensive numerical evaluation, and sensitive to round off errors. However, only the linear part of the collision operator in a power expansion about a uniform equilibrium state of zero velocity and constant density, (say ρ_o), matters for the Chapman–Enskog asymptotic expansion that leads to the incompressible Navier Stokes equations for small Knudsen and Mach numbers.[4] Based on this consideration, a simplified Boltzmann equation was proposed by keeping only the essential elements of Eq. (1) necessary for recovering the first few terms of the double asymptotic expansion. The simplified equation is

$$N_i(\mathbf{x}+\mathbf{c}_i, t+1) = N_i(\mathbf{x}, t) + \Omega_{ij}\left[N_j(\mathbf{x}, t) - N_j^{eq}(\mathbf{x}, t)\right] \qquad (2)$$

where $\Omega_{ij} = (\partial\Omega_i/\partial N_j)_{N=\rho_o/b}$ is the constant symmetric matrix of the linearised collision operator, and

$$N_i^{eq} = \frac{\rho_o}{b}\left\{ 1 + \frac{D}{c^2} c_{i\alpha}v_\alpha + G(\rho_o) Q_{i\alpha\beta}v_\alpha v_\beta + \frac{\Delta\rho}{\rho_o} \right\}, \qquad (3)$$

with

$$G(\rho_o) = \frac{D^2}{2c^4 b} \frac{b - 2\rho_o}{b - \rho_o}, \qquad Q_{i\alpha\beta} = c_{i\alpha}c_{i\beta} - \frac{c^2}{D}\delta_{\alpha\beta}, \qquad c = |\mathbf{c}_i|,$$

is the new equilibrium distribution function for Eq. (2), depending on the density variation $\Delta\rho$ about the reference value ρ_o , and on the velocity $\mathbf{v}$, given by

$$\rho_o + \Delta\rho = \sum_{i=1}^{b} N_i , \qquad \rho_o v_\alpha = \sum_{i=1}^{b} c_{i\alpha}N_i . \qquad (4)$$

It must be emphasized that, although the simplified formulation was suggested by the small Knudsen and Mach number expansions of the complete Boltzmann equation, and the equilibrium solution (3) coincides with the first few terms of the low Mach number expansion of the equilibrium solution for the complete Boltzmann equation, Eq. (3) is exact in the framework of the present formulation, i.e. it does not depends on the order of the Mach number.

All the required information about the collisions is contained into the matrix elements of the linearised collision operator. This is a symmetric $b \times b$ matrix whose element Ω_{ij} represents the change through collisions of the i-th direction population due to a unit excess of the j-th direction population relative to the local equilibrium value given by Eq. (3). As such, Ω_{ij} can depend only on the angle between directions i

and j, that has a limited set of values for a given lattice. See ref.[7] for details on usual lattices.

The number of independent matrix elements is further decreased by the conditions of conservation of mass and momentum, requiring that $D+1$ eigenvalues of the collision operator be zero, with the null space being spanned by the $D+1$ vectors

$$M_0 = (1\ldots 1)\ ,\qquad M_\alpha = (c_i)_\alpha\ ,\qquad i = 1\ldots b\ ,\qquad \alpha = 1\ldots D. \qquad (5)$$

The other eigenvalues of the collision operator must be negative and larger than -2 to ensure the decay through collisions of the non-equilibrium part of the distribution function in the evolution of the discrete system. They were calculated[7] as functions of the independent matrix elements of the collision operator. One of them, say λ, determines the viscosity coefficient in the Chapman-Enskog expansion. Its eigenspace is generated by the $D(D+1)/2-1$ independent elements of the set

$$(Q_i)_{\alpha\beta}\ ,\qquad \alpha,\beta = 1,\ldots,D \qquad (6)$$

The numerical scheme based on Eq. (2) is linearly stable if the eigenvalues of Ω_{ij} verify the conditions $-2 \le \sigma \le 0$, see[7].

OPTIMAL EFFICIENCY FACTOR. ESTIMATIONS OF THE AMOUNT OF WORK

The ratio of the largest (L) to the smallest (δ) scales appearing in a high Reynolds number flow is

$$L/\delta = O(Re^\gamma) \qquad (7)$$

where γ depends on the type of flow. Typically $0 \le \gamma \le 1$.

On the other hand, the Reynolds number attained in a LG simulation is

$$Re = R^* M\ L \qquad (8)$$

in terms of the size of the lattice L, referred to the lattice pitch, the Mach number M, and the efficiency factor R^*, given by[1],

$$R^* = \frac{g(\rho_0)\ c_s}{\nu}\ , \qquad (9)$$

where $c_s = c/D^{1/2}$ is the sound speed, $g(\rho_0) = \dfrac{2\ c^4\ b}{D(D+2)}\ G(\rho_0)$ is the usual factor arising from the lack of Galilean invariance, and

$$\nu = \frac{c^2}{D+2}\left(-\frac{1}{\lambda} - \frac{1}{2}\right) \qquad (10)$$

is the kinematic viscosity. Full control on the efficiency factor is achieved through the simplified form of the Boltzmann equation, Eq. (2), because ν can be varied at will by changing λ, i.e. by tuning a few parameters in the matrix of the collision operator.

Minimization of the lattice size requires that the lattice pitch be only moderately smaller than δ. Relations (7) and (8), (with $\delta = 1$), are sketched in Fig. 1. An optimum value for the product $R^* M$, leading to minimum waste of lattice points while still allowing a complete

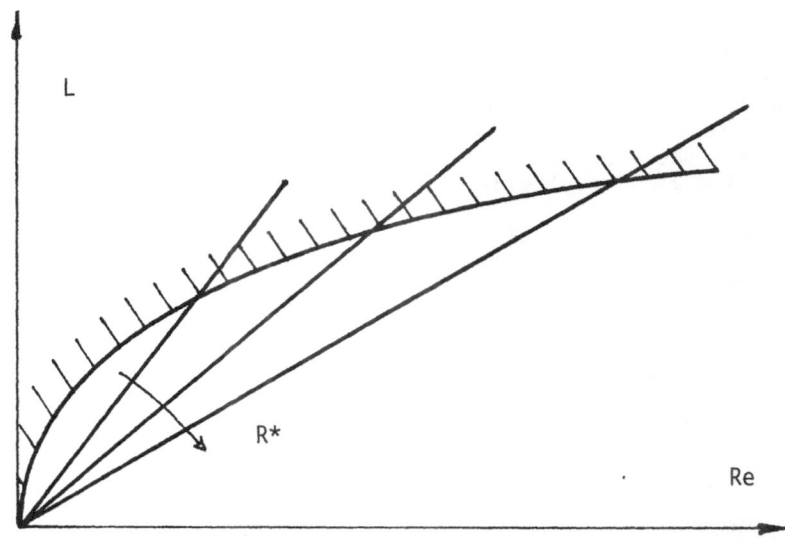

Fig. 1 Optimal efficiency factor and lattice size in terms of the Reynolds number.

description of the flow features, corresponds to the shaded region slightly above curve (7). This imply that R^* should increase with the Reynolds number of the flow. Note that this estimation of the number of lattice sites required for a high Reynolds number simulation is the same as for classical numerical methods. The minimum is dictated by the physical features of the flow and has nothing to do with any limitation of the LG method.

A standard way of saving memory and processing work in the usual cases in which steep changes coexist with regions of low spatial activity is provided by non uniform grids. Some preliminary results on the application of non uniform grids to LG computations are presented in reference[8]. The idea is that, in spite of the fact that the lattice carrying the particles of the gas does not allow for the necessary flexibility, a coarser non uniform grid can be superimposed on the fine lattice in those regions for which the lattice resolution is excessive, and the bulk of the computation can be carried out on the coarse grid. The same populations are assigned to all the lattice sites within the region spanned by a single grid cell and the interaction between neighbouring cells is accounted for, during the streaming phase of the gas evolution, in accordance with the number of lattice links joining the cells along each direction. The exchanged populations are then equidistributed among all the lattice sites in the cell. It was shown[8] that the new "equidistribution" phase increases the entropy of the gas.

Another difficulty encountered in the application of the LG method to incompressible fluid dynamics, shared by many classical numerical methods, is associated to dealing with the incompressibility condition for the velocity field, *div* **v** = *0*. Since the mass and momentum conservation equations for a LG are evolution equations, like for a real gas, a very low Mach number is necessary to rule out compressibility effects by keeping the acoustic time much shorter than the characteristic time for the evolution of the flow, which is the residence time for high Reynolds number flows. In addition, for lattice gases, spurious terms of order M^2 relative to the other terms in the momentum equation must be kept below appropriate limits. This, however, is exceedingly expensive in computational terms, because the time stepping for the LG automatically follows the faster time scale, resulting in a very large number of elementary time steps for the realistic simulation of non steady flows, of the order of the lattice size divided by the Mach number.

Possible modifications of the LG method to improve its efficiency by allowing some reduction in the disparity of time scales involved are presently under study[9,10].

APPLICATIONS

. We summarize here some details on the applications of the LBE to three different problems whose numerical treatment is presently under way at the IBM-ECSEC centre of Rome.

Three-dimensional flows in porous media

One of the mostly advocated merits of the LG method, besides the fact of being an ideal candidate for massive vector and parallel computing, is its flexibility with respect to complex geometries. This is easily understood by recalling that, even though based on a very stylized cellular automaton dynamics, the LG method is basically a particle tracking technique, and as a such, it allows to treat the intricacies associated with complex boundary conditions in simple terms of particle reflections and bounces at appropriate spatial locations flagged as wall sites. These properties point to the LG method as to an excellent candidate to address a long-standing problem in the physics of random media, namely the calculation of transport coefficients in terms of the medium's microscopic geometry. Pioneering work along these lines has been developed by Rothman[11], who applied the six-bit FHP Boolean automaton to the study of two-dimensional flows in porous media.

Very recently, we have extended Rothman's work to three dimensional cases[12]. Following Rothman, prior to the simulation of the random medium, we assessed the range of parameters in which the LBE is indeed a faithful picture of the Navier-Stokes equation. In this respect, a crucial parameter is the size of the open pores. In fact, if the pore size, say h, is too small, then it can become comparable to the mean free path of the particles and no fluid behaviour can be obtained within the pore. On the opposite side, the simulation rapidly becomes too expensive as h increases. A relatively large pore size might, however, prove convenient when simulating flows at non small Reynolds numbers, because otherwise the spatial resolution could be very poor, or the flow speed could grow up to the point where compressibility effects would invalidate the very theory on which the LG method is based.

To test this point, we have considered the Poiseuille flow in a channel with a square cross section of size H^2. The flow in the channel is generated by an uniform body force f acting along the stream direction, (x), with an intensity calculated to produce a given preselected flux, $Q_t = 0.035 \ (h^4/\nu) \ f$, (whose expression arises from the solution of the Poisson equation $\nabla^2 v = -f/\nu$ with Dirichlet boundary conditions at the channel walls). Here ν is the kinematic viscosity as taken from lattice gas theory. The range of validity of the LB method is judged by comparing the numerical flux $Q = \int v \ dy \ dz$ with the theoretical one Q_t for several values of H. The result for the standard FCHC scheme is that at an average flow speed $U = 0.1$, and with $v = 0.03$, a channel as small as 4×4 sites is large enough to reproduce fluid behaviour, in the sense that Q matches Q_t within less than one percent. This value of $H = 4$ is about 3 times smaller than the one found by Rothman. This could somehow be attributed to the fact that in the Boltzmann approach each particle is already an average over several histories and is consequently supposed to carry a certain amount of global information which has no counterpart in the Boolean formulation.

Preliminary simulations of random media were performed with a small resolution 32^3 cubic lattice and $v = 0.03$. The complex medium was modelled as a random sequence of elementary blocks of four units in size distributed in such a way that no free pore smaller than 4×4 lattice units in cross section can appear. On the surfaces of the blocks, as well as on the lateral walls, no-slip boundary conditions were imposed, i.e. particles impinging in a given direction were reflected back into the fluid along the opposite direction. Along the streaming direction periodic conditions were applied.

The first goal of this investigation was to test the validity of Darcy's law,

$$\mathbf{v} = - \frac{k}{\rho_0 v} \nabla P \; , \qquad\qquad (11)$$

for low Reynolds numbers (based on the pore size). Here k is the permeability of the medium and $P = p - \rho_0 f x$ is the total pressure. This test was carried out by varying the total pressure drop via the body force, and measuring the average flux through the porous medium. The results are indicated in Figure 2, where the numerical flux Q is displayed as a function of ∇P for three different values of the porosity $\phi = 0.635$, 0.5, and 0.375

From this figure, we see that Darcy's law is well fulfilled for values of ∇P up to 1.2×10^{-4}, (with $\rho_0 = 7.872$), corresponding to an effective average speed in the range of a few milli-lattice units per step. The data reported in Figure 2 correspond to a maximum Reynolds number of order $0.3 - 0.6$, calculated from to Eq. (8) with $R^* = 7.57$. This seems an appropriate range for viscosity dominated flows. No significant deviations from the linear Darcy's law have been observed for values of the Reynolds number up to about 5. For the present simulation, ($H = 32$ and $v = 0.03$), Eq. (11) and the results in Fig. 2 yield $k \simeq 5, 2$, and 0.75 , for $\phi = 0.635, 0.5$, and 0.375, respectively.

These values are in qualitative agreement with theoretical estimations based on simple pictures of shrinking-tube and grain consolidation models. Quantitative agreement is prevented by the fact that the medium simulated here cannot literally be regarded as a porous medium, on account of the large size of the solid blocks with respect to the total size of the domain. For instance, relatively free channels of size $h \simeq 4 - 8$ lattice units were often detected in the simulations which, according to the Poiseuille law in a square channel, would yield alone a substantial contribution to the global permeability, of the order of $0.5 - 2$ lattice units squared.

The main result of this study was the demonstration of the adherence of the LB model to Darcy's law for a three-dimensional flow through a complex medium. Even though the ultimate confirmation of the validity of

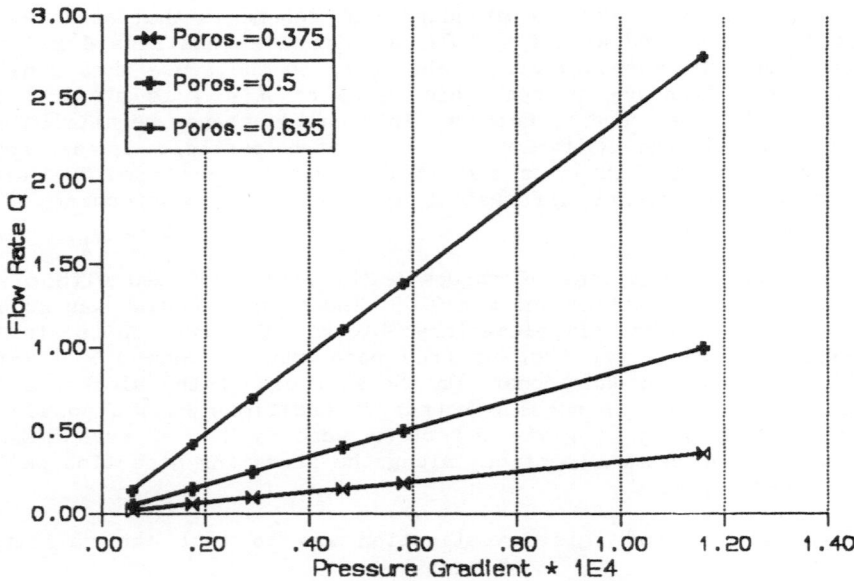

Fig. 2 The volumetric flow Q as a function of the pressure gradient
for three different values of the porosity, ϕ = 0.635, 0.5, 0.375.

the LB method must necessarily come from laboratory experiments, we
believe that LBE can help to gain new insights into rock physics and other
related fields. To this end, higher resolution simulations and extension
to higher Reynolds number flows, beyond the range of Darcy's law, are
certainly needed.

Bifurcations in a two-dimensional Poiseuille flow

It has been recently shown that obstacle induced instable shear
layers can be used to excite the Tollmien-Schlichting waves for the
Poiseuille flow in a channel containing the obstacles, and so promote
instabilities of the flow in the channel flow well below the critical
Reynolds number for the same flow in a clean channel[13]. Apart from the
practical motivation of being a mechanism suitable for sustaining very
large heat and mass transfer at relatively low Reynolds numbers without
incurring, therefore, in the cost due to energy dissipation at higher
Reynolds numbers, these instabilities provide an interesting scenario for
a transition at moderate Reynolds numbers, typically of the order of
hundred. We have studied this problem by considering a plane channel
containing a periodic array of identical plates. The plates, one fifth of
the channel width, are disposed normally to the flow stream and spaced 6.6
times the channel half-width. This spacing is about three times the
wavelength of the channel mode that we intend to destabilize, which is
among the least stable modes of the Poiseuille flow in the range of
Reynolds numbers of interest, with the critical layer by the middle of the
channel.

The primary bifurcation from the steady flow can be described within
the framework of Landau's equation for the complex amplitude $A = \rho\, e^{i\vartheta}$ of
the primary bifurcated mode,

$$\frac{d\rho}{dt} = S_r \rho - L_r \rho^3$$

$$\frac{d\vartheta}{dt} = S_i - L_i \rho^2$$

(12)

where $S = S_r + iS_i$ and $L = L_r + iL_i$ are two complex coefficients characterizing the transition. In the vicinity of the critical point, $(S = 0)$, they can be taken in the form $S_r = \alpha r$ and $S_i = S_0 + \beta r$, where $r = (Re^r - Re)/Re$ is the reduced Reynolds number. $(Re = U H/2\nu$, where H is the width of the channel and U is the maximum velocity of a Poiseuille flow leading to the same flux as the actual flow.) For a supercritical bifurcation, $L_r > 0$, which seems to be the case here, the steady-state solution of the equations (12) above the critical threshold $(r > 0)$ is

$$\rho = \sqrt{S_r/L_r} = O(r^{1/2}) \quad , \qquad \omega \equiv \frac{d\vartheta}{dt} = S_0 + \left(\beta - \alpha \frac{L_i}{L_r} \right) r \qquad (13)$$

which shows the typical square root dependence of the amplitude on the bifurcation parameter. We found that the critical Reynolds number Re_c is about 86, i.e. much smaller than the critical Reynolds number for the same flow in the absence of any obstacle. For $Re > Re_c$ the flow exhibits a stable limit cycle with frequency $\omega = 0.38$, (measured in units of $2U/H$), in close agreement with the results of Karniadakis et al[13]. The streamlines of a typical flow configuration for $Re \simeq 100$ are shown in Figure 3. Beyond $Re = 90$ a secondary bifurcation of much smaller amplitude and higher frequency was observed.

The $r^{1/2}$ scaling of the primary bifurcated mode is roughly recovered even though conspicuous fluctuations are present. On the contrary, the linear correction of ω to ω_c was found to be too small to be detected, if it exists at all. A complete characterization of the transition which requires the numerical determination of the coefficients appearing in the equations (12) is still under way.

Fully developed two-dimensional forced turbulence

We have already noticed that Reynolds numbers as high as allowed by the finite size of the grid can in principle be accessed with the LBE method at the same overall cost as with other classical numerical methods. In this context, homogeneous turbulence provides a particularly severe test, in the sense that all the scales ranging from the macroscopic scale L to the dissipative scale $\delta = L/Re^{0.5}$ do play an important role in the flow evolution. It is therefore of primary interest to test the ability of LBE to reproduce the basic physics of fully turbulent flows as well as its computational efficiency with respect to other conventional techniques. To this purpose, we have studied a two-dimensional flow in a square box of size L with periodic boundary conditions and a periodic long-wave forcing of the type $F = F_0 \sin(2\pi py/L)$ along the x axis on the wave number $p = 4$. The results of the LG method are compared with those of a Fourier pseudo-spectral method. Details on the implementation of the spectral method and on the correspondence between its parameters and those of the LG method (namely number of Fourier nodes/lattice sites, and time step equivalence) will be given elsewhere[14]. As a first instance, we have compared the results for a 64^2 grid, i.e. at a moderately low resolution. The total energy and enstrophy for both LBE and spectral simulations are shown in Figure 4 as a function of time. From this figure we see that indeed the two simulations yield quite similar results, even though the intermittent spikes of enstrophy are somewhat higher in the LBE model.

To get a more refined idea of the agreement, we have increased the

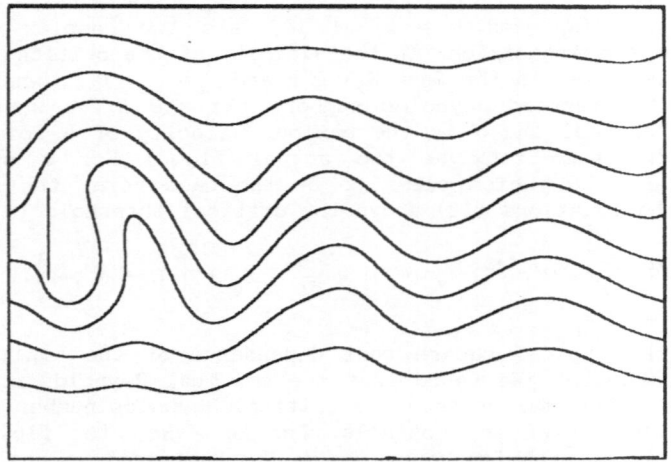

Fig. 3 Streamlines of the unstable flow at *Re ≃ 100*.

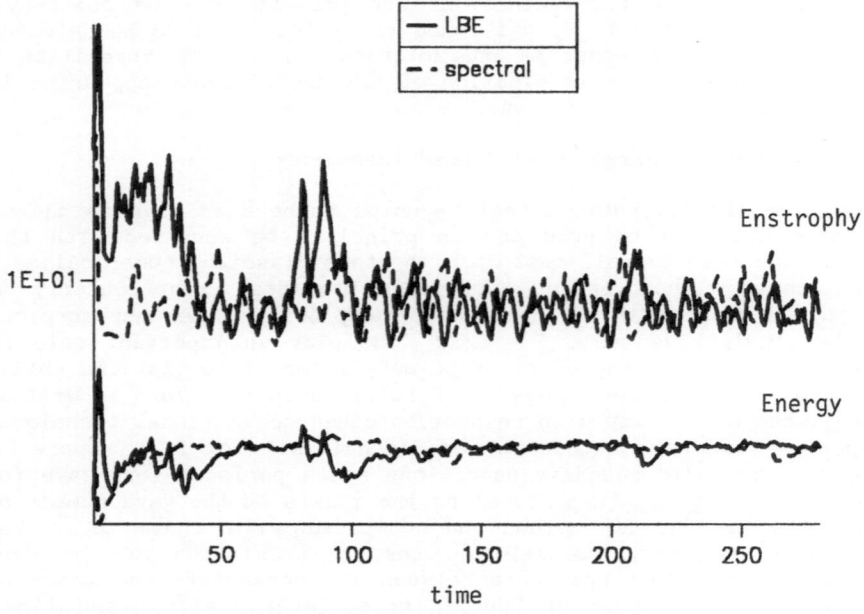

Fig. 4 Time evolution of the total energy and enstrophy for the spectral and LBE models.

resolution by a factor two and repeated the experiment with the same values of the other parameters. The energy spectra corresponding to LBE and spectral simulations after about *150* times the residence time in the computational square (a region of time free of enstrophy spikes) are reported in Figure 5, from which a remarkable agreement is well visible. The flat high-wave number region of these spectra is due to an excess of resolution with respect to the actual Reynolds number of the simulation: the remarkable point is that the flat region begins at the same wave number for both spectra.

However, the hardest challenge for LBE is to test whether it is able to reproduce the statistical properties of two-dimensional fully developed turbulence, namely the enstrophy inertial range at high k and the energy inertial range at low k. To this purpose we have performed a series of high-resolution experiments on a 512^2 grid. A typical series of energy spectra for five almost steady energy spectra is shown in Fig. 6, from which we see that the predicted k^{-3} law is not obtained and much steeper spectra with slope $\simeq$ *4-5* are observed instead. This discrepancy may tentatively be attributed to the absence in the LBE model of any sort of superviscosity effects, i.e. higher order dissipative terms of the form ∇^q, with $q > 2$ that allow coherent structures to survive down to small scales which would otherwise be over damped by normal viscosity ($q = 2$). Indeed, the absence of small-scale structures is well visible by looking at an instantaneous vorticity map (see Figure 7). This figure highlights the presence of strong small-scale agglomerates which are the result of the inverse energy cascade producing energy pile-up at low wave numbers. This energy pile-up is easily removed in pseudo-spectral simulations by adding an inverse Laplacian term[15]. The inclusion of both superviscosity and inverse dissipation is quite straightforward in the Fourier spectral method, since it reduces to a trivial algebraic replacement. (of the type $\nabla^2 \rightarrow -k^2$). Unfortunately, there's nothing equivalently simple in a lattice approach. We may summarize these results by stating that the LBE is indeed a faithful reproduction of the Navier-Stokes equation; however, it doesn't lend itself to the same kind of numerical tricks which are readily available in a spectral context to increase the effective resolution.

Having appreciated the physical picture, we can proceed to examine the computational efficiency of the methods. On a single processor of the IBM 3090 vector multi processor, the LBE required *60 ms/step* and the spectral code (no dealiasing and a 2 step time-marching scheme) *300 ms/step*. Each step of the spectral method requires two units of computational work to advance *0.01* physical units, as opposed to LBE which requires one unit of work to advance *1/382* physical units. Accordingly, the LBE method supersedes the spectral method by a factor *5/2*. Furthermore, the number of floating-point operations (N_{fpo}) per unit of computational work is $N_{fpo} \simeq 150 N^2$ for the LBE and $N_{fpo} \simeq 25 (log_2 N) N^2$ for the spectral method[16]. For $N = 128$, the number of operations is already the same for the two methods, and above this value the logarithmic factor will penalize the spectral method. This might become an important feature as future developments of computer technology push the frontier of large-scale computing further ahead.

CONCLUSIONS

We have reviewed the formulation of the LBE method and presented a variety of fluid-dynamics applications ranging from laminar to fully turbulent flows. Quantitative analysis and comparisons with other well-established numerical techniques seem to indicate that, even though a large margin of improvement is still left, LBE provides a new viable tool

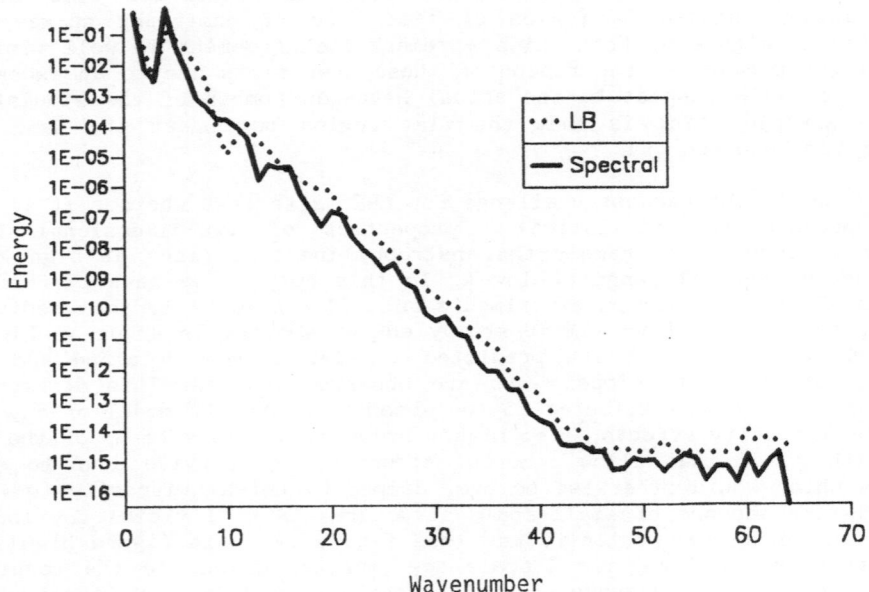

Fig. 5 Energy spectra for the spectral and LBE simulations on a 128^2 grid.

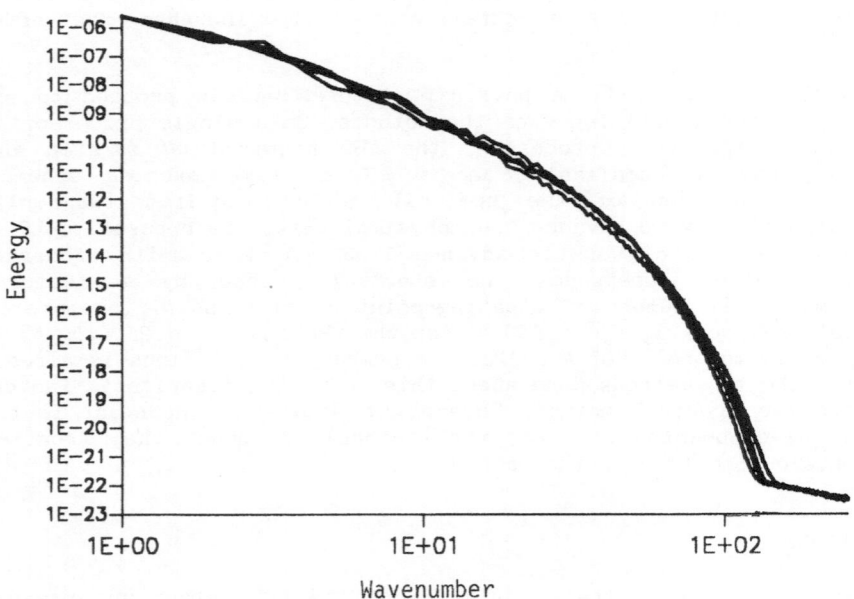

Fig. 6 Energy spectra for the LBE simulation on a 512^2 grid.

Fig. 7. Instantaneous vorticity map for a 512^2 LBE simulation.

for two and three-dimensional computational fluid dynamics.

Acknowledgements: R.B and F.H. wish to thank the support of IBM European Center for Scientific and Engineering Computing where all the numerical simulations were performed.

REFERENCES

1. U. Frisch, D. d'Humieres, B. Hasslacher, Y. Pomeau, and J.P. Rivet, Complex Systems 1, 646 (1987). 6 (1987).
2. S. Wolfram, J. Stat. Phys. 45, 471 (1986).
3. G. Mc Namara and G. Zanetti, Phys. Rev. Lett. 61, 20 (1988).
4. F.J. Higuera and J. Jiménez, Europhys. Lett. to appear (1989).
5. J.P. Rivet, Ph.D. Thesis, Univ. de Nice, June 1988, and references therein.
6. F.J. Higuera, Proc. Workshop on Lattice Gases, Torino (1988). World Scientific 1989, Singapore. ingapore.
7. F. Higuera, S. Succi and R Benzi, Europhys. Lett., to appear (1989).
8. F.J. Higuera and S. Succi, 5th Int. Symp. Num. Eng. Lausanne, July (1989).
9. C. Treviño and F.J. Higuera, 10th Brazil. Congress Mech. Eng. Rio de Janeiro, Dec. 1989.
10. F.J. Higuera, in preparation.
11. D. Rothman, Geophys., 53, 509 (1988).
12. S. Succi and E. Foti, Europhys. Lett., submitted.
13. G.E. Karniadakis, B.B. Mikic, and A.T. Patera, J. Fluid Mech. 192, 365 (1988).
14. R. Benzi and S. Succi, in preparation.
15. B. Legras, P. Santangelo and R. Benzi, Europhys. Lett. 5, 37 (1988).
16. G.S Patterson and S. Orszag, Phys. Fluids 14 (1971) 2538.

FLOWS (II): ILLUSTRATIONS

PART II

SIMULATION AND SYMMETRY OF SIMPLE AND COMPLEX FLOWS

M. W. Evans and D. M. Heyes

Department of Chemistry
Royal Holloway and Bedford New College
University of London, Egham, Surrey TW20 OEX, UK

1. INTRODUCTION

In this study we interpret the microscopic simulation of flows in a new way, making use of the flow and property symmetry as embodied in the *new* principles of group theory statistical mechanics.

The hydrodynamics of complex flows is a subject which has been based historically on constitutive equations derived from the fundamental equations of motion in the classical approximation.[1,2] This approach "washes out" the microscopic structure of the fluid, and is forced to ignore the fact that this is generally molecular in nature. D.J. Evans has shown that the constitutive equations for a molecular (structured) fluid such as carbon tetrachloride become numerous, complicated, and insoluble within the reasonable future.[3] The many body dynamics are traditionally kept under control by approximations which not only ignore the existence of molecules but which also lead to severe contemporary controversy over frame indifference, a problem arising *purely* from the constitutive equation methodology.[4]

The consequencies of ignoring the details of molecular dynamics in fluids under elongational and shear stress include the following.

1) The constitutive approach remains empirical, *i.e.*, 'Newtonian' or 'non-Newtonian' according to whether the stress / strain relation appears linear or not.

2) Empiricism, as usual, means that the traditional approach is not predictive, but descriptive, in nature. A general solution of flow in atomic or molecular ensembles requires the fundamental equations of motion in the classical approximation applied to the many body problem through computer simulation.

3) The constitutive approach misses some fundamental details of the response of fluids to shear stress.[5-7] This happens even in the 'simplest' of fluids, composed of atoms. Examples of this basic weakness in the constitutive approach are given later in this article.

4) The computer power needed to implement the constitutive approach is about the same as that needed in the more fundamental approach made possible by

recent advances in computer simulation. So we might as well use the latter whenever possible.

5) The fundamental approach attacks problems at the fundamental level, and once these are solved, a new world of possibilities is there for the exploration. It becomes possible, for example, to implement group theory and symmetry to reveal the presence of indicator functions which appear in a fluid in response to shear. Their presence leads directly to new types of spectra which are characteristic of the rheology of a sheared liquid. An example is depolarised light scattering.[8]

Recent work, using nonequilibrium molecular dynamics (NEMD) and Brownian dynamics (BD) simulation has shown that the response of an atomic ensemble to shear and elongational stress is a rich landscape of interwoven patterns, a picture which can be drawn in terms of cross correlation functions (c.c.f.'s) governed by the three principles of group theoretical statistical mechanics (g.t.s.m.).[9-15] Time correlation functions in the transient (non-equilibrium) state can be used to generalise the Green-Kubo relations and linear response theory in one simple and elegant equation.[16] The third principle of g.t.s.m. applies also to these transient c.c.f.'s, and is ideal for the description of shearing and elongational stress in terms of symmetry. In molecular liquids considerations of symmetry apply not only in the laboratory frame (X, Y, Z) but also in the molecule fixed frame (x, y, z) as defined in the standard point group character tables.[17,18] This implies that the response of a molecular fluid under complicated external stress can be analysed precisely with a variety of new indicator c.c.f.'s in both frames. Not only are these symmetry-specific to the type of stress (shear, elongation, compression, or any combination, [19-21]) but also to the point group of the molecule in frame (x, y, z). The effect of stress can be studied in both frames (X, Y, Z) and (x, y, z) by NEMD and Brownian Dynamics, BD simulation,[22,23] implementing the SLLOD equations, profile unbiased thermostatting (PUT) or any other numerical technique.

This represents a significant advance over the constitutive approach to rheology, which is not able to describe flow at the atomic or molecular level, and in consequence is not able to describe or anticipate the existence of indicator c.c.f.'s.

Another major advantage of the fundamental approach is that it reveals phenomena on the picosecond/angstrom scale of a computer simulation which remain valid on the second/metre scale of laboratory and industrial flow processes. This follows quite clearly from the fact that the classical equations of motion are valid on both scales. A one-to-one relation can be developed straightforwardly between simulation and observation, and it follows that the former is capable of predicting experimental observables on the second/metre scale. Several master curve and extrapolative techniques for doing just this have been developed recently by Heyes and co-workers.[23,24] It follows that in the design of specific materials with specific and predictable response to imposed stresses of various kinds, fundamental computer simulation is the obvious answer.

The fundamental technique has been applied with notable success by Rapaport and co-workers,[25] and by Clementi and co-workers[26-29] to flows around objects on the picosecond/angstrom scale, revealing on this scale the presence of vortices, eddies, and other flow phenomena whose behaviour extrapolates linearly to the second/metre scale. These phenomena obey symmetry principles that can be used to relate them to indicator c.c.f.'s at the fundamental level and thus to close the gap between traditional hydrodynamics and molecular dynamics. Some of these symmetries are discussed later, in the context of flow past objects, hydrodynamic instability (Rayleigh-Benard phenomena), and in other less well-known contexts such as rotating electric fields as first tackled by Born,[30] in the early twenties of this century and demonstrated experimentally by Lertes,[31], Grossetti,[32] and Dahler,[33]. This elegant work shows that spinning electric fields can physically rotate a liquid at up to MHz frequencies, and is the historical precursor of electrorheology, an important contemporary technique. G.t.s.m. can be applied in all these contexts, and in combination with NEMD and BD simulation, has already produced indications of the presence of explanations and phenomena unknown to the constitutive approach. Examples described later include 1) the first explanation for the Weissenberg effect in terms of pressure tensor correlation functions; 2) the discovery by combined simulation/symmetry of thermal conductivity produced by simultaneous shear and elongational stress in atomic ensembles.

2. SYMMETRY AND FLOWS

Symmetry is simple and powerful, and its application to complicated flow is no exception. Of key importance in this application are the three principles of g.t.s.m.,[11−15] recently developed and stated in terms of contemporary point group theory. The first principle is the Neumann/Curie Principle applied with group theory, and states that the ensemble average over a product of scalars, vectors or tensors exists in the laboratory frame if its product of representations contains at least once the totally symmetric irreducible representation of $R_h(3)$, the point group of all rotations and reflections. If the ensemble is chiral, the relevant point group becomes $R(3)$ of all rotations. The second principle applies principle (1) to the molecule fixed frame (x, y, z), where the relevant totally symmetric irreducible representation is that of the molecular point group, e.g. C_{2v} for water or T_d for carbon tetrachloride etc. The third principle is a simple but very powerful cause effect theorem which states that if an external field of force is applied to an ensemble of atoms or molecules, that ensemble develops transient or steady state ensemble averages with the same symmetry as that of the applied field itself. In other words the symmetry of cause and effect is identical.

Principle (3) is found to be particularly useful for NEMD and BD simulation of complicated flows. It applies across the whole of mechanics, classical, statistical, quantum and grand unified, and also in other contexts, such as classical, quantum, and relativistic electromagnetic field theory, both in linear and non-linear contexts.

In order to implement these principles it is necessary to define the irreducible representations of scalar, pseudoscalar, vector and tensor quantities in the point group of interest. In the context of principle three, the field symmetry is defined through these irreducible representations in the point group of the ensemble (i.e. $R_h(3)$ or $R(3)$). For $R_h(3)$ the set of irreducible representations consists of the the D representations, $D_g^{(0)}, \dots\dots\dots\dots\dots D_g^{(n)}$ and $D_u^{(0)}, \dots\dots\dots\dots\dots, D_u^{(n)}$. In $R(3)$ the set consists of, $D^{(0)}, \dots\dots\dots\dots\dots, D^{(n)}$. Thus, scalar, vector, or tensor quantities in $R_h(3)$ can be either positive (g) or negative (u) to the parity inversion operation, which takes (X, Y, Z) to (-X, -Y, -Z) and which is defined as the operation, $\hat{P} : (r, p) \to (-r, -p)$, where r and p denote position and linear momentum respectively. Of additional interest is the time reversal operator, defined as $\hat{T} : (r, p) \to (r, -p)$. In $R(3)$ the g and u subscripts are undefined because the $\hat{P}$ operator takes the ensemble to the opposite enantiomer, a distinctly different entity, and is therefore not a valid group theoretical operation.

In $R_h(3)$ a scalar quantity such as mass is $D_g^{(0)}$; a pseudoscalar is $D_u^{(0)}$; a polar vector such as linear momentum or position is $D_u^{(1)}$; an axial vector such as angular velocity or vortex is $D_g^{(1)}$; and higher order tensors are u or g. The totally symmetric irreducible representation is $D_g^{(0)}$. In $R(3)$ the same definitions apply without subscripts, so that the symmetries of a scalar and pseudoscalar are identical in $R(3)$; as are those of polar and axial vectors and so on. The totally symmetric irreducible representation is $D^{(0)}$. In order to build up the symmetries of ensemble averages over products of two or more quantities, such as time c.c.f's of the generic type $< A(0)B(t) >$, we use the Clebsch Gordan Theorem,

$$D^{(n)}D^{(m)} = D^{(n+m)} + D^{(n+m-1)} + \ldots\ldots\ldots + D^{(|(n-m)|)}, \tag{1}$$

with the rule for subscript multiplication, $gg = uu = g; ug = gu = u$, in $R_h(3)$. If we are dealing with the nine element tensor $< v(t)v(0) >$, where v is atomic or molecular centre of mass velocity, for example, the product of representations in $R_h(3)$ is,

$$R_h(3) : \Gamma(v)\Gamma(v) = D_u^{(1)}D_u^{(1)}$$
$$= D_g^{(0)} + D_g^{(1)} + D_g^{(2)}, \tag{2}$$

so that the signature of the complete c.c.f. tensor is the sum of scalar, vector and second rank tensor symmetries. The complete tensor is overall positive to $\hat{P}$ and $\hat{T}$. Similarly, the signature of the c.c.f. $< v(t)\omega(0) >$, where v and ω are respectively the linear and angular velocities of a diffusing molecule, is

$$R_h(3) : \Gamma(v)\Gamma(\omega) = D_u^{(1)}D_g^{(1)}$$
$$= D_u^{(0)} + D_u^{(1)} + D_u^{(2)}, \tag{3}$$

which is negative to $\hat{P}$ and positive to $\hat{T}$.

In the point group $R(3)$ of chiral ensembles the representation of both c.c.f.'s is the same,

$$R(3) : \Gamma(v)\Gamma(v) = \Gamma(v)\Gamma(\omega) = D^{(0)} + D^{(1)} + D^{(2)}, \tag{4}$$

with positive time reversal symmetry. Principle (1) shows therefore that the scalar component of $< v(t)v(0) >$ exists in $R_h(3)$ but all elements of $< v(t)\omega(0) >$ vanish. In $R(3)$ the $D^{(0)}$ element of the latter c.c.f. exists as a pseudoscalar quantity. To apply principle (3) we need to examine the symmetries of flow fields in both point groups. Some examples of cause and effect relations obeying principle (3), together with their $\hat{P}$, $\hat{T}$ and D symmetries are as follows,

Linear force: Acceleration, - , +, $D_u^{(1)}$

Torque: Angular acceleration, + , + , $D_g^{(1)}$

Electric field (E): Polarisation, -, +, $D_u^{(1)}$

Electric field (E): $< v(t)\omega(0) >$, -, +, $D_u^{(1)}$

Magnetic Field (B): Magnetization, + , -, $D_g^{(1)}$

Shear Stress: Off diagonal $< v(t)v(0) >$, + , +, $D_g^{(1)} + D_g^{(2)}$

Elongational stress: Diagonal $< v(t)v(0) >$, $+$, $+$, $D_g^{(0)}$

Rotating E: Bulk rotation, $+$, $-$, $D_g^{(1)}$

Above, the first entry records Newton's second law of motion, whereby force results in a net acceleration of an N body ensemble with the given symmetries. An intrinsically time independent electric field of force results in dielectric polarisation, which is an ensemble average of the same symmetry over permanent electric dipole moments. Less obviously, the electric field also causes the vector part of the c.c.f. matrix $< v(t)\omega(0) >$ to appear in frame (X, Y, Z), a result first noted in the mid-eighties using conventional computer simulation in molecular liquids.

Stress is the negative of the pressure tensor, which is force divided by a scalar unit volume cross section. Stress therefore has the symmetry of its equivalent force field, and principle three may be used to find the symmetry of the ensemble averages set up by stress. These are the same symmetries as strain rate multiplied by a scalar viscosity, in general the nine element tensor product of velocity and inverse position,

$$\Gamma(v)\Gamma(1/r) = D_g^{(0)} + D_g^{(1)} + D_g^{(2)}, \tag{5}$$

with negative $\hat{T}$. Shear stress and strain rate symmetry is the off-diagonal part, made up of a combination of symmetry $D_g^{(1)}$ (vorticity) and $D_g^{(2)}$ (deformation). Elongational stress and strain rate symmetry is the diagonal part, $D_g^{(0)}$. Combined shear and elongational stress (complex stress) produces a response with the complete D symmetry on the right hand side of eqn. (5).

In the context of complex stress we arrive at the important conclusion that ensemble averages of the same symmetry are set up by principle (3) in frame (X, Y, Z) of the laboratory. Among these are a new and useful class of indicator c.c.f.'s exemplified by $< P_{XY}(0)P_{YZ}(t) >$ for the pressure tensor components and $< v(t)v(0) >$, whose characteristics have been described recently by Evans and Heyes using $SLLOD$, PUT, and BD simulation.[34,35]

3. INDICATOR C.C.F'S, THE SIGNATURES OF COMPLEX FLOW

These are available only from the fundamental $NEMD$ and $NEBD$ computer simulation methods developed in the past few years, and appear to be inaccessible to the traditional constitutive description of stress induced flow. The work of Evans and Heyes has shown that the indicator c.c.f. $< v(t)v(0) >$ for shear stress has the unique property of being neither symmetric nor antisymmetric to index reversal (or the equivalent time displacement),[9]

$$< v_X(t)v_Z(0) > \neq < v_Z(t)v_X(0) >, \tag{6}$$

in the steady state under applied shear. (Here the shear rate is $\dot{\gamma} = \partial v_X/\partial Z$.) This property appears to be the first counter-example to the Onsager/Casimir Reciprocal Principle, which demands symmetry or antisymmetry, but cannot account for asymmetry, in X and Z. The same time asymmetry arise in the c.c.f.'s of the pressure tensor (zero in the absence of shear). Illustrations of this new and unique property are provided from $SLLOD$ and PUT simulations in figs. 1 and 2 in the two and three dimensional fluid states.

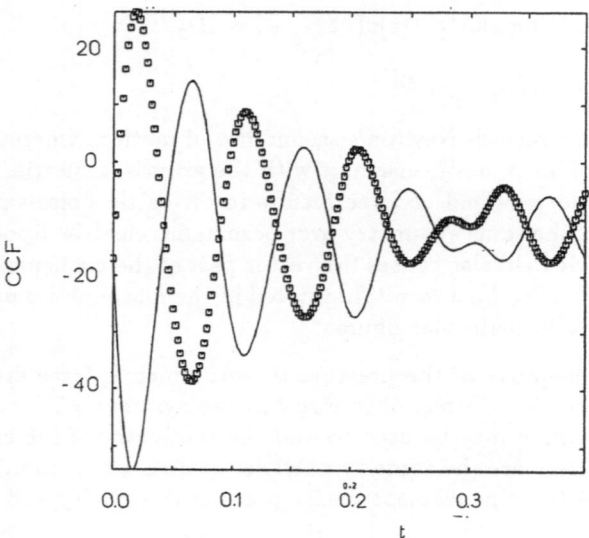

Fig. 1 The time cross-correlation function $(V/k_BT) < P_{XY}(0)P_{YZ}(t) >$, solid line and $(V/k_BT) < P_{YZ}(0)P_{XY}(t) >$, squares using the PUT algorithm ($r_p = 2\sigma$) at the $3D$ LJ $N = 500$ state $\rho = 0.8442$, $T = 0.722$ and $\dot\gamma = 20$.

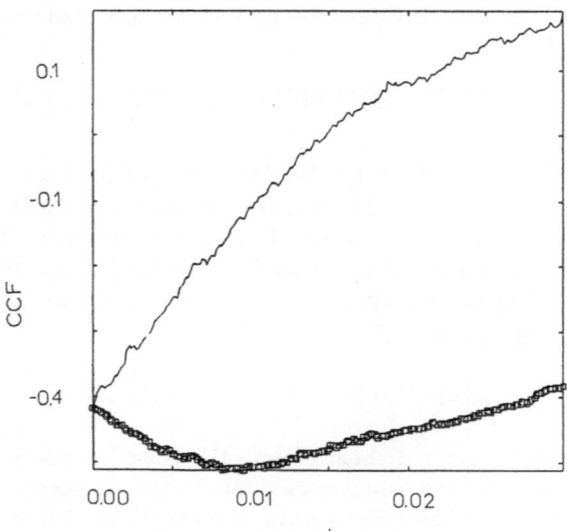

Fig. 2 The time correlation function, $< v_X(0)v_Z(t) >$, squares, and $< v_Z(0)v_X(t) >$, solid line for $2D$ soft-disk states at $\rho = 0.733, T = 1.0$ and $\dot\gamma = 17.8$ using the PUT algorithm.

90

The indicator c.c.f. in eqn (6) has the overall D, $\hat{P}$ and $\hat{T}$ symmetries of the above list from principle three, $i.e.$, a *weighted* combination of $D_g^{(1)}$ and $D_g^{(2)}$. This is a weighted combination of,

$$D_g^{(1)} :< v_X(t)v_Z(0) >= - < v_Z(t)v_X(0) >, \tag{7}$$

and

$$D_g^{(2)} :< v_X(t)v_Z(0) >= < v_Z(t)v_X(0) >, \tag{8}$$

$i.e.$, of antisymmetric shear induced vortex flow and symmetric shear induced deformational flow. The combined effect (eqn. (7)) is neither symmetric nor antisymmetric, and is therefore called "asymmetric" (non-Onsager/Casimir) flow. The complete D symmetry of the macroscopic stress field downstream of the disk is therefore,

$$D_g^{(0)} + D_g^{(1)} + D_g^{(2)}, \tag{9}$$

implying the existence of all elements of the indicator c.c.f. tensor $< v(t)v(0) >$. The explorative work and conclusions of Evans and Heyes on combined shear and elongational flow,[19] may therefore be applied to the region downstream of the disk in two dimensions, or cylinder in three. For example, we expect the presence downstream of asymmetric, non-Onsager/Casimir, c.c.f.'s of velocity, position, mixed velocity and position, and various asymmetric c.c.f.'s of pressure tensor components coming from the shearing part of the complete downstream stress field symmetry. We expect symmetric diagonal components of these fundamental indicator correlation functions due to the elongational part of the complete downstream stress field. These functions would be the microscopic signature of the macroscopic flow phenomena discussed by Clementi and co-workers,[25-28] $i.e.$, eddies, vortices, and so on, the conventional hydrodynamic description. On a more fundamental level, these are described by indicator cross correlation functions.

As shown by Evans and Heyes, the shear and elongational parts of the complete stress field are mutually interdependent. The time evolution of the respective indicator c.c.f.'s would likewise depend on each other, but *not* the symmetry, which is an ineluctable signature of the elongational, or alternatively of the shearing, component of the complicated macroscopic flow. Interestingly, the flow combined downstream stress field symmetry is expected to produce thermal conductivity, defined as the Green-Kubo (or the more general Morriss-Evans) integral over the heat flux tensor of Irving and Kirkwood. The D symmetry of the downstream stress field remains the same for more complicated objects, such as an ellipse, or an object shaped like a car body, aircraft wing, or ship's hull. Downstream turbulent behaviour, lift, and so on are characterised therefore by these indicator c.c.f.'s on a fundamental level, and this should be of direct use in industrial design processes, based on the flow of air or water past macroscopic objects in NEMD or NEBD simulations.

4. SYMMETRY OF MACROSCOPIC FLOW CREATED BY A ROTATING DISK

A rotating macroscopic disk in an N body ensemble creates a macroscopic flow consisting of vorticity and elongational shear, the symmetry of whose strain rate tensor is, $D_g^{(1)}(-) + D_g^{(2)}(-)$ in $R_h(3)$. The time reversal symmetry in brackets is negative. The equivalent stress field tensor has the same D symmetry, but is positive to $\hat{T}$, generated by the fact that stress is a $\hat{T}$ negative strain rate tensor multiplied by a scalar $\hat{T}$ negative viscosity. By principle (3), the stress field causes an effect of the same symmetry, which may be measured through the same asymmetric indicator c.c.f.'s, positive to $\hat{T}$, as described by Evans and Heyes in the context of couette flow.[9]

The flow created by a rotating disk therefore has the same non-Onsager Casimir characteristics as those of couette flow. These fundamental, microscopic, indicator c.c.f.'s are functions of the visible, macroscopic, non-Newtonian rheology of the fluid. In some cases, stirring (rotating macroscopic rod) causes the fluid to solidify (corn-flour paste), a response which is a non-Newtonian in the extreme. In this case the asymmetric indicator c.c.f. in the shear-induced solid state would be highly oscillatory and very asymmetric. In the context of stirred liquid crystals, the same non-Newtonian characteristics dominate, some liquid crystals having the property of acting like clock-work springs, once wound up (stirred) they unwind again at a different pace after the stirrer is switched off. In this case the fundamental, indicator c.c.f.'s are going to be highly asymmetric.

In fig. (3) we show the velocity flow lines around two counter-rotating solid disks in a $2D$ LJ fluid. Note the symmetry breaking aspects of the flow lines. In fig. (4) we show the associated density variation in the configuration. The major density variations at different points in the flow field reveal the power of the microscopic simulation technique in accounting for the many-body properties *in situ*.

5. FLOW FROM ROTATING ELECTRIC AND ELECTROMAGNETIC FIELDS

Mechanical shear rates cannot exceed about a megahertz. This limits experimental investigation to these frequencies. In contrast, the strain rates of a typical NEMD or NEBD simulation are in the megahertx (MHz) to terahertz (THz) range. It would be of interest to stir a fluid in the mechanically inaccessible MHz to GHz range to complete the investigation of its non-Newtonian response. This is easily possible in principle with contemporary technology by using the Born/Lertes effect.[30,31] Among Max Born's earliest contributions was a theoretical paper published in 1920,[30], which showed that a liquid suspended between rotating electric fields in a thin walled glass vessel on a torsion wire produces a measurable torque. Born analysed the effect in terms of the Debye relaxation time, showing that the effect maximises at the Debye peak frequency corresponding to the Debye relaxation time. This was confirmed almost immediately by Lertes.[31] Further exparimental work was reported more than thirty years later by Grossetti,[32] and a masterly teatment by Dahler appeared in 1965.[33] The Born/Lertes effect seems not to have been utilised to investigate the non-Newtonian rheology of fluids, but allows such work to proceed with rapidly spinning electric fields or circularly polarised lasers, in which the electric field spins about the propagation axis. The complete D symmetry of the macroscopic stress field downstream of the disk is therefore, $D_g^{(0)} + D_g^{(1)} + D_g^{(2)}$, implying the existence of all elements of the indicator c.c.f. tensor $< v(t)v(0) >$. The original Born/Lertes effect has the D, $\hat{P}$ and $\hat{T}$ symmetries of a rotating electric field. The electromagnetic equivalent, which seems not to have been described in the literature, involves the same symmetry, that of the rotating electric field component of the circularly polarised electromagnetic field. The nearest equivalent to the electromagnetic Born/Lertes effect is the discovery by Beth,[36] that a circularly polarised laser rotates an optically birefringent crystal suspended on a torsion wire. This was used as one experimental proof of the fact that a photon has spin, i.e. is a chiral particle travelling at the speed of light.

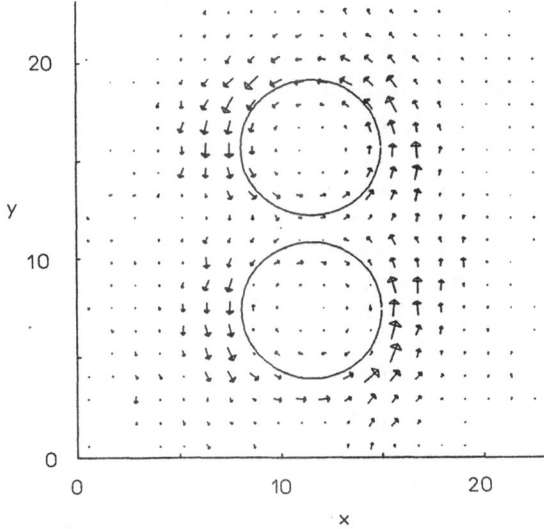

Fig. 3 Grid velocities around two counter-rotating 2D disks in a 2D Lennard-Jone fluid. The average density is $\rho = 0.75$ and $T = 0.6$. $N = 400$, $\delta t = 0.015$, thermostatted by velocity rescaling. The sidelength is $S = 23.1$ $LJ\sigma$. The radius of each solid LJ disk is $0.15 \times S$. The length and direction of each arrow indicates the relative magnitude and direction of the the flow velocity in the grid domain. The maximum velocity shown os 0.33 LJ reduced units. Note the symmetry breaking of the flow lines around the two disks.

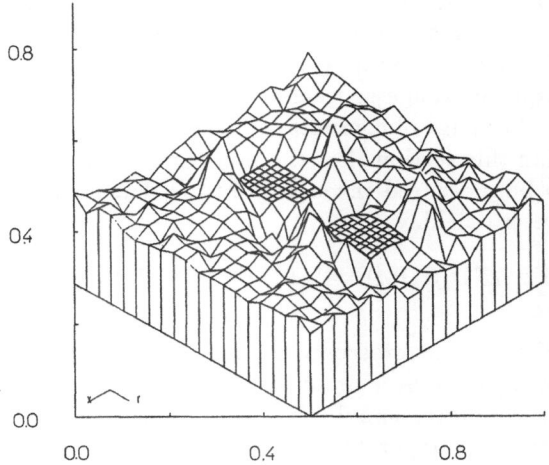

Fig. 4 The fluid density contours for the state of Fig. 3. For clarity the two disks are assigned zero density, therefore appearing as voids in the middle of the figure (as finer grids). The peak density in the figure is 1.44 reduced units. Notice the enhanced fluid density around the fluid disk peripheries.

The tip of the rotating electric field vector draws out a circle, and the overall symmetry is the same as that of a shear stress field, $i.e.$, $D_g^{(1)}(+) + D_g^{(2)}(+)$, where $D_g^{(1)}$ refers to $\nabla \times E$, the curl, which through Maxwell's equations is minus the time derivative of the magnetic field, and the $D_g^{(2)}$ refers to the off-diagonal part of the dyadic ∇E. Dahler,[33] has clearly analysed the macroscopic flow set up by the electric field with analytical hydrodynamics. The above rotating electric field symmetry is imparted by principle three to the molecular ensemble contained in the suspended thin walled sphere, provided that the molecules have a permanent electric dipole moment. We expect off diagonal, asymmetric, indicator c.c.f.'s of the type $< A(t)A(0) >$ where A is molecular centre of mass velocity, dipole moment, angular velocity, and so on to appear in a computer simulation analysis of this effect. The electric field gradient has the same symmetry as a stress tensor, and sets up a shear in the ensemble.

The D symmetry of the rotating electric field includes that of angular acceleration, or torque, $D_g^{(1)}(+)$, which causes the entire vessel to accelerate rotationally against the torsion wire in the Born/Lertes experiment, following the simple cause/effect principle three. This torque is a direct measure of non-Newtonian effects and is dependent on an effective viscosity in the same way as mechanical stress is viscosity multiplied by mechanical strain rate. The strain rate due to the rotating electric field has, $D_g^{(1)}(-) + D_g^{(2)}(-)$, symmetry, which includes the $D_g^{(1)}(-)$ symmetry of macroscopic angular momentum, angular velocity, vortex, or magnetic field. The advantage of using the Born/Lertes effect to study non-Newtonian macroscopic flow is that the rotating electric field frequency can be varied over the Hz to upper MHz range using four electrodes arranged around the sample suspended in the thin walled vessel from the torsion wire. The macroscopic torque depends on the Debye relaxation time as shown by Born, and confirmed by Dahler's extensive analysis. The Debye relaxation time is dependent on the viscosity as is well known in dielectric relaxation. From the MHz range upwards into the far infra red (to about 100 THz) the rotating electric field can be generated by circularly polarised electromagnetic radiation, and here again we expect the appearance of asymmetric indicator c.c.f.'s as the beam becomes elliptically polarised due to sample absorption through the molecular permanent dipole moment.[37] Here again, the maximum torque on the torsion wire appears at the inverse of the Debye relaxation time, which under certain circumstances occurs in the microwave frequency range just below the far infra red. This brings the experimental study within the range of computer simulation of shear stress phenomena. We mention in concluding this section that the non-Newtonian nature of the response of the sample can also be measured through the time dependent magnetization set up by the curl of the electric field.

6. THE SYMMETRY AND COMPUTER SIMULATION OF CHANNEL FLOW

The characteristics of macroscopic channel flow have recently been studied by Lie and co-workers using computer simulation and stochastic hydrodynamics.[38,39] Using two dimensional computer simulations with thermal walls, these authors have described the velocity and temperature characteristics of very dilute gases in channel flow, and have begun to investigate the effects of density variation across a temperature gradient in a three dimensional system. They have also simulated Rayleigh Benard phenomena and hydrodynamic instabilites and cell patterns using upwards of a quarter of a million particles and specialised boundary conditions. This work shows clearly the advantages of the fundamental approach, and shows a number of features which are again inaccessible to the constitutive approach.

The D, $\hat{P}$ and $\hat{T}$ symmetry of the stress field in channel flow is, $D_g^{(0)}(+) + D_g^{(1)}(+) + D_g^{(2)}(+)$, and the indicator function, $< v(t)v(0) >$, contains all nine elements in general by principle three. The elongational stress field dominates near the channel centre away from the walls and the shear stress field dominates near the walls. Lie and co-workers describe the resulting velocity profile for various Knudsen numbers, which on average shows a well defined parabolic character. This much is known hydrodynamically, but the fundamental approach goes deeper, and may examine the root cause of this in terms of indicator c.c.f. symmetries at the flow applied steady state. It is significant that Lie et al. measure a heat dissipation by viscous dissipation as the density of the ensemble decreases, one symptom of which is increased velocity near the thermal walls of the flow channel. This appears to support the conclusions of Evans and Heyes in another context, described already, where combined elongational and shear flow results in heat flux and thermal conductivity. There is also a temperature profile across the flow channel which is a kinetic energy profile related to the square of the velocity. In channel flow we may therefore expect indicator c.c.f.'s of non-Onsager/Casimir symmetry, and also diagonal elements of the indicator c.c.f.'s. The off-diagonal elements dominate near the walls and the diagonal elements near the channel centre, thus explaining the observed velocity profile on a fundamental level.

7. HYDRODYNAMIC INSTABILITY AND RAYLEIGH-BENARD PATTERNS

The symmetries of hydrodynamic instability arise from convection when, for example, a fluid is heated from below. These Rayleigh-Benard patterns have recently appeared from a fundamental computer simulation.[40] Hydrodynamical instability and the Rayleigh-Benard problem are usually treated with an infinite layer of fluid in which a steady adverse temperature gradient is maintained, usually by heating from below. This implies a pressure and density distribution as well as a thermal gradient. Rayleigh analysed the early observations of Benard of cell structures due to this hydrodynamic instability by building up triangular, quadrilateral, and hexagonal cells using classical hydrodynamic theory. He showed that the component of horizontal velocity perpendicular to the cell walls observed by Benard disappears.

A recurring D symmetry in the Rayleigh Benard problem is, $D_u^{(1)} D_u^{(1)} D_u^{(1)}$, i.e., the cube of $D_u^{(1)}$. This occurs in such quantities as the gradient of the pressure tensor (positive $\hat{T}$), the pressure distribution caused by hydrodynamic instability (heating a liquid from below). It also occurs with negative $\hat{T}$ symmetry in the equations governing velocity within the Rayleigh Benard cells, a typical scalar component of which is, $(n.\nabla)v = 0$, where n is a unit vector perpendicular to the cell wall. The presence of a temperature gradient implies a heat flux tensor J, which in the Irving/Kirkwood definition has the same D symmetry as the gradient of the pressure tensor but is negative to $\hat{T}$.

The cell patterns emerge from Rayleigh's analysis by considering equations which contain scalar components of the second gradient of velocity and the gradient of the vorticity, which both have the same general D, $\hat{P}$ and (negative) $\hat{T}$ symmetries as the Irving/Kirkwood heat flux tensor J. The thermal conductivity is the Green-Kubo, or more generally, the Morriss-Evans, integral over the correlation function of J. With these symmetries we note that principle (3) applies to the Rayleigh-Benard problem and its complicated macroscopic cellular flow patterns as follows. The cause (external field) in principle (3) is a heat flux, which has the units of force gradient per unit volume cross section (the same units as the gradient of the pressure tensor); and one of the effects is represented by the scalar element $(n.\nabla)v$ which vanishes at the cell walls. This is the same $\hat{P}, \hat{T}$ and D symmetry, and governs the Rayleigh-Benard cell patterns. Principle (3) relates the symmetry of the heat flux tensor directly to the symmetry of the Rayleigh-Benard macroscopic flow cells. We have reduced the complicated Rayleigh-Benard macroscopic flow problem to a $(D_u^{(1)})^3$ symmetry. This can be rewritten as the product, $D_u^{(1)}(D_g^{(0)} + D_g^{(1)} + D_g^{(2)})$, one of whose two components is the combined symmetry of the indicator c.c.f. $< v(t)v(0) >$. In the Rayleigh-Benard cells, therefore, we expect non-Onsager Casimir off-diagonal elements of $< v(t)v(0) >$ and symmetric diagonal elements, the former predominating near the cell walls and the latter near the centre of the macroscopic flow cells. Interestingly, the Rayleigh-Benard D symmetry also allows all scalar elements in general of the triple indicator correlation function $< vvv >$. The latter is unique to Rayleigh-Benard flow and computer simulation should reveal many of its interesting features.

8. CONCLUDING REMARKS

Several different macroscopic flow problems recently treated by microscopic computer simulation have been analysed and reduced to combinations of D symmetries of the point group $R_h(3)$. The elongational stress field has symmetry $D_g^{(0)}(+)$; the shearing stress field is a linear combination of the vorticity component of $D_g^{(1)}(+)$ symmetry, and the deformational component of $D_g^{(2)}(+)$ symmetry. The combined elongation and shear stress field is a linear combination of these three D components. In channel flow, $D_g^{(0)}(+)$ dominates near the channel centre, and $D_g^{(1)}(+) + D_g^{(2)}(+)$ near the walls. In the Born/Lertes effect, the rotating electric field sets up $D_g^{(1)}(+) + D_g^{(2)}(+)$ macroscopic flow; and Rayleigh Benard hydrodynamic instability has the symmetry signature $(D_u^{(1)})^3$. Principle three of g.t.s.m. applies to all these complicated macroscopic flows, and shows the presence of indicator c.c.f.'s of each type of flow. Some components of the indicator c.c.f.'s break Onsager/Casimir symmetry (vortex and deformational shear flow), others are symmetric diagonal (elongational flow). The Rayleigh Benard cells allow both types, together with scalar components of the triple indicator c.c.f $< vvv >$.

Acknowledgements

D.M.H. gratefully thanks *The Royal Society* for the award of a *Royal Society 1983 University Research Fellowship*. M.W.E. thanks RHBNC for the award of an Hon. Research Fellowship. We thank S.E.R.C. for the award of computer time at the University of London Computer Centre, and the RHBNC Computer Center for use of their VAX 11/780 computer facilities.

References

1. J. Harris, "Rheology of non-Newtonian Flow", Longman, London (1977).

2. B. C. Eu, J. Chem. Phys., 82:3773 (1985).

3. D. J. Evans, Mol. Phys. 42:1355 (1981).

4. A. S. Lodge, R. B. Bird, C. F. Curtiss, and M. W. Johnsson, J. Chem. Phys., 85:2341 (1986).

5. M. W. Evans and D. M. Heyes, Mol. Phys., 65:1441 (1988).

6. M. W. Evans and D.M. Heyes, Physica Scripta, in press (1988).

7. M.W. Evans and D.M. Heyes, J. Mol. Liq., in press (1988).

8. M. W. Evans, Mol. Phys., in press (1988).

9. M.W. Evans and D.M. Heyes, Mol. Phys. 65:1441 (1988).

10. D. M. Heyes, Computer Phys. Rep., 8:71 (1988).

11. M. W. Evans, Chem. Phys. Lett., 152:33 (1988).

12. M. W. Evans, Phys. Lett. A, 134:409 (1989).

13. M. W. Evans, Chem. Phys., 127:413 (1988); 132:1 (1989).

14. M. W. Evans, Physica B, 154:313 (1989).

15. M. W. Evans, Phys. Rev. A, 39:6041 (1989).

16. G. P. Morriss and D. J. Evans, Phys. Rev. A, 35:792 (1987).

17. R. L. Flurry, Jr., "Symmetry Groups, Theory and Chemical Applications.", Prentice Hall, Englewood Cliffs (1980).

18. F. A. Cotton, "Chemical Applications of Group Theory", Wiley Interscience, New York (1963).

19. M. W. Evans and D. M. Heyes, Mol. Phys., submitted for publication.

20. L. D. Landau and E. M. Lifshitz, "Fluid Mechanics", Pergamon, Oxford (1959).

21. A. Garcia, Phys. Rev. A, 34:1454 (1986).

22. D. M. Heyes, J. Chem. Soc., Faraday Trans. 2, 82:1365 (1986).

23. D. M. Heyes and R. Szczepanski, J. Chem. Soc., Faraday Trans. 2, 83:319 (1987).

24. D.M. Heyes and J.R. Melrose, Mol. Sim. 2:281 (1989).

25. D. Rapaport and E. Clementi, Phys. Rev. Lett., 57 (1986)

26. L. Hannon, G. C. Lie and E. Clementi, J. Stat. Phys., 51:965 (1988).

27. ibid., Phys. Lett. A, 119:174 (1986).

28. A. L. Garcia, M. M. Mansour, G. C. Lie, and E. Clementi, Phys. Rev. A, 36:4348 (1987).

29. M. W. Evans, G. C. Lie, and E. Clementi, Phys. Rev. A, 37:2551 (1988).

30. M. Born, Z. Phys., 1:221 (1920).

31. P. Lertes, ibid., 4:315 (1921); 6:56 (1921); 22:261 (1921).

32. E. Grossetti, Il Nuovo Cimento, 10:193 (1958); 13:621 (1959).

33. J. S. Dahler, in "Research Frontiers in Fluid Dynamics", (ed. R. J. Seeger and G. Temple, Interscience, New York, (1965).

34. M. W. Evans and D.M. Heyes, J.C.S. Faraday Trans 2, submitted.

35. M.W. Evans and D.M. Heyes, Mol. Sim. submitted.

36. R.A. Beth, Phys. Rev., 50:115 (1936).

37. M.W. Evans, Chem. Phys. Lett., (1989) in press.

38. M.M. Mansour. A.L. Garcia, G.C. Lie and E. Clementi, Phys. Rev. Lett., 58:874 (1987).

39. D.K. Bhattacharya and G.C. Lie, Phys. Rev. Lett, 62:897 (1989).

40. J.A. Given and E. Clementi, J. Chem. Phys., 90:7376 (1989).

MOLECULAR DYNAMICS OF FLUID FLOW AT SOLID SURFACES[#]

Joel Koplik[*], Jayanth R. Banavar[**], Jorge F. Willemsen[***]

Schlumberger-Doll Research

Old Quarry Road, Ridgefield, Connecticut 06877-4108, USA

ABSTRACT: We use molecular dynamics techniques to study the microscopic aspects of several slow viscous flows past a solid wall, where both fluid and wall have a molecular structure. Systems of several thousand molecules are found to exhibit reasonable continuum behavior, albeit with significant thermal fluctuations. In Couette and Poiseuille flow of liquids we find the no-slip boundary condition arises naturally as a consequence of molecular roughness, and that the velocity and stress fields agree with the solutions of the Stokes equations. At lower densities slip appears, which can be incorporated into a flow-independent slip-length boundary condition. We examine the trajectories of individual molecules in Poiseuille flow, and also find that their average behavior is given by Taylor-Aris hydrodynamic dispersion. An immiscible two-fluid system is simulated by a species-dependent intermolecular interaction. We observe a static meniscus whose contact angle agrees with simple estimates and, when motion occurs, velocity-dependent advancing and receding angles. The local velocity field near a moving contact line shows a breakdown of the no-slip condition and, up to substantial statistical fluctuations, is consistent with earlier predictions of Dussan.

I. INTRODUCTION

The equations of continuum fluid mechanics are incomplete without specification of boundary conditions. For over a century, the no-slip condition that fluid velocity vanishes at a solid surface has successfully accounted for

[#] This article is reprinted from Physics of Fluids **A1** (5), May 1989

[*] present address: Benjamin Levich Institute and Department of Physics, City College of the City University of New York, New York, NY 10031

[**] Department of Physics and Materials Research Laboratory, Pennsylvania State University, University Park, PA 16802

[***] RSMAS - AMP, 4600 Rickenbacker Causeway, University of Miami, Miami, FL 33149-1098

the experimental facts, but several problems remain. The first is that there is no compelling theoretical argument for no-slip. As we shall review below, kinetic theory provides some insight into the question, but cannot deal with the realistic situation - an interacting many-body problem of a dense liquid adjacent to a nearly- rigid molecular solid. Indeed, the no-slip condition is usually presented as an empirical result of 19th Century experiments[1,2], and has not been questioned until recently. For most purposes it is not necessary to do so, but a serious problem arises when a contact line separating two immiscible fluids moves along a solid surface[3-5]. The straightforward hydrodynamic analysis of this situation predicts a divergent energy dissipation rate, and slip boundary conditions, among other explanations, have been proposed to deal with the problem. Lastly, closely related questions involving the connection between the microscopic scale and the continuum arise in the description of fluid phenomena in very small systems where the number of molecules is not that large. Examples of current interest include the study of very thin films and surface forces[6] and flow in microporous solids[7].

In this paper we address some of these issues using molecular dynamics (MD) simulations of several cases of fluid flow near a solid boundary. In MD one explicitly computes the detailed molecular motion from the instantaneous positions andvelocities and an interaction potential[8-10]. A key feature of ourwork is the use of a wall with molecular structure. We are thus able to set up a controlled and unprejudiced numerical experiment on a system with specified reasonable microscopic properties, examine the resulting average or large scale behavior, and correlate the results to the continuum description. Compared to other methods there are both advantages and disadvantages to MD. In contrast to laboratory experiments, there are no difficulties with contaminant effects, optical distorsion, or microscopic flow visualization. Compared to analytic methods, no assumptions about correlations in intermolecular distribution functions are required. The negative side of MD is its current practical restriction to small systems (up to about 100 Å) and short times (10^{-9} sec or less), with concomitantly large thermal and statistical fluctuations. In consequence, we have taken some pains to verify that continuum behavior is observed in our simulations.

The utility of MD in the study of some fluid mechanical problems has been recognized earlier by a number of authors. In addition to work on channel flows which we shall describe in detail below, interesting calculations of flow past obstacles[11] and Rayleigh-Benard convection[12] have appeared recently. The particular stress of the present paper is rather on using MD as a bridge between the microscopic and macroscopic levels of description. In this sense, our closest precursor is the work of Alder and collaborators[13] on velocity correlations and Brownian motion in hard sphere systems, where extrapolation of 'micro-hydrodynamics' to the continuum is examined.

The outline of the paper is as follows. In Section 2 the background of the no-slip condition is discussed in more detail. We recall the kinetic theory arguments of Maxwell[14] and others[15] for the no-slip condition, raise some questions that are not answerable at the purely continuum level, and then consider the problem of the moving contact line singularity. Section 3 reviews the molecular dynamics algorithm and the particular system studied here, and discusses various methods for confining a fluid system between walls. Dense fluids exhibit the layering effects known from experiments and other simulations, while at low temperature and density capillary condensation is observed. In Section 4, we present our simulation results for Poiseuille and Couette flow of a single fluid in a channel. In the liquid phase, the no-slip condition emerges with no explicit assumptions, and the velocity and stress fields are found to agree with solutions of the Stokes equations. For both flows we compute a consistent fluid viscosity in agreement both with experiment and other simulations. At lower densities we observe slip, and find that a single slip-length boundary condition of Maxwell form applies to either Poiseuille or Couette flow. We also examine the behavior of individual particle trajectories, which are seen to essentially correspond to weakly perturbed Brownian motion, and show that the average behavior is given by Taylor-Aris[16] hydrodynamic dispersion. Section 5 considers immiscible two-fluid systems, which are simulated by an additional species-dependent interaction. An approximately predictable static contact angle is found, and when motion occurs distinct velocity-dependent advancing and receding angles are observed. In this situation the flow field is two-dimensional, which entails an average over many fewer molecules per bin than in the previous cases. In consequence, the thermal fluctuations prevent us from obtaining the local velocity with any precision, and the stress field is still noisier.The approximate velocity field shows a breakdown of no-slip near the contact line, and appears to have a jet and eddy structure consistent with that proposed by Dussan[17]. Section 6 presents our conclusions. A brief account of our results, with color illustrations, has appeared previously[18].

II. THE NO-SLIP CONDITION

The no-slip condition states that the velocity of a viscous fluid at a solid surface vanishes. The vanishing of the normal component is not controversial, as this is essentially the definition of a solid surface as one which confines the fluid. The vanishing of the tangential component is not obvious at all, however. While one would naturally expect the solid to exert a frictional force on fluid moving past and tend to slow it down, it is unclear why such friction would bring the fluid exactly to rest.

The first quantitative theoretical study of the no-slip condition is that of Maxwell[14], who considered a dilute gas of molecules in the presence of an idealized solid wall. A molecule colliding with the wall was assumed to undergo diffuse reflection with probability f or specular reflection with complimentary probability $1-f$. In specular reflection the colliding molecule reverses its velocity normal to the wall, but preserves its tangential velocity, as in an elastic collision with a very massive solid. Diffuse reflection was an idealization of a multiple collision process between a gas molecule and the wall molecules, wherein the former emerges from the wall after adsorption and evaporation with a velocity appropriate to a Maxwellian temperature distribution, as if the wall acted as a heat bath. Maxwell showed that the average tangential velocity u at the wall was related to the shear rate $\partial u / \partial z$, z being the coordinate normal to the wall, by

$$u = \lambda \frac{\partial u}{\partial z} \text{ (wall)} \tag{1}$$

where the 'slip length' is given by

$$\lambda = \frac{2}{3} \left(\frac{2}{f} - 1 \right) l \tag{2}$$

where l is the mean free path of the gas molecules. Now in a dilute gas, the mean free path can become appreciable and this argument then predicts slip at a solid surface, a phenomenon known to be occur for dilute gases in the Knudsen regime[19]. On the other hand, if one (heuristically) applied the argument to a liquid, where the mean free path is comparable to the molecular size,the tangential velocity would be proportional to a microscopic number and, on the macroscopic scale, would effectively be zero.

While Maxwell's argument provides physical insight into the origin and limitation of the no-slip condition, it is not quantitatively useful because of twostrong assumptions. The first is the restriction to a dilute gas whose intermolecular interactions are neglected except insofar as they lead to thermal equilibrium. More recent work[15] has removed the latter restriction byusing a molecular distribution function for the gas obtained by solving the Boltzmann equation in various approximations. By this means at least two-body interaction effects may be incorporated. The second limitation is in the modeling of the walls, where Maxwell's boundary condition of a mixture of specular and diffuse reflection is retained. There is really little alternative in a kinetic theory approach, short of incorporating multiple scattering from the individual wall molecules. The latter is not feasible analytically, but presents no difficulty in MD.

The no-slip condition is a statement about the continuum velocity, the time and space average velocity of fluid molecules occupying a sampling region at the wall. One may also enquire into the motion of individual molecules, which could be a question of practical importance, for example, if the fluid reacted with the wall. The (continuum, Eulerian) no-slip assumption provides no molecular information, and several microscopic possibilities exist. Conceptually, the simplest mechanism is that the molecular velocity vanishes, as if the fluid molecules near the wall were bound to it. An equally good alternative, kinematically, is a rolling or caterpillar-like motion of the molecules. This is analogous to a ball rolling on a plane, where the part of the ball touching the plane is instantaneously at rest but has non-zero acceleration and rolls off. Other possibilities can be devised, and in any case the thermal motion of the molecules must be added. A purely continuum theory cannot discriminate among the various molecular mechanisms, and we will use MD to settle the question in Section 4.

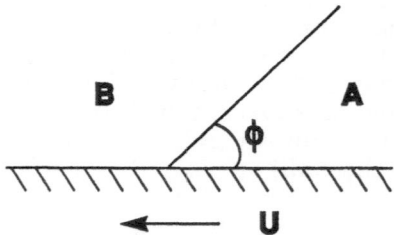

FIGURE 1. Definition sketch for a moving contact line.

While for most applications the no-slip condition may be accepted uncritically as a phenomenological rule, recent work on moving contact lines has raised a serious question of principle. The difficulty appears most simply in an analysis due to Huh and Scriven[3], which we now outline in order to illustrate the basic nature of the problem. Consider the steady state displacement of fluid A by fluid B along a plane solid surface, shown in Fig. 1 in the rest frame of the meniscus where the solid moves backward with velocity U. In the immediate vicinity of the contact line it is plausible to assume that the meniscus appears flat with some contact angle ϕ, the flow field is two-dimensional, and the relative velocity of fluid and solid is sufficiently small for the Stokes equations to apply. In a polar coordinate system centered at the contact line, the velocities are given by

$$u_r = - \frac{1}{r} \frac{\partial \psi}{\partial \theta}$$

$$u_q = \frac{\partial \psi}{\partial r} \tag{3}$$

where the stream function $\psi (r,\theta)$ satisfies the biharmonic equation

$$\nabla^4 \psi = 0 \tag{4}$$

A general solution corresponding to finite velocity at $r \to 0$ and $r \to \infty$ is

$$\psi_\alpha = r \, (a_\alpha \sin\theta + b_\alpha \cos\theta + c_\alpha \theta \sin\theta + d_\alpha \theta \cos\theta) \tag{5}$$

where $\alpha = A, B$ refers to the two fluid regions. The boundary conditions assumed by Huh and Scriven are vanishing of the normal velocity at the wall and the meniscus:

$$u_{A\theta} = 0, \text{ for } \theta = 0 \, , \, \phi$$

$$u_{B\theta} = 0, \text{ for } \theta = \phi \, , \, \pi \tag{6a}$$

continuity of tangential velocity at interfaces (no-slip):

$$u_{Ar} = -U, \text{ for } \theta = 0$$

$$u_{Br} = +U, \text{ for } \theta = \pi$$

$$u_{Ar} = u_{Br}, \text{ for } \theta = \phi \tag{6b}$$

and continuity of tangential stress at the interface:

$$\mu_A \frac{\partial u_{Ar}}{\partial \theta} = \mu_B \frac{\partial u_{Br}}{\partial \theta} \, , \, \theta = \phi \tag{6c}$$

where the μ_α are the viscosities. The normal stress condition at the fluid-fluid interface is ignored. Now equations (6a-c) give 8 conditions, sufficient to determine the 8 constants in (5); their values and some examples of the resulting flows are given in the original reference. The critical problem is that the stress tensor

$$T_{\alpha ,ij} = - \delta_{ij} p_\alpha + \mu_\alpha (\partial_i u_{\alpha j} + \partial_j u_{\alpha i}) \tag{7}$$

has an r^{-1} divergence as $r \to 0$, because $\psi \sim r^1$, $u \sim \partial \psi \sim r^0$ and $T \sim \partial u \sim r^{-1}$. Therefore, a measureable quantity, the energy dissipation per unit time per unit length of contact line,

$$\dot{E} = \frac{1}{2} \int d^2\bar{x} \, m(\bar{x}) \, T_{ij}(\bar{x}) \, T_{ij}(\bar{x}) \qquad (8)$$

has a logarithmic divergence at small r. If for purposes of discussion the integration is cut off at a small value r=λ , then

$$\dot{E} \sim \log(R/\lambda) \qquad (9)$$

where R is some outer scale of the flow. While the details of this argument may be criticized[3,4], e.g., the normal stress boundary condition is not satisfied, the conclusion that a singularity arises at a moving contact line survives. A fairly general argument for this assertion is given by Dussan and Davis[20], who by experiment and kinematic analysis show essentially that a moving contact line is consistent with the no-slip condition when the fluids undergo a rolling motion (see Section 5 below for further evidence). They further show that this motion necessarily gives a multiple-valued velocity which implies a singularity in the force exerted by the fluid on the solid.

A divergence in a measurable quantity usually represents an inconsistency in the assumptions that produced it, and many proposals have been advanced to remedy the moving contact line singularity. In view of the earlier discussion in this paper and the results to be given, we favor relaxing the no-slip condition, but other possibilities exist. First of all, the contact line motion may be unsteady, exhibiting 'stick and slip' motion[21], and perhaps may be occasionally rapid enough for inertial effects to enter. Secondly, as alluded to above, the interface may not be locally flat and some curvature may be present down to microscopic length scales giving an oscillatory interface[22]. In addition, van der Waals forces should enter at short distances and perhaps remove the singularity[23].More generally, van der Waals forces may give rise to 'precursor films' advancing ahead of the bulk meniscus. The no-slip condition at the contact line is then replaced by a matching to the precursor film[5], but it is not clear what happens at the tip of the film.

If one allows slip at a solid surface, the first issue is the precise functional form of the boundary condition. This may not be critical, as Dussan[24] has shown that the macroscopic flow resulting from a variety of particular forms of slip conditions is the same, all giving a finite stress and an energy dissipation rate of the form (9). The question is now the magnitude of the slip length. The simplest possibility is that λ is of the order of a molecular size, but another is that the slip length is related to roughness of the solid surface. Various calculations[25] have shown how a slip-free flow over a rough surface can lead to effective slip, perhaps associated with stick-slip motion of the meniscus over the irregularities. One might then say λ is of order the roughness size, but other calculations[26] find a more complicated dependance.

To summarize this situation, there is no consensus on the correct resolution of the moving contact line singularity problem. As with all of the issues discussed in this section, macroscopic modeling, kinetic theory, or laboratory experiments do not yield a definite conclusion, and the MD technique seems ideally suited to their elucidation.

III. MOLECULAR DYNAMICS WITH WALLS

We have used a standard molecular dynamics algorithm[8] in which a pair of molecules separated by distance r interacts through a 12-6 Lennard-Jones potential

$$V_{LJ}(r) = 4\varepsilon \left[(r/\sigma)^{-12} - (r/\sigma)^{-6} \right] + r \, \delta V \tag{10}$$

characterized by energy and distance scales ε and σ, respectively. The potential has a repulsive core at $r \sim \sigma$, and the depth of the attractive well is ε. It is computationally convenient to cut off the potential and treat purely short-range interactions, and we do so at a value $r_c = 2.5 \, \sigma$. The additional correction term $r \, \delta V$ is inserted so that the force vanishes at the cutoff. The chosen interaction is known to successfully reproduce the properties of liquid Argon, if the parameters are chosen as $\sigma = 3.4$ å and $\varepsilon / k_B = 120°K$, along with a molecular mass $m = 40$ A.U. The natural time unit in the calculations is then $\tau = \sigma \sqrt{m/\varepsilon} = 2.16 \times 10^{-12}$ sec. For later discussion, we show a schematic phase diagram of the system[27] in Fig. 2. In the remainder of this paper, all dimensionalquantities given as pure numbers will be understood to be multiplied by an appropriate combination of σ, ε and m.

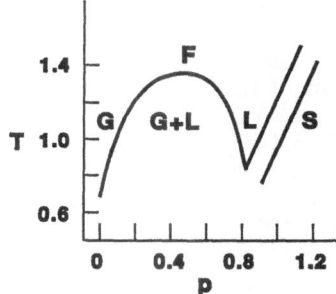

FIGURE 2. Phase diagram of a 12-6 Lennard-Jones system, after Ref. 27.

In our computer simulations, the molecular positions are specified to give the instantaneous potential, and Newton's laws numerically integrated using a fifth order predictor-corrector scheme, with a time step of $0.005 \, \tau$. The molecules are initialized on an fcclattice whose spacing is chosen to obtain the desired density, with initial velocities randomly assigned subject to a fixed temperature. The substance is then allowed to melt. To simulate an infinite system, the system would be placed in a box with periodic boundary conditions on all sides, so that a molecule exiting the box is reinserted on the opposite face. Periodic copies of the molecules are used inthe force computation. To study flow in a channel geometry,we have modified the algorithm to provide constraining walls on two sides, while retaining periodicity in the other two directions.

A number of different schemes for providing walls in MD have been implemented in the literature. The simplest of these has perfectly reflecting walls[28], analogous to Maxwell's specular reflection,and leads to slip at the walls. Likewise, one may implement diffuse or 'thermal' walls[29,30], which would correspond to a heat bath maintained at a fixed temperature. Molecules which collide with the thermal walls are reinjected into the system with new velocities chosen from a Boltzmann distribution characteristic of the wall temperature. One would at first expect this choice to build in the no-slip boundary condition,because by construction molecules leaving the wall have their velocity components parallel to thewall equal to zero on average. In fact, slip is found to occur, both in numerical simulations and in the kinetic theory calculations referred to above. Presumably, this occurs because the average velocity near the wall has a contribution from molecules that have not yet or will not enter the heat bath, whose tangential velocity is likely to be in the direction of average flow.In order to obtain exact no-slip in this context of artificial walls, one must use 'reverse reflection', wherein the tangential velocity is reversed upon collision with a wall[31]. None of these algorithms is applicable to the problem of elucidating the boundary conditions, because they essentially assume the answer.

A more realistic approach is to incorporate into the wall the molecular structure of a solid. There has been extensive work on MD simulation of the structure of the solid-liquid interface. In a pioneering study, Abraham[32] modeled the constraining walls in four different ways:

(1) Solid molecules were put at the lattice sites of the first two planes of the <100> surface of an fcc solid. During the simulation the solid molecules were constrained to remain at their lattice sites.

(2) The discrete set of centers of interaction in each solid plane were replaced by a continuous distribution of molecules with uniform planar density. The interaction of a liquid molecule a distance z away from the plane is obtained by integrating the potential along the x and y coordinates, leading to a 10-4 Lennard-Jones wall. The interaction of the entire infinite series of struc-tureless planes with a given liquid particle can also then be determined and leads to a Lennard-Jones 9-3 wall.

(3) The 10-4 wall was replaced by a Boltzmann weighted wall whose potential is proportional to the probability that a single molecule is at a distance z from the <100> surface, irrespective of its x and y coordinates. This leads to a potential deeper, broader and softer than the 10-4 potential.

(4) The wall potential is assumed to be infinite when a liquid molecule strays into the wall and zero otherwise. Further studies of the static fluid properties of the 10-4 wall system have been made by Toxvaerd[33] and Magda, Tirrell and Davis[34].In addition, the crystal-fluid interface was simulated by Broughton and Gilmer[35] by joining bulk crystal and bulk liquid systems with the liquid and crystal systems in equilibrium at two different temperatures, while Tallon[36] has studied the layer structure of a soft sphere fcc <111> solid-liquid interface by creating a crystal-liquid-crystal sandwich with roughly equal quantities of liquid and crystal.

Our walls most resemble the first scheme of Abraham. We assign the molecules in the top and bottom two layers of the initial fcc lattice a heavy mass, $m_H = 10^{10} m$, but allow these to move in accord with the equations of motion. In this way collisions between fluid and wall molecules conserve energy, and the walls retain their integrity over the duration of the simulation, although eventually they would disintegrate. In the simplest case, the wall-fluid and fluid-fluid interactions are assumed to be the same.

The phase space trajectory of the collection of the molecules is obtained by monitoring their positions and velocities. The internal energy, the displacement of each molecule from its starting position, the running average temperature of the two walls and of the fluid and the total energy are all monitored as a function of the time. The region between the walls is divided into bins along the z direction (for the immiscible two-fluid simulations below, the channel is binned across the gap as well) to obtain running averages for the density, the mean x, y and z displacements and velocities and their squares, and the nine components of the stress tensor in each of the bins. Typically the number of bins was equal to the twice the number of unit cells of the initial fcc lattice, giving a bin width $O(\sigma)$.

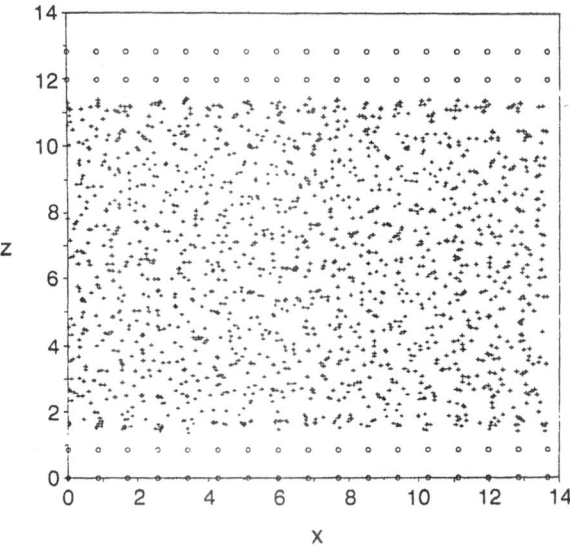

FIGURE 3. Instantaneous molecular positions for a liquid in equilibrium between walls at T=1.2 and ρ =0.8, where molecular positions for all values of y are superposed.

For our initial studies, the density was chosen to be $\rho = 0.8$ and temperature to be 1.2 ε/k_B. From the phase diagram of Fig. 2, this should place us in the liquid region. Note that the phase diagram refers to a bulk (periodic) system, and is not precisely correct for a small system with walls[7]. Most of our runs were carried outwith 1536 fluid molecules and 256 molecules for each of the two walls. The system was divided into 12 bins in this case, having an average of 128 molecules per bin. The fluid melts and achieves equilibrium (defined here as a steady state in temperature, pressure, energy, etc.) after several thousand time steps. In Fig. 3 we show an x-z snapshot of instantaneous molecular positions after 5000 time steps of constant-temperature equilibration,in which molecules at all values of y are superposed.Note first that the wall molecules remain very close to their original lattice sites and that the fluid molecules are indeed confined.Near the walls distinct molecular ordering can be seen: the ordering normal to the walls is exhibited quantitatively in Fig. 4, where the solid line gives the time-averaged density profile as a function of z. This layering effect is consistent both with experiment[37,38] and with earlier MD studies of fluids confined by both artificial or molecular boundaries[32-36], although we believe the precise structure to be model-dependent. In addition, there is molecular ordering in the layer adjacent to the wall, with the most probable location of a fluid molecule being at an fcc lattice site with respect to the wall molecules. This x- y ordering has not been directly observed in experiments as yet.

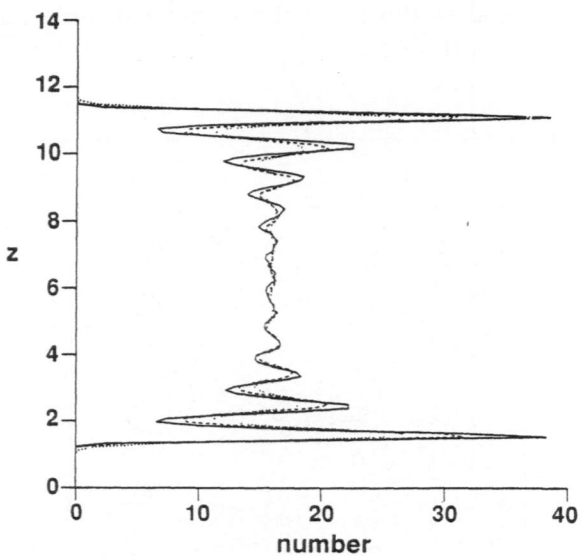

FIGURE 4. Average density profile associated with Fig. 3.

The discussion above concerned the liquid phase of the Lennard-Jones system, but we have also examined lower density phenomena. Referring to the phase diagram in Fig. 2, there are two cases to distinguish. At relatively low temperatures, below the critical point, there is a two-phase coexistence region, where one might expect liquid to condense along the walls in thermodynamic equilibrium with gas in the gap. We have studied this by fixing the densities of the walls, while systematically lowering the fluid density. In Fig. 5 we show typical instantaneous snapshots of molecular positions at temperature T = 0.5 and densities ρ = 0.2, 0.4 and 0.6, respectively. At the lowest density the configuration is as described above, but as the density increases a bridge of fluid spans the gap --capillary condensation[39] has occurred. In addition, as the density variesthe amount of liquid near the walls also varies from approximately one layer at ρ = 0.2 to two at 0.4 to three at 0.6. The continuum origin of capillary condensation is that the equilibrium vapor pressure of a gas over a concave mensicus of liquid is reduced, so that gas condenses at lower pressure if the confining geometry allows it to form such a mensicus.The phenomenon is outside themain thread of this paper.

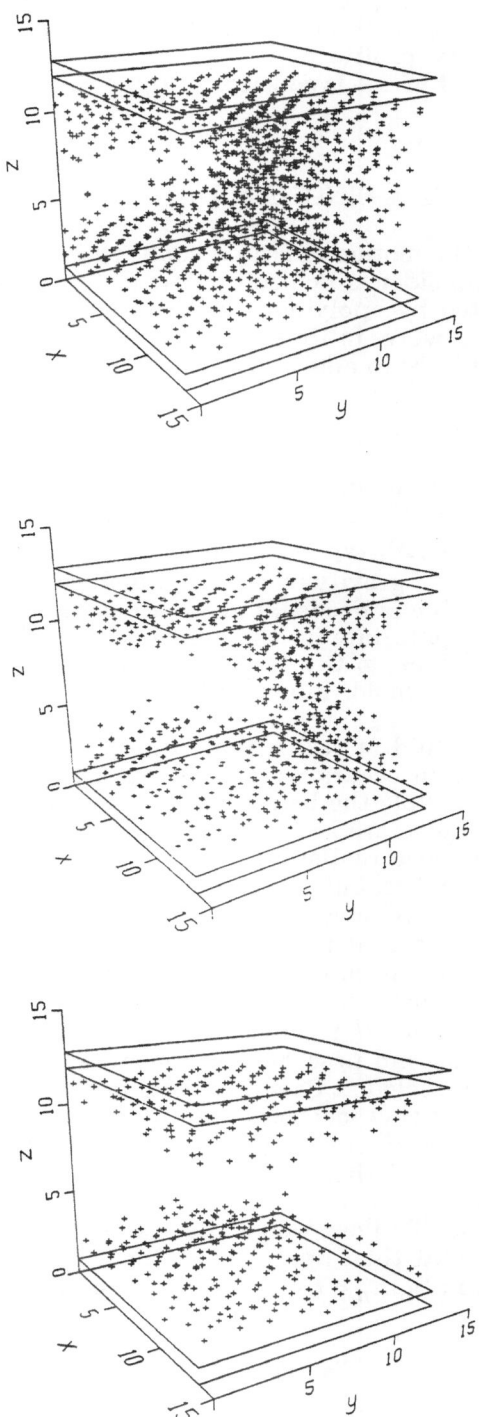

FIGURE 5. Instantaneous molecular positions showing capillary condensation in the two-phase coexistence region: T=0.5 and ρ = (a) 0.6, (b) 0.4, and (c) 0.2 .

and we have not yet examined it in detail. In contrast at a higher temperature above the critical point, T = 2.5, where there one should have only 'fluid' rather than separate solid and gas phases, we find at all densities a fairly uniform density profile across the gap. There are some density oscillations near the wall of the type seen in Fig. 4, but significantly reduced in magnitude.

IV. CHANNEL FLOWS OF A SINGLE FLUID

We now turn to the motion of a fluid confined between molecular walls, and the first case we consider is Poiseuille flow in the liquid phase. This flow is most easily generated by applying a uniform acceleration parallel to the walls, as if the fluid were falling under gravity. Quantitatively, each molecule experiences a force in addition to that arising from (10) of the form

$$\delta \mathbf{F} = mg\mathbf{x} \tag{11}$$

Because we are dealing with small systems and short times, some care must be taken in order to extract a signal from substantial thermal noise. In Lennard-Jones units (ε, σ, m), typical velocities and forces are $O(1)$, whereas the real, laboratory force of gravity is $O(10^{-13})$, and would have no discernible effect. A much larger value is needed, but we are constrained to simulate low Reynolds number, as appropriate for the neighborhood of a contact line. After some trial and error, values $g = O(0.1)$ were selected, which produce average velocities $u = O(0.1)$ and Reynolds numbers $< O(1)$.

Whereas an MD fluid at rest usually has a stable temperature, up to fluctuations, when motion occurs work is being done on the system and it tends to heat up. In principle the heat could escape through the walls, but in our case the wall molecules are so heavy that the wall is effectively an insulator. (In a somewhat more realistic treatment of a wall, one might assign the molecules a normal mass, but tether them to fixed positions with Hookes' Lawsprings[40]). We have followed either of two procedures with regard to temperature control. The first is to simply allow the fluid temperature to rise, which has the general drawback that the transport coefficients would vary as the temperature does. In our runs the temperature can rise by as much as 65 %, but the viscosity is known[41] to have a weak variation over the parameter range of interest. The second procedure is continued equilibration, in which all fluid velocities are rescaled at fixed intervals to maintain the mean kinetic energy; the drawback of this method is its unphysical nature. Fortuitously, in our work the statistical fluctuations are sufficiently large that we see the same results using either method.

To simulate Poiseuille flow, we let the system 'melt' for 5000 time steps without acceleration, and then apply (11) with a non-zero value of g. The velocity in a given z-bin $\{ B_n: z_n < z < z_{n+1} \}$ is computed as

$$\bar{u}_n = \frac{1}{N_n} \sum_{i \in B_n} \frac{d\bar{x}_i}{dt} \tag{12}$$

where N_n is the number of molecules in bin-n , and its x-component is fit to a parabolic profile

$$u(z) = (\rho \, g/2 \, \mu) \, (z - z_1)(z_2 - z) \qquad (13)$$

The fit determines the viscosity μ, as well as the locations z_i where the velocity vanishes. In Fig. 6 we show the longitudinal velocity profiles at g=0.1 for nine different sets of data points, corresponding to different intervals of 5000 time steps and different initial velocity distributions. The solid dots show the location and (zero) velocities of the two wall layers on each edge, and the solid line is an overall fit. The other components of **u** are consistent with zero. Although there is considerable spread in the data, each of the nine individual sets is well-fit by a parabola, in that a least-squares fit typically accounts for 99 % of the variation. We see first that the fitted velocity vanishes at a distance approximately one layer spacing from each innerwall layer, so that the fluid molecules adjacent to the wall are on average at rest. In other words, the no-slip condition has emerged naturally from a simulation based on a simple intermolecular interaction and walls with molecular roughness. We have repeated the calculation for different values of g, and we find global average values z_1/σ =1.6 $\pm$ 0.3 and z_2/σ =11.4 $\pm$ 0.3, with a slight tendency for z_1 to decrease and z_2 to increase with g. The fixed wall layers are located at z/σ =0, 0.86, 11.97, and 12.83, and so z_1 and z_2 are approximately the positions of the fluid layers of closest approach to the walls.

We have seen no significant compressibility effects in the simulations, at least in the liquid phase of the system: the density in each of the velocity bins varied by no more than 1 % or so. The Mach number in these flows can be estimated to be in the range 0.01-0.1, based on the observed average velocity and an experimentally determined sound speed for liquid Argon[42].

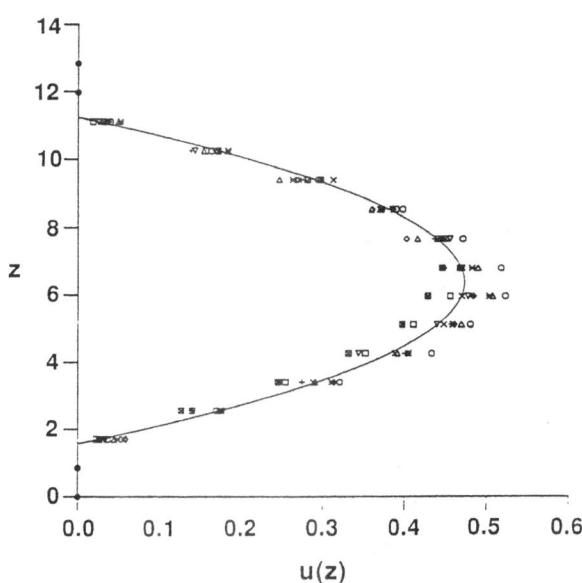

FIGURE 6. Velocity profile in simulations of Poiseuille flow of a liquid at g=0.1. The solid dots give the velocities and positions of the wall layers.

From an examination of the trajectories of the molecules (see below), we have found no evidence for adhesionof a layer of fluid molecules to the wall for the potential we have used, but it is instuctive to examine the effect of different wall interactions.We have examined two variant cases, wherein the attractive part of the LJ potential between fluid and wall is either doubled in strength or removed. In the former case there is a greater tendencyfor adhesion, and the location of the zeroes of the velocity profile moves deeper into the fluid. It is as though the trapped liquid layer had become part of the solid surface, and the location of no-slip occurs within σ of the trapped layer. With no wall attraction, on the other hand, the velocity in the bin adjacent to the walls is rather larger, and the fitted locations of zero velocity move out into the walls a distance $O(\sigma)$. For all macroscopic purposes, however, in both of these variant cases the no-slip condition is still appropriate.

TABLE I. Measured viscosity, average velocity, and longitudinal dispersivity for various values of acceleration, all at T=1.2 and ρ =0.8.

g	μ	U	$D_{||}$
0.00	...	0.00	0.065 ±0.02
0.025	2.00±0.54	0.078±0.01	0.14 ±0.02
0.05	2.21±0.54	0.14 ±0.02	0.29 ±0.03
0.075	2.14±0.54	0.21 ±0.03	0.58 ±0.1
0.1	2.00±0.27	0.25 ±0.03	0.96 ±0.3
0.2	1.94±0.16	0.50 ±0.05	3.1 ±1.0
0.4	1.94±0.16	0.99 ±0.10	9.0 ±3.0

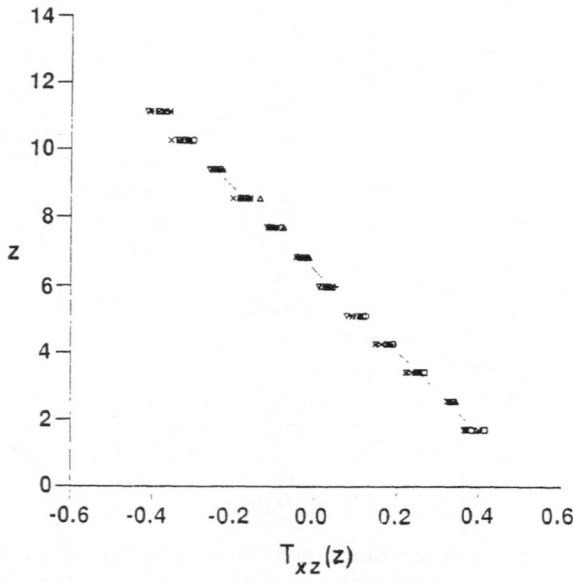

FIGURE 7. Shear stress profile corresponding to Fig. 6.

Returning to the viscosity, upon repeating the simulation and fit for different values of g, we obtain Table I[43], which gives a stable value μ= 2.0 ± 0.3 . Simulations using different system sizes, from 6 to 16 layers or 576 to 3200 fluid molecules, are consistent with this value.Using parameter values for Argon, this gives a value of approximately 0.2 cp, consistent with values obtained in experiment[44] or in other simulations using different non-equilibrium MD techniques[45].Other authors[46,47] have observed a variation of viscosity with shear rate, but at even higher values of shear than those studied in this paper. In the free heating runs, the temperature is expected to be aquartic function of z, so that the thermal conductivity could be extracted from the parameters of a fit to this dependance[29,30,48]. However, our data was too noisy to make a meaningful fit.

As a further test of Newtonian behavior, and of the viscosity computation, the stress tensor can be calculated from the molecular trajectories using the Irving-Kirkwood expression[49]

$$\mathbf{T} = \sum_i m_i \left(\frac{d\bar{x}_i}{dt} - \bar{u}\right)\left(\frac{d\bar{x}_i}{dt} - \bar{u}\right) + \frac{1}{2}\sum_j \bar{r}_{ij}\vec{F}_{ij}$$

(14)

where $\mathbf{u}(\mathbf{x})$ is the local average velocity and where r_{ij} is the separation between molecules i and j, and $\mathbf{F}_{ij}$ is the force between them. A binned version of (14) is computed as a running average during the calculation, with the result for the shear stress T_{xz} shown in Fig. 7,for the same runs whose velocity is given in Fig. 6. The shear stress is linear, as expected from (13), and the other components of the tensor are in accord with continuum Poiseuille flow. Note that Newtonian behavior occurs despite the fact that in physical units the magnitude of the shear rate in this simulation is enormous, $O(10^{11}$ sec$^{-1})$. An independent calculation of the viscosity follows by

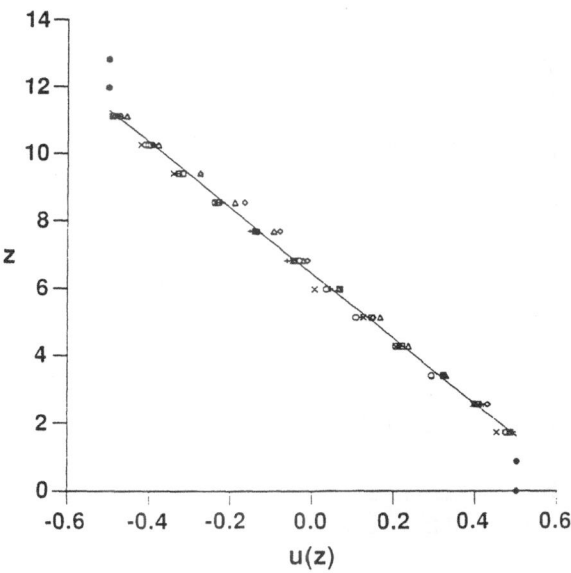

FIGURE 8. Velocity profile in Couette flow of a liquid, with the wall velocities and positions shown as solid dots.

dividing T_{xz} by the derivative of the velocity profile, $\partial u/\partial z$, and again gives a value $\mu \approx 2.0$.

Yet another test of ourconclusions is obtained by simulating plane Couette flow, where the two walls are given opposing constant velocities $u_w = \pm 0.5$ (with g=0). The resulting velocity profile is shown in Fig. 8, where different runs and time intervals are superposed and the wall velocities are shown as solid circles. The profile is linear as expected and exhibits a no-slip boundary value,the shear stress is found to have a z-independent value $T_{xz} = 0.20 \pm 0.02$, and dividing it by the velocity gradient gives the same value $\mu \approx 2.0$.

While no-slip has emerged from the simulation of liquids, a smooth transition to slip is found in the low-density, high-temperature 'fluid' regime. In Fig. 9, we show sample velocity profiles obtained at T=2.5 and a sequence of densities ranging from 0.8 to 0.08: the velocity at the wall systematically increases as the density decreases. We have also computed the shear stress in these runs, and extracted a slip length using the Maxwell definition (1) and extrapolations of the fitted velocity and stress to the wall -- defined here as the center of the inner wall layer.A similar calculation was done for Couette flow at the same temperature and densities, and the results

TABLE II. Slip length vs. density in the fluid phase at T=2.5.

ρ	μ	λ(Poiseuille)	λ(Couette)
0.8	2.09 ± 0.40	0.081 ± 0.014	. . .
0.6	1.04 ± 0.15	0.49 ± 0.09	0.48 ± 0.19
0.4	0.67 ± 0.08	2.1 ± 0.4	1.8 ± 0.9
0.2	0.37 ± 0.06	9.2 ± 1.1	10.0 ± 1.9
0.08	0.18 ± 0.05	56.0 ± 29.0	. . .

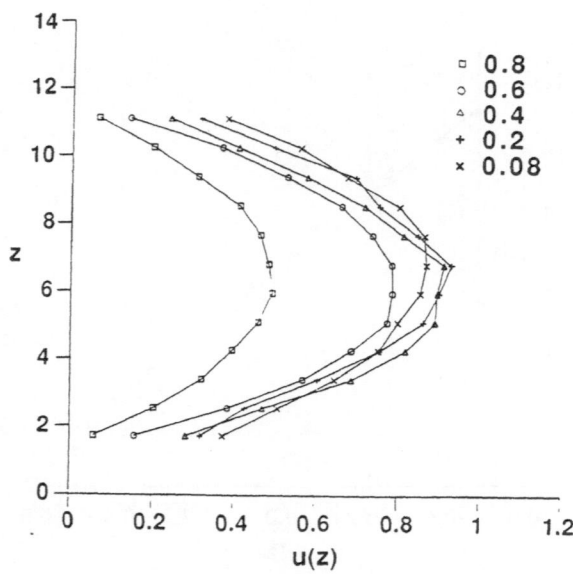

FIGURE 9. Velocity profile in Poiseuille flow at T=2.5 and decreasing densities as labeled.

are collected in Table II. We see first that the slip length increases as the density drops, and further that the same slip length applies to two different flow configurations. This result supports the use of (1) as a genuine continuum boundary condition for gases, applicable to different flows.

Having examined the Eulerian properties of channel flow, we turn to the molecular motion. Figure 10 shows several individual molecular trajectories in Poiseuille flow; note the different horizontal and vertical scales.The typical motion might be described as Brownian, with a superimposed drift in the direction of the applied acceleration, and with some temporary localization near the fixed attracting wall molecules. As the fluid-wall interaction varies the degree of wall localization changes in the obvious way. Note that there is no significant difference between molecules that were initially near the wall and those initially in mid-channel, in that the thermal motion eventually obliterates the initial bias.A last remark is that in a laboratory experiment with a realistic value of g and a macroscopic channel width, the relative amount of drift and wall localization is vastly smaller, and the molecular motion differs only slightly from Brownian.

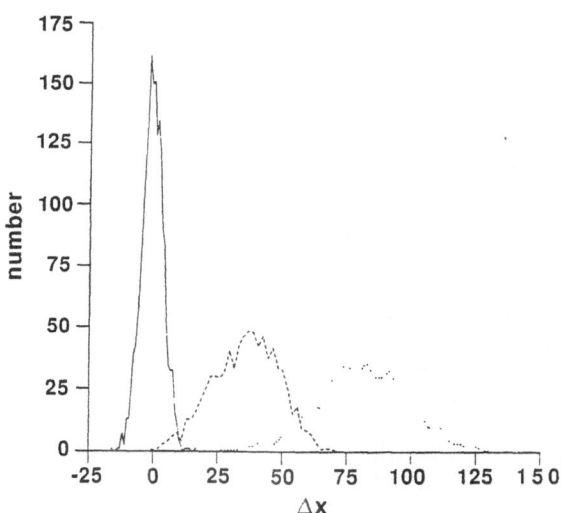

FIGURE 10. Molecular trajectories in Poiseuille flow; note the differing horizontal and vertical scales. The horizontal solid lines give the positions of the wall layers.

The molecular motion can be nicely quantified in terms of diffusivities. First, consider an equilibrium, periodic (wall-free) system. The motion of molecules is diffusive, i.e.,

$$<\Delta x_i> = 0$$
$$\lim_{t \to \infty} < \Delta x_i \, \Delta x_j > = 2 \, D_{mol} \, t \; \delta_{i,j} \, , \, g=0 \qquad (15)$$

where the Δx_i are the displacements from the starting position and where D_{mol} is the bulk molecular diffusivity. In fact, the latter is usually calculated in an MD simulation by monitoring the mean square displacements as a function of time. Here the situation is complicated by the presence of the walls: the z motion is confined and cannot grow indefinitely with time, while the x,y motion is unbounded and still diffusive but impeded by the attracting walls. The solid curves in Figs. 11-13 are histograms of the three components of displacement Δx_i in the presence of walls at g = 0, for a single 25000 step run of 1536 molecules. The x and y results are consistent with Gaussian, but z is not. An acceleration g = 0.1 was then applied, and the same histograms are obtained at time step 50000 (dashed curves) and 75000 (dotted curves). The Δz-histogram covers the same interval in z but becomes less peaked. The Δy-histogram continues approximately Gaussian, but its width grows linearly with time.In Δx, there is both a translation of the peak in the direction of flow, and an enhancement of its broadening compared to Δy. The broadening can be quantified by defining an effective longitudinal diffusivity as

$$D_{||} = \lim_{t \to \infty} < (\Delta x - <\Delta x>)^2 > / 2 t \tag{16}$$

and the resulting values are shown in Table I, along with the average velocity U.

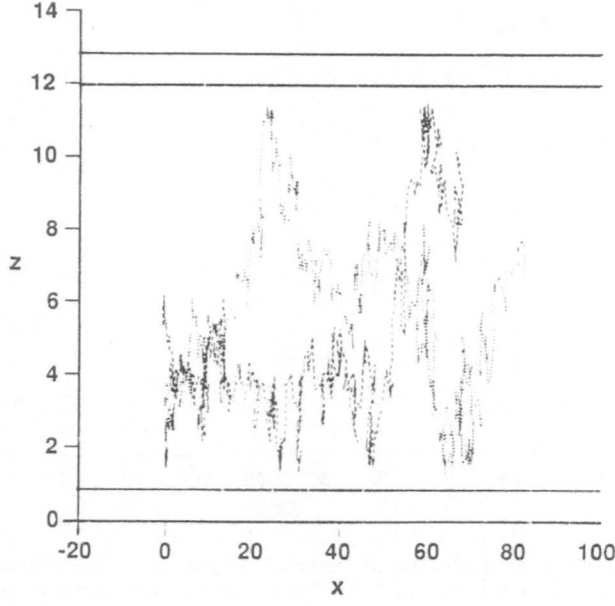

FIGURE 11. Histogram of x-displacements (a) in equilibrium after 25000 time steps, (b) after 25000 further steps at g=0.1, and (c) after another 25000 steps at g=0.1.

At the continuum level, one expects the spread in molecular positions in the direction of flow to be described by Taylor-Aris[16] hydrodynamic dispersion. These authors showed the the behavior of any initial distribution of a passive tracer in a channel asymptotically tends to a Gaussian moving with the average velocity whose variance, or effective longitudinal dispersivity, varied as the square of the Peclet numberin the form

$$D_{||}(U) = D_{mol} + U^2 w^2 / \xi D_{mol} \tag{17}$$

Here w is the width of the channel and the pure number ξ can be calculated for any given channel flow from the velocity profile; for the particular case of flow in a slot Aris[50] gives the value 210. Our numerical results are consistent with the quadratic velocity dependance in (17), and with the values $D_{mol}=D_{||}(0)=0.065 \pm 0.02$ and $\xi = 140 \pm 100$ .Because of the attraction to the nearby walls, our value of D_{mol} is less than the bulk value 0.077, obtained in simulations using fully periodic boundary conditions by other authors[34,51].

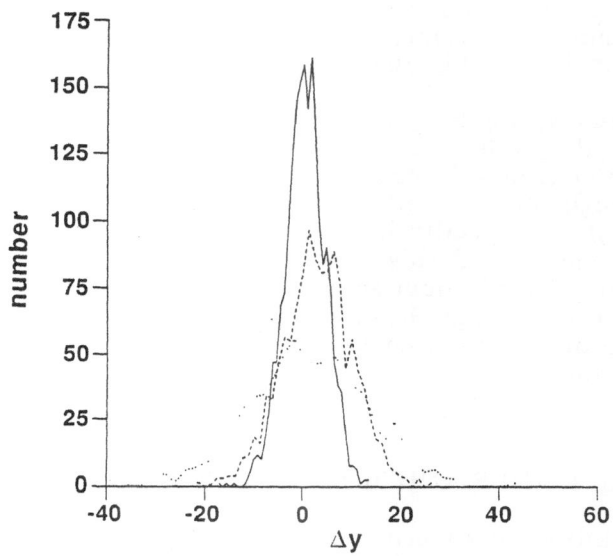

FIGURE 12. Histogram of y-displacements, as in Fig. 11.

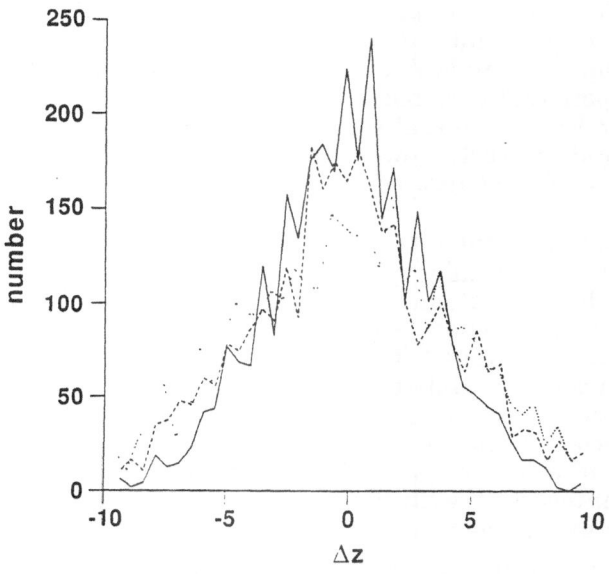

FIGURE 13. Histogram of z-displacements, as in Fig. 11.

This latter discrepancy makes the comparison to (17) slightly ambiguous, since its derivation does not incorporate such small-system walleffects. Further, for the higher values of g, the temperature of the fluid increases from 1.2 to 8.0 in long runs of 50000 time steps, leading to large systematic errors. Such long runs are needed to obtain an accurate value of the diffusivity, since we require the Gaussian asymptotic behavior to have set in. In obtaining the diffusion coefficient for g equalto 0.1, 0.2 and 0.4, we carried out temperature equilibration duringthe entire run. For g = 0.05, we carried out both the normal and equilibrated runs to ensure that the diffusion coefficient and the velocity were substantially independent of whether the equilibration was carried out or not. To summarize, we regard the comparison with the Taylor-Aris formula as satisfactory, and one further piece of evidence that our MD system is described by the usual continuum equations.

Another interesting aspect of the molecular motion is the persistence of layering near the walls. The dashed and dotted curves in Fig. 4 give the density distribution across the channel after motion occurs, at 25000 time step intervals. Although their magnitude drops with time (perhaps due to a rise in temperature), the oscillations are still evident. Several previous studies[30,46,48] have noted this dynamic layering, which is of course not accounted for in the continuum equations. Nonetheless, once the system is larger than the range of significant density oscillations, (which decay after 3 or so molecular diameters from a wall, here and in Ref. 13) the average behavior is macroscopic.

V. IMMISCIBLE FLUIDS AND MOVING CONTACT LINES

We first introduce a mechanism for segregating immiscible fluids. A simple choice is to add an additional repulsive interaction between species,

$$V_{\alpha\beta}(r) = V_{LJ}(r) + (c_\alpha - c_\beta)^2 \, 4\varepsilon \, (r/\sigma)^{-6} \qquad (18)$$

where c_α is an adjustable 'pseudo-charge' associated with species α. The bulk interaction of pure fluid is thus unchanged, but we can vary the interaction between the two fluids A and B and with the wall W. The resulting intermolecular potential does not represent any specific pair of fluids that we are aware of, and more generally one could choose interactions specific to a realistic system of interest. However, this choice respects the physics of immiscibility and is effective in producing the desired separation of the fluids.

As an example, consider the case $c_A = +0.5$, $c_B = -0.5$ and $c_W = 0$, where initially 1536 molecules of each fluid occupy half an FCC lattice with a plane boundary between them, and the lattice is allowed to come to equilibrium. In Figs. 14a-c we show several snapshots of the instantaneous molecular positions, at time steps 5000, 7500 and 10000; the molecules of the two fluids are displayed separately to make the interfacial region visible. Outside of a thin transition zone, the fluids retain their initial segregation. The position of the interfaces between the fluids showssome fluctuation with time, over a distance of a few molecular diameters. Since in this case both fluids have the same interaction with the wall, one expects a 90° contact angle and this is the case on average.Next taking the value $c_W = 0.35$ along with the previous $c_{A,B}$, we obtain Fig. 15, where the angle $\phi \approx 45°$ reflects the larger attraction of A (the lower fluid in the figure) to the wall. The value of ϕ is in

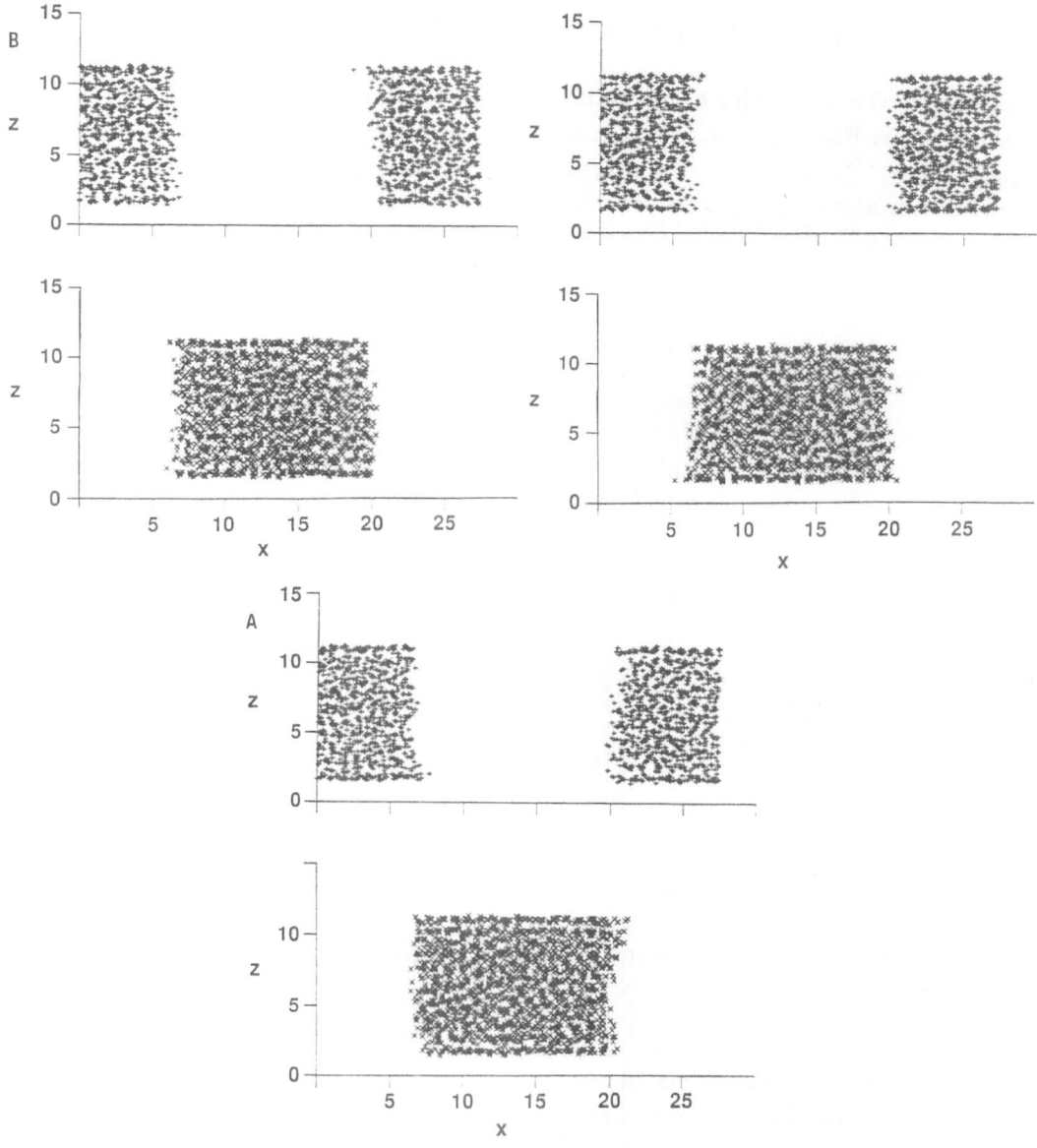

FIGURE 14. Meniscus fluctuations in a symmetric two-fluid system at rest; the fluids are displayed separately for ease of visualization although they in fact are in contact. Fluid A is below, and B above.

rough agreement with the following simple argument given by Israelachvili[38]. Consider Young's equation relating the solid-liquid and liquid-liquid interfacial free energies per unit area for the configuration shown in Fig. 1 :

$$\gamma_{AW} + \gamma_{AB} \cos\phi = \gamma_{BW} \tag{19}$$

The interface arises from the additional interspecies interaction (18), and it is plausible to assume that the interfacial free energy is simply proportional to its strength, $\gamma_{\alpha\beta} \sim (c_\alpha - c_\beta)^2$. Substituting into (18) gives

$$\cos \phi = [(c_B - c_W)^2 - (c_A - c_W)^2] / (c_A - c_B)^2 \qquad (20)$$

and using the values for c's given above, (20) predicts $\phi = 45$. Another example is shown in Fig. 16a, where $c_W = 0.1$, the observed angle is $\phi \approx 80°$, and (19) predicts 79°. Needless to say, this simple argument is not the last word on the calculational aspects of contact angle[52] but due to the small-simulation fluctuations illustrated in Fig. 14a more precise comparison is inappropriate.

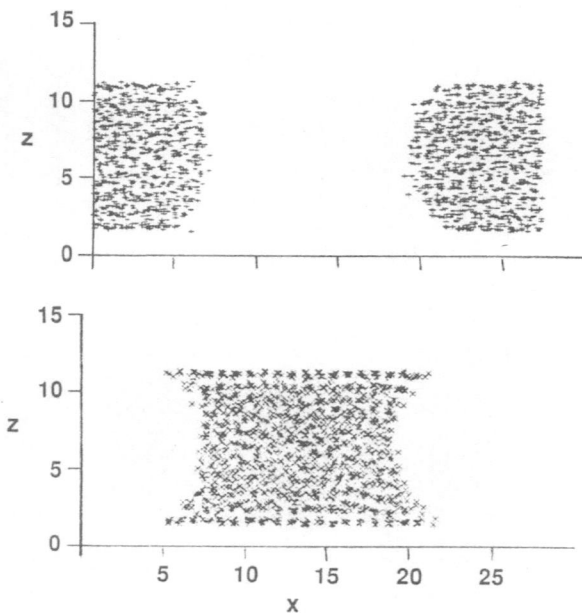

FIGURE 15. Meniscus profile for parameters $c_A = -c_B = 0.5$ and $c_W = 0.35$.

We have attempted to use the observed contact angle to determine the value of the surface tension, using the Laplace relation between the pressure difference in the fluids and the curvature κ of the interface. For the geometry at hand, the principal radii of curvature are $w(2\cos\phi)^{-1}$ across the gap, and ∞ along it, so

$$p_A - p_B = \gamma\kappa \rightarrow 2\,\gamma_{AB} \cos \phi / w \qquad (21)$$

To calculate the pressure we computed the local stress tensor from (14), using a 32 x 16 rectangular array of x-z bins, and then took one-third of the trace of the stress tensor averaged separately over each fluid region. Unfortunately this procedure gives values of the pressure such as 4.0 ± 0.2 in each fluid, with a difference of order 0.2 - 0.4 between the two. Thus the difference is barely distinguishable from the fluctuations, and a reliable estimate is impossible. If we ignore this problemand use the nominal difference of the means, along with a similarly crude estimate of ϕ (obtained by applying a protractor to the instantaneous snapshots, or with comparable accuracy to a contour plot of average fluid densities, or even equivalently through the use of (20)), we obtain $\gamma \sim 2.5$ for $c_A = -c_B = 0.5$ and a range of c_W. Using physical parameters for Argon, this value corresponds to 34 dynes/cm, which is in the right ballpark for a surface tension, but as noted above this is an artificial two-fluid system and there is no data to compare to.

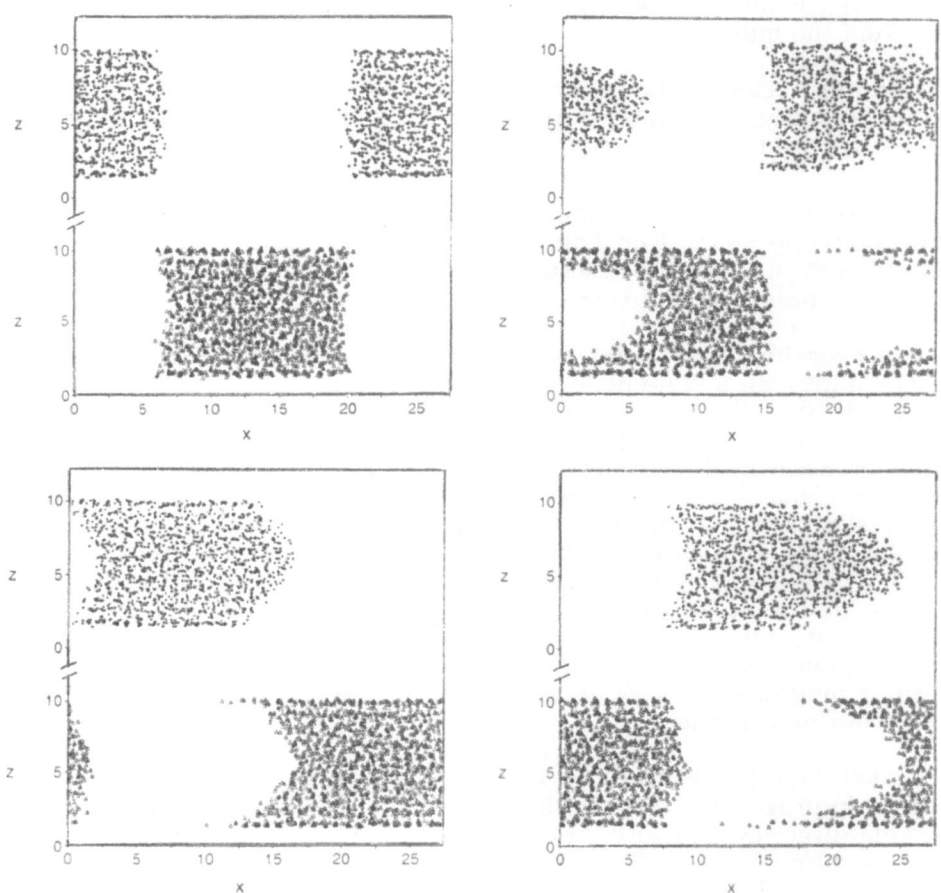

FIGURE 16: Time sequence of two-fluid immiscible displacement and moving contact lines, as described in the text.

Next, we study the motion of a contact line, by applying an acceleration to each fluid particle as before. A typical time sequence is shown in Figs. 16b-d at intervals of 8000 time steps after melting, for the choices $c_W = 0.1$ and $g = 0.1$. First we see that the two fluids maintain their phaseseparation even after two complete tranversals of the periodic channel. Secondly, note the distinct advancing and receding contact angles at the two A-B interfaces. The two angles differ both from the static angle and each other, and vary systematically with the velocity of the fluid. As in experiments[4,5], with increasing velocity the advancing angle increases and the receding angle decreases. Unfortunately, we have found it difficult to quantify the latter variation because of the meniscus fluctuations at low velocity, and the limited range of higher velocities available in the low Reynolds number regime. Lastly, note the thin film of fluid A left behind along the walls. The film contains the fluid more strongly attracted to the wall, as one would expect, and has the additional consequence thatthe apparent receding contact angle effectively varies with time.When the bulk motion halts, on setting g to zero, the film begins to retract very slowly into the bulk, but apparently on a time scale long compared to the duration of the simulation. We have seen no evidence for a *precursor* film in the cases we have examined, but this absence could be due to such factors as insufficient wall material to develop a reasonable van der Waals force, or simply not allowing enough time for a film to extend outwards from the bulk.

At high velocity the interface motion is fairly steadywith a roughly constant velocity, for example $u_{int} \sim 0.2$ for the case shown in Fig. 16. In contrast, at very low acceleration and velocity, the bulk motion of the fluids is rather erratic and jumpy. For example at g= 0.001 the fluids remain more or less in place for 60000 time steps, then advance a distance $O(\sigma)$ over the next 10000, then stop for (at least) another 30000 steps. Aside from being a clear manifestation of stick-slip motion,this behavior is suggestive of the phenomenon of contact angle hysteresis[4,5].

In experiment one often observes a range of static angles where, for example, a drop on a plate distorts without bulk motion as the plate is tilted, until a critical inclination is reached and the drop begins to slide. The effect is commonly attributed to the roughness of the solid surface: even if the microscopic contact angle is unique, when it lies on a rough surface of variable orientation it may have a variety of apparent contact angles. Thus, in the sliding drop situation a small tilt may just produce a tiny motion of the interface up to the next bump on the surface, and one might say that the interface is 'pinned' by the bumps on the surface (see, e.g., Schwartz and Garoff[53]). The irregular motion described just above suggests that the mensicus is *almost* pinned by the molecular scale roughness of the walls. It is conceivable that at values of g even lower than 0.001, we might find a meniscus truly pinned by molecular roughness, but we do not have unbounded computer time available to perform this experiment.

Next we consider the flow field in the moving contact line system. In trying to improve the statistics beyond our earlier work[18], we have used a somewhat larger system with 8000 molecules (3200 per fluid and 800 per wall) and use a 40 x 20 array of x-z bins moving with the velocity of the interface, adjusting this velocity periodically. In Fig. 17a-c we show the average velocities of molecules occupying each bin over an interval of 12000 time steps, The three figures represent different initial distributions of molecular velocity, and have several important qualitative features in common.The walls are moving to the left in this reference frame, and away from the contact line the fluid velocity near the wall coincides with the wall velocity. Near the contact lines the no-slip condition appears to fail, however, and we observe a jet of fluid into the interior from the contact line. In addition, there is a return flow in the center of the channel, and hints of closed eddies of fluid behind the interface. While the flow fields in the three runs are not identical and all quite noisy, they each exhibit the qualitative features just cited.We have also simulated immiscible flow in a Couette geometry, as suggested by Thompson and Robbins[40]; the results show the same qualitative behavior as well as the same regrettable lack of quantitative detail.

To gain some insight into the velocity field, it is useful to ask what one might expect to see. The discussion to follow is suggested by the work of Dussan and collaborators[17,20,54]. First consider a viscous fluid advancing through a slot into vacuum, illustrated in Fig. 18a. In the rest frame of the meniscus, the velocity of the wall should agree with the wall velocity, except perhaps near the contact line, and by incompressibility there must be a return flow which can most economically be located in the center of the channel. This reasoning leads to a fountain-like motion of fluid up the center and back along the walls, which has been observed in experiment. Now suppose two immiscible fluids are in motion; the simplest possible flow field is obtained by simply duplicating the previous flow in each fluid, which leads to Fig. 18b. While perfectly sensible from a kinematic point of view the latter flow field has the feature that the velocity must go to zero along the meniscus, because the velocities to either side of it are in opposite directions. This behavior can be ruled out by experiment, simply by placing some dye on the meniscus and observing that it moves outwards to the wall. The simplest means of

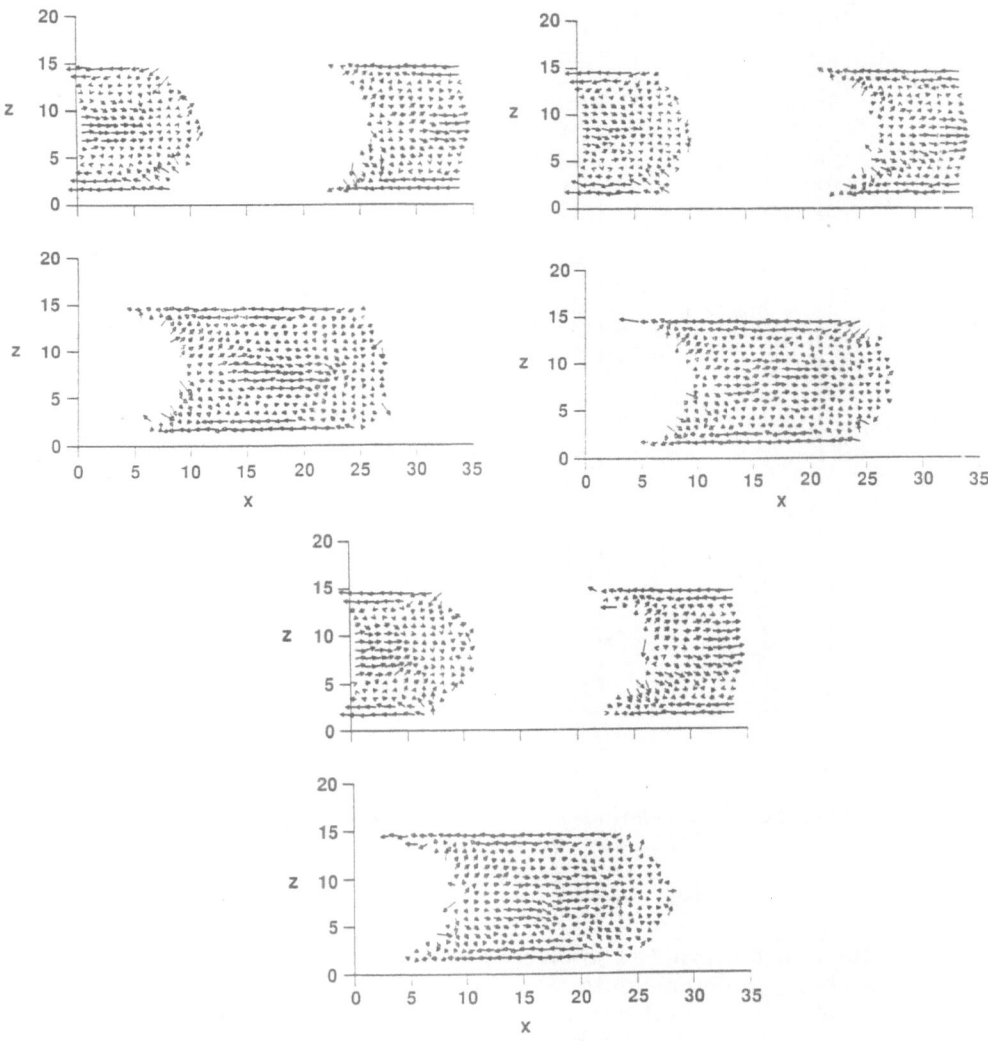

FIGURE 17. Examples of the velocity field in immiscible displacement, obtained in different runs.

allowing a non-zero velocity on the meniscus is to introduce a closed eddy in one fluid which reverses that fluid's velocity near the meniscus, as in Fig. 18c. The choice of which fluid has the closed eddy is not dictated by kinematics, but one might anticipate that it would occur as drawn in the wider of the two wedges, because this choice minimizes the bending of streamlines and would tend to lower the energy dissipation, in accord with both the detailed calculation of Huh and Scriven[3,55] and the laziness principle of Joseph et al.[56] Finally, if this reasoning is extended to the specific geometry of the present simulation, one obtains the flow field of Fig. 18d. While our simulation results in Fig. 17 do not exhibit this detailed streamline pattern, they are *consistent* with it, in that the reproducible features noted in the previous paragraph are all present.

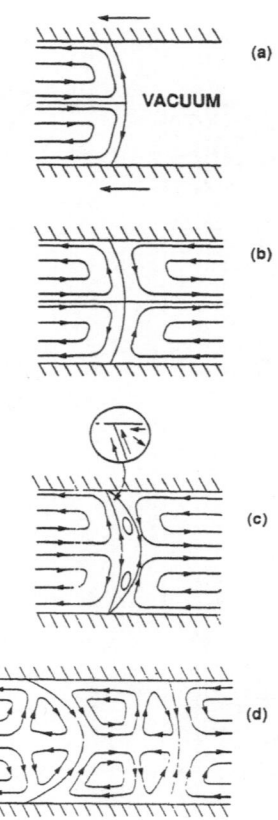

FIGURE 18. Qualitative streamlines expected in various displacement situations, as described in the text.

We have also computed the local stress field using the Kirkwood-Irving expression (14). Analogously to the static pressure computation described above, the thermal fluctuations due to the small number of molecules in an averaging bin mask the systematic variation. There is some evidence for an enhancement of the shear stress near the contact line, presumably the remnant of the divergence associated with no-slip, but we have not been able to describe even qualitatively the spatial variation of the stress.

VI. CONCLUSIONS

We have shown that molecular dynamics calculations can successfully reproduce the continuum properties of low Reynolds number flow near realistic solid boundaries, as well as answer questions that cannot be settled from a purely macroscopic viewpoint. In channel flows of a single fluid, where the interesting variation occurs only in one (spanwise) dimension, MD produces both the correct variation of the continuum velocity and stress fields and the Brownian and dispersive properties of the molecular motion. More complicated flows, involving two fluids or a two (or three) dimensional flow field, are difficult to fully resolve with system sizes currently acessible to MD, but anticipated advances in computing power should remove this restriction in the near future. Nonetheless, we have been able to draw significant qualitative conclusions for a particularly interesting two-fluid problem.

The no-slip boundary condition appears to be a natural property of a dense liquid interacting with a solid wall with molecular structure and interactions. There is some unavoidable ambiguity in assigning the precise location where the tangential velocity should vanish at a boundary, due to both a finite molecular size and to a quite reasonable variation with the details of the fluid-wall interaction. The fluid-wall interface is inherently ill-defined on the length scale of individual molecules, although any such imprecision is imperceptible at the continuum scale. Indeed, whenever such effects are important the continuum equations are insufficient.

The no-slip condition appears to fail at a moving contact line between immiscible fluids. One may describe this situation heuristically by saying that each fluid 'tries' to stick to the wall, but the high shear rates required by the kinematics of a moving contact line overcomes the natural adherence. If no-slip is imposed, the shear rate tends to diverge, but the molecules are compliant enough to effect a compromise with some slip and high but finite shear[57]. A hint of the latter is seen in the MD simulations. Our results, with limited statistical resolution, do not suggest the precise form of the correct continuum boundary condition, but presumably further simulations of larger systems over longer times will settle this issue.

The more general conclusion is that even surprisingly small systems of several thousand molecules exhibit continuum behavior, and that MD is therefore a viable tool in continuum fluid mechanics. For most purposesMD involves far too much microscopic detail to be an efficient computational method, but in appropriately chosen problems it is a valuable addition to the repertoire of techniques. A particularly promising area is micro-hydrodynamics[58], where one considers flows with length scales around a micron, plus or minus a decade or two, and in particular the derivation of effective transport properties of systems with non-trivial microscopic structure. A host of other problems may be studied with MD,ranging from the flow of more complicated (e.g., non-Newtonian or composite) fluids, to the effects of convection on reaction and phase change, to flows over irregular surfaces. We hope to address such problems soon.

ACKNOWLEDGEMENTS

We have benefitted from discussions with and the encouragement of Berni Alder, Stephen Garoff, Jacob Israelachvili, Mark Nelkin, John Ullo and Sidney Yip. We particularly thank Elizabeth Dussan for her continued interest in this work and her thoughtful comments on the manuscript.

REFERENCES

1   S. Goldstein, ed., <u>Modern Developments in Fluid Dynamics</u> (Oxford University Press, Oxford, 1938), Appendix. We thank G. M. Homsy for bringing this reference to our attention.

2   G. K. Batchelor, <u>An Introduction to Fluid Dynamics</u> (Cambridge University Press, Cambridge, 1967)

3   C. Huh and L. E. Scriven, J. Colloid. Interface. Sci. **35**, 85 (1971).

4   E. B. Dussan V., Ann. Rev. Fluid Mech. **11**, 371 (1979).

5   P. G. De Gennes, Rev. Mod. Phys. **57**, 827 (1985).

6   J. N. Israelachvili, P. M. McGuiggan and A. M. Homola, Science **240**, 189 (1988).

7   D. D. Awschalom and J. Warnock, Phys. Rev. B **35**, 6779 (1987)

8   M. P. Allen and D. J. Tildesley <u>Computer Simulation of Liquids</u>, (Clarendon Press, Oxford, 1987).

9   G. Coccotti and W. G. Hoover, eds., <u>Molecular-Dynamics Simulation of Statistical Mechanics Systems</u> (North-Holland, Amsterdam, 1986).

10 J. M. Haile, <u>A Primer on the Computer Simulation of Atomic Fluids by Molecular Dynamics</u> (Clemson University, Clemson, SC, 1980).

11 E. Meiburg, Phys. Fluids **29**, 3107 (1986); D. C. Rapoport, Phys. Rev. **A 36**, 3288 (1987).

12 M. Mareschal and E. Kestemont, Nature **329**, 427 (1987); J. Stat. Phys. **48**, 1187 (1987); D. C. Rapaport, Phys. Rev. Lett. **60**, 2480 (1988).

13 B. J. Alder and T. E. Wainwright, Phys. Rev. **A 1**, 18 (1970); W. E. Alley and B. J. Alder, ibid. **A 27**, 3158 (1983); W. E. Alley and B. J. Alder, Phys. Today **37** (1), 56 (1984).

14 J. C. Maxwell, Phil. Trans. Roy. Soc. **170**, 231 (1867); see also R. Jackson, <u>Transport in Porous Catalysts</u> (Elsevier, Amsterdam, 1977).

15 C. Cercignani, <u>Mathematical Methods in Kinetic Theory</u> (Plenum Press, New York, 1969).

16 G. I. Taylor, Proc. Roy. Soc. **A 219**, 186 (1953); R. Aris, ibid. **235**, 67 (1956).

17 E. B. Dussan V., AIChE J. **23**, 131 (1977).

18 J. Koplik, J. R. Banavar and J. F. Willemsen, Phys. Rev. Lett. **60**, 1282 (1988).

19 G. N. Patterson, <u>Molecular Flow of Gases</u>, (Wiley, New York, 1956).

20 E. B. Dussan V. and S. H. Davis, J. Fluid Mech. **65**, 71 (1974).

21 See, e. g., G. E. P. Elliot and A. C. Riddiford, J. Colloid. Interface. Sci. **23**, 389 (1967).

22 Y. Pomeau and A. Pumir, C. R Acad. Sci. **299 II**, 909 (1984).

23 A. Nir, private communication (1988).

24 E. B. Dussan V., J. Fluid Mech. **77**, 665 (1976).

25 S. Richardson, J. Fluid Mech. 59, 707 (1973); L. M. Hocking, J. Fluid Mech. **76**, 801 (1976).

26 K. M. Jansons, J. Fluid Mech. **154**, 1 (1985); Phys. Fluids **31**, 15 (1988).

27 J. P. Hansen and I. R. McDonald, Theory of Simple Liquids, (Academic Press, New York, 1976).

28 A. W. Lees and S. F. Edwards, J. Phys. C **5**, 1921 (1972).

29 A. Tenenbaum, G. Ciccotti and G. Gallico, Phys. Rev. **A 25**, 2778 (1982); C. Trozzi and G. Ciccotti, ibid. **29**, 916 (1984).

30 L. Hannon, G. C. Lie, E. Clementi and V. Yakhot, IBM Kingston preprint (1988).

31 J. Schnute and M. Shinbrot, Can. J. Math. **25**, 1183 (1973); D. Rothman, Geophysics **53**, 509 (1988).

32 F. F. Abraham, J. Chem. Phys. **68**, 3713 (1978); see also F. F. Abraham, Rept. Progr. Phys. **45**, 1113 (1982).

33 S. Toxvaerd, J. Chem. Phys. **74**, 1998 (1981).

34 J. J. Magda, M. Tirrell and H. T. Davis, J. Chem. Phys. **83**, 1888 (1985).

35 J. Q. Broughton and G. H. Gilmer, J. Chem. Phys., **64**, 5759 (1986), and earlier references therein.

36 J. Tallon, Phys. Rev. Lett. **57**, 1328 (1986).

37 J. N. Israelachvili and G. E. Adams, J. Chem. Soc. Faraday Trans. I 74, 975 (1978).

38 J. N. Israelachvili, Intermolecular and Surface Forces (Academic, London, 1985).

39 S. J. Gregg and K. S. W. Sing, Adsorption, Surface Area and Porosity , 2nd ed. (Academic, New York, 1982).

40 P. A. Thompson and M. O. Robbins, to be published (1988).

41 W. T. Ashurst and W. G. Hoover, Phys. Rev. **A 11**, 658 (1975)

42 The sound speed in liquid Argon at 44.4 MHz at 85 K and vapor pressure is 853 m/s, as cited in A. I. P. Handbook of Physics, 3rd ed. (McGraw-Hill, 1972).

43 This Table corrects an error in a similar one given in Ref. 18, where the wrong value of ρ was inadvertently used.The values of μ in the earlier Table should all be scaled upwards by a factor 13/12.

44 S. A. Mikhailenko, B. G. Dudar and V. A. Schmidt, Sov. J. Low Temp. Phys. 1, 109 (1975).

45 W. G. Hoover, D. J. Evans, R. B. Hickman, A. J. C. Ladd, W. T. Ashurst and B. Moran, Phys. Rev. A 22, 1690 (1980).

46 D. M. Heyes, J. J. Kim, C. J. Montrose and T. A. Litovitz, J. Chem Phys. 73, 3987 (1980).

47 J.-P. Ryckaert, A. Bellemans, G. Cicotti, and G. A. Paolini, Phys. Rev. Lett. 60, 128 (1988).

48 I. Bitsanis, J. J. Magda, M. Tirrell and H. T. Davis, J. Chem Phys. 87, 1733 (1987).

49 J. H. Irving and J. G. Kirkwood, J. Chem. Phys. 18, 817 (1950).

50 R. Aris, Proc. Roy. Soc. A 252, 538 (1959).

51 S.-H. Suh and J. M. D. MacElroy, Mol. Phys. 58, 445 (1986).

52 J. S. Rowlinson and B. Widom, Molecular Theory of Capillarity (Clarendon Press, Oxford, 1982).

53 L. W. Schwartz and S. Garoff, J. Colloid. Interface. Sci. 126, 422 (1985).

54 F. Y. Kafka and E. B. Dussan V., J. Fluid Mech. 95, 539 (1979)

55 Although their solution cannot be correct down to the contact line, it can represent an outer or matching solution when combined with a slip boundary condition in the inner region near the wall; see Ref. 54.

56 D. D. Joseph, K. Nguyen and G. S. Beavers, J. Fluid Mech. 141, 319 (1984).

57 A contrasting situation occurs when a single fluid flows past a sharp corner, as discussed by K. Moffatt, J. Fluid Mech. 18, 1 (1964). The geometry requires the streamlines to bend sharply at the corner, but the velocity vanishes there, and there is no potentially divergent shear stress.

58 G. K. Batchelor, in Theoretical and Applied Mechanics, W. T. Koiter, ed. (North Holland, Amsterdam, 1976).

HYDRODYNAMIC INTERACTIONS AND TRANSPORT COEFFICIENTS IN A SUSPENSION OF SPHERICAL PARTICLES

Anthony J.C. Ladd

Lawrence Livermore National Laboratory, Livermore, California 94550

Particulate suspensions of solids in liquids can be understood in terms of the microstructure of the solid phase by using the well-known techniques of numerical statistical mechanics. The major problem with such an approach has been the incorporation of the long-range, many-body hydrodynamic forces between the suspended particles. In this paper I describe a general computational method for calculating the forces and torques exerted by slowly moving spheres suspended in an incompressible fluid. The method can be used to determine bulk constitutive properties of solid-fluid suspensions or particulate porous media. Numerical results for the sedimentation velocity and high frequency viscosity of monodisperse suspensions have been obtained, and the results are shown to compare very well with experimental measurement.

1. INTRODUCTION

Transport processes in simple liquids can be described by linear constitutive laws, characterized by transport coefficients that depend only on the thermodynamic state of the system. This is because there are large separations of length scale and time scale between the macroscopic fluxes and the underlying microscopic processes. For particulate suspensions these scale separations are not so large and the macroscopic constitutive laws can be non-local in time and non-linear in the applied fields. These complicated constitutive relations can be obtained directly from a statistical analysis of the microscopic interactions [1]. This involves first determining what the microscopic interactions are for a particular configuration of solid particles and then averaging over an appropriate ensemble of configurations to obtain macroscopic transport coefficients such as viscosity, sedimentation velocity (mobility), self-diffusion coefficients, and permeability. This is a well-defined procedure which requires as input only the distribution of solid particle sizes, the viscosity of the suspending fluid (assuming it is a simple Newtonian fluid), and a boundary condition at the solid-fluid surface, usually the hydrodynamic "no-slip" condition.

The first application of a statistical theory of particulate suspensions was contained in Einstein's thesis work around 1905. He obtained an expression, now well known, relating the viscosity of a dilute suspension of spheres η to the viscosity of the pure fluid η_0 and the solid volume fraction ϕ [2],

$$\eta/\eta_0 = 1 + \frac{5}{2}\phi. \tag{1.1}$$

In the dilute limit it is sufficient to treat the suspension as an assembly of non-interacting spheres, but at higher solid densities the particles interact via viscous forces transmitted through the fluid. These hydrodynamic interactions cannot be written down in closed form except in special cases; the

Microscopic Simulations of Complex Flows
Edited by M. Mareschal, Plenum Press, New York, 1990

difficulties involved in calculating them in general led to the use of oversimplified, pair-wise additive approximations in early computer simulations of solid-fluid suspensions [3-5]. None of these calculations addressed the crucial issue of the many-body contributions to the hydrodynamic interactions, the importance of which has been emphasized by Beenakker and Mazur [6]. Recently however, attempts have been made to calculate the many-body hydrodynamic interactions numerically [7-10], using a multipole-moment expansion of the force density on the surface of each particle [11]. In Refs. [7,8] an exact calculation of the pairwise-additive hydrodynamic interactions was combined with a force-torque-stresslet approximation to the many-body interactions. This is insufficient at high packing fractions as will be seen later. In Refs. [9,10] I described a method for calculating the hydrodynamic interactions between spherical particles that included as many force multipoles as were needed for convergence, at least in principle. However convergence was very poor when particles were close together, because the short-range lubrication forces were not explicitly included. It became clear that the general multipole expansion of the force density [9,10] had to be combined with an exact calculation of the pairwise-additive lubrication forces [7]. This has now been done and a full account of this work will appear elsewhere [12]. In this paper I summarize the computational method, developed in Refs. [9,10,12], for calculating the forces and torques exerted by slowly moving spheres suspended in an incompressible fluid, and indicate how one can then determine the bulk constitutive properties of solid-fluid suspensions such as viscosity and sedimentation velocity. Recent numerical results for the viscosity and sedimentation velocity of monodisperse suspensions are shown to compare very well with experimental measurement.

2. HYDRODYNAMIC INTERACTIONS

The motion of solid particles suspended in a fluid is a complex problem, primarily because of the difficulties involved in calculating the long-range, many-body hydrodynamic interactions. If the relative velocity of the solid and fluid phases is small, the inertial terms in the Navier-Stokes equation describing the motion of the fluid can be ignored. The fluid flow is then stationary in time and incompressible, leading to the so-called "creeping-flow" equations for the velocity field v and the pressure p [1],

$$\nabla \cdot \mathbf{v} = 0, \tag{2.1a}$$

$$\nabla p = \eta_0 \nabla^2 \mathbf{v}, \tag{2.1b}$$

where η_0 is the shear viscosity of the pure fluid. Finite Reynolds number corrections are generally small for particle sizes up to 1 mm. However, even with these important simplifications of the full non-linear Navier-Stokes equation, the calculation of the resulting forces on an assembly of moving particles is far from trivial.

Since the creeping-flow equations are linear in the velocity and pressure fields, the forces **F** and torques **T** exerted on the particles by the fluid can be linearly related to the particle velocities **U** and angular velocities Ω via the configuration dependent friction-coefficient matrices ζ,

$$\mathbf{F}_i = -\sum_{j=1}^{N}[\zeta_{ij}^{TT} \cdot \mathbf{U}_j + \zeta_{ij}^{TR} \cdot \Omega_j], \tag{2.2a}$$

$$\mathbf{T}_i = -\sum_{j=1}^{N}[\zeta_{ij}^{RT} \cdot \mathbf{U}_j + \zeta_{ij}^{RR} \cdot \Omega_j], \tag{2.2b}$$

where the superscripts T and R refer to the translational and rotational components of ζ, and the sums extend over all N particles in the suspension. Equations (2.2) can be inverted to find the corresponding mobility matrices μ defined by,

$$\mathbf{U}_i = -\sum_{j=1}^{N}[\mu_{ij}^{TT} \cdot \mathbf{F}_j + \mu_{ij}^{TR} \cdot \mathbf{T}_j], \tag{2.3a}$$

$$\Omega_i = -\sum_{j=1}^{N}[\mu_{ij}^{RT} \cdot \mathbf{F}_j + \mu_{ij}^{RR} \cdot \mathbf{T}_j]. \tag{2.3b}$$

The friction and mobility coefficients are complicated functions of all the particle coordinates;

calculating them for arbitrary assemblies of spherical particles is the main objective of the method described in this work. Specifically, it will be assumed that the spheres are of uniform size, and that hydrodynamic stick boundary conditions apply at the solid-fluid surfaces. Then the fluid velocity field on the surface of a particle is given by

$$\mathbf{v}(\mathbf{r}) = \mathbf{U} + \mathbf{\Omega} \times (\mathbf{r} - \mathbf{R}) \quad \text{for} \quad |\mathbf{r} - \mathbf{R}| = a, \tag{2.4}$$

where $\mathbf{U}$, $\mathbf{\Omega}$, and $\mathbf{R}$ are the velocity, angular velocity, and location of the particle, and a is the particle radius. Equations (2.1) and (2.4), together with periodic boundary conditions applied to a set of N spheres in a cubic unit cell of volume V, completely specify the problem, which can be solved by the method of induced forces [11].

The stick boundary conditions are satisfied by introducing an induced-force density on the surface of each particle, which is to be chosen so that the fluid velocity field matches the surface velocity of each particle [Eq. (2.4)] at all points on the particle surface. The fluid velocity field anywhere in the system can by written in terms of this induced force density $\mathbf{F}_{ind}$,

$$\mathbf{v}(\mathbf{r}) = \mathbf{v}_0(\mathbf{r}) + \int_{V_M} \mathbf{T}(\mathbf{r} - \mathbf{r}') \cdot \mathbf{F}_{ind}(\mathbf{r}') d\mathbf{r}', \tag{2.5}$$

where $\mathbf{v}_0$ is an externally imposed velocity field. The Green's function $\mathbf{T}$ is the Oseen tensor, describing the velocity field arising from a point force in an infinite medium,

$$\mathbf{T}(\mathbf{r}) = \frac{1}{8\pi\eta_0 r}[\mathbf{I} + \hat{\mathbf{r}}\hat{\mathbf{r}}]; \tag{2.6}$$

the vector $\hat{\mathbf{r}}$ denotes the unit vector $\mathbf{r}/r$, and $\mathbf{I}$ represents the second rank unit tensor. The integral in Eq. (2.5) extends over the macroscopic sample volume V_M, and includes all the periodic images of the unit cell. Because the Oseen tensor is long-range this integral can diverge, and it is therefore necessary to take careful account of the macroscopic boundary conditions. In general there can be no net force on the fluid; otherwise it will accelerate without bound. A gravitational force, for instance, must always be balanced by a compensating pressure gradient. With this constraint, we can rewrite Eq. (2.5) in a form consistent with periodic boundary conditions [10],

$$\mathbf{v}(\mathbf{r}) = \mathbf{v}_0(\mathbf{r}) + \frac{1}{V} \sum_{\mathbf{k} \neq 0} \frac{e^{i\mathbf{k}\cdot\mathbf{r}}}{\eta_0 k^2}[\mathbf{I} - \hat{\mathbf{k}}\hat{\mathbf{k}}] \cdot \mathbf{F}_{ind}(\mathbf{k}), \tag{2.7}$$

where $\mathbf{F}_{ind}(\mathbf{k})$ is a finite Fourier transform over the volume V of the unit cell containing the N spheres,

$$\mathbf{F}_{ind}(\mathbf{k}) = \int_V e^{-i\mathbf{k}\cdot\mathbf{r}} \mathbf{F}_{ind}(\mathbf{r}) d\mathbf{r}. \tag{2.8}$$

The sum in Eq. (2.7) includes all k-vectors commensurate with the unit cell. The $\mathbf{k} = 0$ term is omitted from the summation, as it is assumed that there is no net ($\mathbf{k} = 0$) force on the system.

The periodic Oseen equation [Eq. (2.7)] can be solved by a multipole expansion of the induced force density on the surface of each particle. Force multipoles are defined as surface integrals involving the induced traction $\mathbf{t}_i$ (i.e. the induced force per unit area on the particle surface) and irreducible tensor products of the unit vector $\hat{\mathbf{r}}_i$, which denotes a point on the surface of sphere i relative to its center [9]. Thus the p-th order force multipole of particle i is a tensor of rank p+1

$$\mathbf{F}_i^{p+1} = \frac{1}{p!} \int_{S_i} \overline{\hat{\mathbf{r}}_i^p} \mathbf{t}_i(\mathbf{r}_i) d\mathbf{r}_i, \tag{2.9}$$

where S_i indicates an integral over the (spherical) surface of particle i. The notation $\overline{\mathbf{r}^p}$ is used to indicate an irreducible tensor of rank p. The forces, torques, and stress exerted by the fluid on a particular particle are related to the first two force moments, *i.e.*

$$\mathbf{F}_i = -\mathbf{F}_i^1 \quad ; \quad \mathbf{T}_i = a\,\boldsymbol{\varepsilon}{:}\mathbf{F}_i^{2a} \quad ; \quad \sigma_{\text{ind}}V = -a\sum_{i=1}^{N}\mathbf{F}_i^{2s}, \tag{2.10}$$

where $\boldsymbol{\varepsilon}$ is the Levi-Civita tensor and $\mathbf{F}^{2a}$ indicates the part of $\mathbf{F}^2$ that is antisymmetric with respect to permutation of the cartesian indices ($\boldsymbol{\varepsilon}{:}\mathbf{F}^2 \equiv \varepsilon_{\alpha\beta\gamma}F_{\gamma\beta}^2$). The traceless symmetric part $\mathbf{F}_i^{2s}$ is the contribution to the induced stress tensor σ_{ind} arising from the force density on particle i; it is related to the shear viscosity, as will be seen in section 3. The trace of $\mathbf{F}^2$ is related to the induced pressure and has no effect on the fluid motion because of the incompressibility condition. All the properties of the suspension that are of physical interest can be derived from the first two force moments, but the higher moments are necessary to give an accurate description of the fluid flow in densely packed suspensions.

Irreducible moments of the fluid velocity on the surface of each sphere can be defined in an analogous manner to the force moments

$$\mathbf{V}_i^{p+1} = \frac{(2p+1)!!}{4\pi a^2}\int_{\mathbf{S}_i}\overline{\hat{\mathbf{r}}^p}[\mathbf{v}(\mathbf{r}_i) - \mathbf{v}_0(\mathbf{r}_i)]d\mathbf{r}_i. \tag{2.11}$$

These moments are related to the particle velocities, angular velocities, and external shear rate

$$\mathbf{U}_i - \mathbf{u}_0 = \mathbf{V}_i^1 \quad ; \quad \boldsymbol{\Omega}_i - \boldsymbol{\omega}_0 = -(2a)^{-1}\boldsymbol{\varepsilon}{:}\mathbf{V}_i^{2a} \quad ; \quad \dot{\boldsymbol{\varepsilon}}_0^s = -a^{-1}\mathbf{V}_i^{2s}, \tag{2.12}$$

where $\mathbf{u}_0$, $\boldsymbol{\omega}_0$, and $\dot{\boldsymbol{\varepsilon}}_0^s$ describe the velocity, angular velocity, and traceless-symmetric strain rate of a homogeneous velocity field,

$$\mathbf{v}_0(\mathbf{r}) = \mathbf{u}_0 + \boldsymbol{\omega}_0\times\mathbf{r} + \dot{\boldsymbol{\varepsilon}}_0^s{\cdot}\mathbf{r}. \tag{2.13}$$

Because the stick boundary conditions allow no relative solid-fluid motion on the surface of the particles all higher velocity moments are zero.

It follows from Eqs. (2.7), (2.9) and (2.11), that the force moments and velocity moments are linearly related,

$$\mathbf{V}_i^{p+1} = \sum_{j=1}^{N}\sum_{p'=0}^{\infty}\mathbf{G}_{ij}^{p+1,p'+1}\odot\mathbf{F}_j^{p'+1}. \tag{2.14}$$

where $\odot$ indicates a $p'+1$-fold contraction between the the tensors $\mathbf{G}_{ij}^{p+1,p'+1}$ and $\mathbf{F}_j^{p'+1}$, contracting the last index of $\mathbf{G}_{ij}^{p+1,p'+1}$ with the first index of $\mathbf{F}_j^{p'+1}$, etc.. After substituting the spherical-harmonic expansion of the plane wave $e^{i\mathbf{k}\cdot\mathbf{r}}$ into the expression for the velocity field [Eq. (2.7)], and using the definitions of force and velocity moments [Eqs. (2.9) and (2.11)], it can be shown that [9]

$$\mathbf{G}_{ij}^{p+1,p'+1} = (i)^{p-p'}(2p+1)!!(2p'+1)!!\sum_{\mathbf{k}\neq0}e^{i\mathbf{k}\cdot(\mathbf{R}_i-\mathbf{R}_j)}\frac{j_p(ka)j_{p'}(ka)}{\eta_0 k^2 V}\overline{\hat{\mathbf{k}}^p}[\mathbf{I}-\hat{\mathbf{k}}\hat{\mathbf{k}}]\overline{\hat{\mathbf{k}}^{p'}}. \tag{2.15}$$

By restricting the number of force moments, that is by restricting p and p' to be less than or equal to some p_{max}, Eqs. (2.14) are reduced to a finite set of linear simultaneous equations. The calculation of the matrix elements $\mathbf{G}$ was described in Appendix A of Ref. [9].

The set of linear simultaneous equations represented by Eqs. (2.14) can be solved by triangular decomposition, followed by the appropriate backsolves to obtain the friction coefficients $\boldsymbol{\zeta} = \mathbf{G}^{-1}$,

$$\boldsymbol{\zeta}_{ij}^{TT} = \boldsymbol{\zeta}_{ij}^{1,1} \quad ; \quad \boldsymbol{\zeta}_{ij}^{TR} = a\,\boldsymbol{\zeta}_{ij}^{1,2a}{:}\boldsymbol{\varepsilon} \quad ; \quad \boldsymbol{\zeta}_{ij}^{RT} = -a\,\boldsymbol{\varepsilon}{:}\boldsymbol{\zeta}_{ij}^{2a,1} \quad ; \quad \boldsymbol{\zeta}_{ij}^{RR} = -a^2\boldsymbol{\varepsilon}{:}\boldsymbol{\zeta}_{ij}^{2a,2a}{:}\boldsymbol{\varepsilon}. \tag{2.16}$$

The effects of an external shear rate can be calculated from the dipole friction tensors $\boldsymbol{\zeta}^{1,2s}$, $\boldsymbol{\zeta}^{2a,2s}$, and $\boldsymbol{\zeta}^{2s,2s}$, as discussed in section 3.

132

When a pair of particles is close to contact the hydrodynamic forces diverge as s^{-1} and $\ln s^{-1}$, where $s = (R_{12}-2a)/a$ is the gap between the particles relative to the radius. Under these circumstances the multipole moment expansion described above converges very poorly, or not at all [10]. However, since these lubrication forces are dominated by the interactions between contact points rather than boundary surfaces, they are pairwise additive. Thus we can replace the approximate pairwise-additive friction coefficients in Eqs. (2.16) with the exact ones [7], which are already known from other work [13-15]. The computational overhead is negligible, since the two-body friction coefficients are just scalar functions of the interparticle separation multiplied by appropriate spherical harmonics to describe the angular variation. The scalar functions, computed from the difference between the exact and approximate two-particle friction coefficients, were stored in tabular form at 1000 points between $R = 2a$ and $R = 4a$. The interpolation procedure used ensures that the s^{-1} and $\ln s^{-1}$ singularities are handled exactly.

3. TRANSPORT COEFFICIENTS

The transport properties of a random dispersion of hard spheres have been a subject of theoretical investigation for a long time [1]. In this paper we consider only the sedimentation velocity (mobility) and the high-frequency viscosity; other transport coefficients are considered elsewhere [12]. An underlying assumption is that the distribution of configurations of spheres is the equilibrium one, unaffected by the hydrodynamic interactions. This assumption is strictly valid only at short times or high frequencies.

The transport coefficients we are interested in are related to ensemble averages of the appropriate elements of the friction and mobility tensors. The collective mobility (sedimentation velocity) is obtained from the velocity response of the N-particle suspension to an applied force, and thus from Eq. (2.3a)

$$\mu = N^{-1}\frac{1}{3}\mathrm{tr}<\sum_{i,j=1}^{N}\mu_{ij}^{TT}>,\tag{3.1}$$

where

$$\begin{bmatrix}\mu^{TT} & \mu^{TR}\\ \mu^{RT} & \mu^{RR}\end{bmatrix} = \begin{bmatrix}\zeta^{TT} & \zeta^{TR}\\ \zeta^{RT} & \zeta^{RR}\end{bmatrix}^{-1}.\tag{3.2}$$

Our results are expressed in terms of the dimensionless quantity μ/μ_0, where $\mu_0 = (6\pi\eta_0 a)^{-1}$ is the mobility of an isolated sphere of radius a.

In section II it was shown how to calculate the dissipative hydrodynamic forces acting on suspended particles that are moving under the influence of external forces or fluctuating Brownian forces. We are also interested in the response of the suspension to a homogeneous external shear rate. In this work we will ignore the effects of the imposed flow on the suspension microstructure and assume that it is an equilibrium distribution of hard spheres, freely moving with the flow without hydrodynamic forces and torques. Such a situation can be realized experimentally by using an oscillating strain rate with a sufficiently high frequency that the solid particles cannot follow the imposed fluid flow. The theory developed in Ref. [9], summarized here, applies to that situation; low frequency transport coefficients will be considered in future work.

The stress in a suspension subjected to a homogeneous external shear rate $\dot{\varepsilon}_0^s$ contains a contribution from the velocity gradients in the fluid and a contribution from the forces exerted by the fluid on the solid-particle surfaces. This latter contribution is related to the external shear rate by the ensemble-averaged dipole-dipole friction tensor [Eqs. (2.10), (2.12), and (2.14)]

$$<\zeta^{2s,2s}> = N^{-1}<\sum_{i,j=1}^{N}\zeta_{ij}^{2s,2s}>\tag{3.3}$$

where the superscript 2s indicates that the fourth-rank tensor $\zeta^{2,2}$ has been symmetrized with respect to the 2 pairs of cartesian suffixes; the anti-symmetric part $\zeta^{2a,2a}$ is related to ζ^{RR} [Eq. (2.16)]. Because of the zero-force and zero-torque conditions, we require a modified form of the dipole-dipole friction tensor, $<\zeta^{2s,2s}>$, which incorporates these conditions,

$$\zeta^{2s,2s} = \zeta^{2s,2s} - \left[\zeta^{2s,T}\zeta^{2s,R}\right] \cdot \begin{bmatrix} \zeta^{TT} & \zeta^{TR} \\ \zeta^{RT} & \zeta^{RR} \end{bmatrix}^{-1} \cdot \begin{bmatrix} \zeta^{T,2s} \\ \zeta^{R,2s} \end{bmatrix}, \tag{3.4}$$

where $\zeta^{2s,T}$ and $\zeta^{2s,R}$ are defined by analogy with Eq. (2.16). In Ref. [10] the lubrication contribution to the dipole-dipole friction tensor for freely moving pairs of spheres was added directly to $\zeta^{2s,2s}$, which incorrectly forces a zero-force zero-torque condition on both the many-body and pairwise-additive contributions to $\zeta^{2s,2s}$. The correct procedure [7], adopted in this work, involves adding the appropriate lubrication forces to all the friction-coefficient matrices on the right-hand side of Eq. (3.4) and then imposing the zero-force zero-torque condition to construct $\zeta^{2s,2s}$. This leads to a much more rapid convergence of $\zeta^{2s,2s}$ with increasing number of force moments.

Combining the fluid stress and induced stresses we find for the total stress tensor [9]

$$<\sigma> = \eta_0(2<\dot{\varepsilon}^s> + \frac{5}{2}\phi<\gamma^{2s,2s}>:\dot{\varepsilon}_0^s), \tag{3.5}$$

where $\gamma^{2s,2s}$ is the ensemble-averaged dipole-dipole friction tensor, normalized by its low-density limit $\zeta_0^{2s,2s} = (10/3)\pi\eta_0 a$ (i.e. $\gamma^{2s,2s} = \zeta^{2s,2s}/\zeta_0^{2s,2s}$). The fluid stress is proportional to the ensemble-averaged shear rate in the presence of the suspended particles $<\dot{\varepsilon}^s>$, which in general is different from the imposed shear rate $\dot{\varepsilon}_0^s$ [9,16]. However for periodic boundary conditions $<\dot{\varepsilon}^s> = \dot{\varepsilon}_0^s$ [9,17] and, if the system is isotropic, the viscosity is independent of the direction of shear so that

$$\frac{\eta}{\eta_0} = 1 + \frac{5}{2}\phi<\gamma>, \tag{3.6}$$

where γ is a component of the normalized dipole-dipole friction tensor. There are in general five independent components of the ensemble-averaged dipole-dipole friction tensor, coupling the five irreducible shear stresses and shear rates. An ensemble average of the viscosity in a cubic unit cell has two independent values, which in the large system limit become identical as the system tends to isotropy. The difference between the two values is generally small, vanishing as N^{-1}.

Numerical results were obtained by averaging over 100 statistically independent configurations of hard spheres. The configurations were obtained from snapshots, equally spaced in time, along a molecular dynamics trajectory. The time spacing was large, of the order of the time taken for a particle to diffuse over a distance $2a$. The calculated transport coefficients have small statistical errors; the 95% confidence limits are within 1% of the calculated values in all cases, and in general are closer. Simulations were run for 16, 32, 54, and 108 particles using moment approximation from $p_{max} = 1$ to $p_{max} = 7$, corresponding to between 12 and 192 equations per particle. The packing fractions range from $\phi = 0.01$ to $\phi = 0.45$, near the solid-liquid freezing transition of hard spheres at $\phi = 0.49$.

Hydrodynamic transport coefficients at the highest packing fraction, $\phi = 0.45$, are shown in Table 1. Our results show that low-order moment approximations to the force density are inadequate at high packing fractions even if lubrication forces are explicitly included. Although the viscosity converges quite rapidly, with little change beyond a $p_{max} = 3$ approximation, the collective mobility converges rather poorly; it requires a p_{max} of at least 6 for convergence at high packing fractions, or about 150 equations per particle. By analyzing the convergence of the transport coefficients with successively higher force-moment approximations (Table 1), it can be seen that our results for 16 particles are within 1% of the fully converged result. Owing to the neglect of higher force moments, the collective mobility coefficient calculated in earlier work [8] is too large by about 50% at $\phi = 0.45$.

It is not computationally feasible at present to run the larger systems with the same number of force moments as the 16-particle systems; we used force moments up to $p_{max} = 5$, $p_{max} = 3$, and $p_{max} = 2$ for the $N = 32$, $N = 54$, and $N = 108$ particle systems respectively. Fortunately the higher force moments are only important for clusters of particles that are relatively close together, and there-

Table 1. *Dependence of the mobility* (μ) *and viscosity* (η) *on the number of force moments.*

ϕ	N	p_{max}	μ/μ_0	η/η_0
0.45	16	1	0.0982(5)	5.41(4)
		2	0.0480(3)	5.29(4)
		3	0.0439(2)	5.48(4)
		4	0.0418(2)	5.41(4)
		5	0.0415(2)	5.54(5)
		6	0.0412(2)	5.55(5)
		7	0.0410(2)	5.50(4)
	32	1	0.1012(2)	5.70(2)
		2	0.0499(2)	5.53(2)
		3	0.0453(1)	5.58(2)
		4	0.0433(2)	5.63(2)
		5	0.0427(1)	5.65(2)
	54	1	0.1014(2)	5.69(2)
		2	0.0505(1)	5.55(2)
		3	0.0460(1)	5.61(2)
	108	1	0.1018(2)	5.73(1)
		2	0.0507(2)	5.52(2)

fore their effects are more or less independent of system size. This can be verified from Table 1 by noting that the difference between mobility coefficients derived from successive moment approximations is essentially independent of N. Thus we can estimate the large p_{max} results for all the system sizes considered, assuming the effects of the higher force moments are the same as for $N = 16$. This leads to consistent results for the $p_{max} \rightarrow \infty$ mobility coefficients even at $\phi = 0.45$ where the high-moment effects are most important. The fully converged transport coefficients for the different size systems, over a range of packing fractions, are shown in Table 2.

Table 2. *Variation of the mobility* (μ) *and viscosity* (η) *with system size. Finite-size corrections for the mobility coefficient are shown in parentheses, followed by the resulting estimate for $N \rightarrow \infty$.*

ϕ	N	μ/μ_0	η/η_0
0.01	16	0.800 (+ 0.137 = 0.937)	1.02547
	32	0.829 (+ 0.109 = 0.938)	1.02550
	54	0.839 (+ 0.091 = 0.930)	1.02549
	108	0.868 (+ 0.073 = 0.941)	1.02549
0.05	16	0.570 (+ 0.164 = 0.734)	1.1383
	32	0.600 (+ 0.130 = 0.730)	1.1387
	54	0.621 (+ 0.109 = 0.730)	1.1389
	108	0.640 (+ 0.087 = 0.727)	1.1391
0.25	16	0.162 (+ 0.039 = 0.201)	2.17
	32	0.167 (+ 0.031 = 0.198)	2.17
	54	0.174 (+ 0.027 = 0.201)	2.17
	108	0.178 (+ 0.021 = 0.199)	2.17
0.45	16	0.0410 (+ 0.0041 = 0.0451)	5.5
	32	0.0423 (+ 0.0034 = 0.0457)	5.6
	54	0.0431 (+ 0.0029 = 0.0460)	5.6
	108	0.0437 (+ 0.0023 = 0.0460)	5.6

The viscosity is insensitive to system size, and the difference between results for different size systems is not statistically significant (Table 2). This is because the friction coefficients are screened by the many-body interactions [18]. By contrast, the mobility coefficient is not screened [6]; it shows a strong system-size dependence, deviating from the thermodynamic limit by terms of order $N^{-1/3}$. We can understand and quantitatively predict these deviations by first considering the sedimentation of a single particle in a periodic system, with a large but finite unit cell. At low-density, the sedimentation velocity of a simple-cubic lattice is given by [19]

$$\mu_{sc}/\mu_0 = 1 - 1.7601\phi^{1/3} + \phi - \cdots. \tag{3.7}$$

We can superpose the effects of the neighboring particles and the periodic-image particles to relate the N-particle periodic diffusion coefficient to its thermodynamic limit [8], and thus for a very dilute suspension,

$$\mu(N) = \mu - (6\pi\eta_0 a)^{-1} [1.7601(\phi/N)^{1/3} - \phi/N]. \tag{3.8}$$

This correction works well at low solids packing (Table 2), but at higher packing fractions the neighboring particles partially screen the effects of the periodic images, reducing the magnitude of the correction without changing the functional form of the hydrodynamic interactions at large separations [6]. Since the distances between a sedimenting particle and its periodic images are large compared with the particle size we can account for this screening by replacing the pure-fluid viscosity η_0 with the suspension viscosity η. Moreover the sedimenting particles are no longer independent of each other as they are at low density. However the N-particle cluster can be treated as a single particle of effective radius a_{eff}, where

$$a_{eff}/a = \mu_0/\mu. \tag{3.9}$$

Then, by replacing η_0 by η and a by a_{eff} in Eq. (3.8), our expression relating the infinite system mobility to the N-particle mobility is

$$\mu = \mu(N)\left\{1 + \frac{(\eta_0/\eta)[1.7601(\phi/N)^{1/3} - \phi/N]}{1 - (\eta_0/\eta)[1.7601(\phi/N)^{1/3} - \phi/N]}\right\}. \tag{3.10}$$

These corrections are shown in parentheses in Table 2, and it can be seen that the estimates of the thermodynamic limit of the mobility obtained from the various size systems are consistent.

Results for the collective mobility μ are shown in Fig 1. Experimentally μ can be measured from the sedimentation velocity of the solid phase [20,21] or from long-wavelength light scattering [22]. In both cases the experiments measure only the long-time transport coefficients; the sedimentation experiments are steady state and the relaxation times probed by the light scattering experiments, proportional to k^{-2}, are long compared with the relaxation time for Brownian motion. There has been some question as to the difference between the long-time and short-time mobilities [23], but our results show that the differences are in fact small. Comparison of simulation and experimental results suggest that there is little structural rearrangement during the sedimentation of high-density microsphere suspensions.

It has been claimed that the pairwise-additive Rotne-Prager form for the hydrodynamic interactions accurately predicts sedimentation velocities over the whole density range [24]. This conclusion was based on comparisons with simulations using low-order moment approximations to the force density [8]. However, we find significant discrepancies at all densities between the expression for the sedimentation velocity based on Rotne-Prager hydrodynamics [24],

$$\mu/\mu_0 = \frac{(1-\phi)^3}{(1+2\phi)}, \tag{3.11}$$

and our simulation results. At $\phi = 0.45$ Eq. (3.11) is too large by a factor of 2.

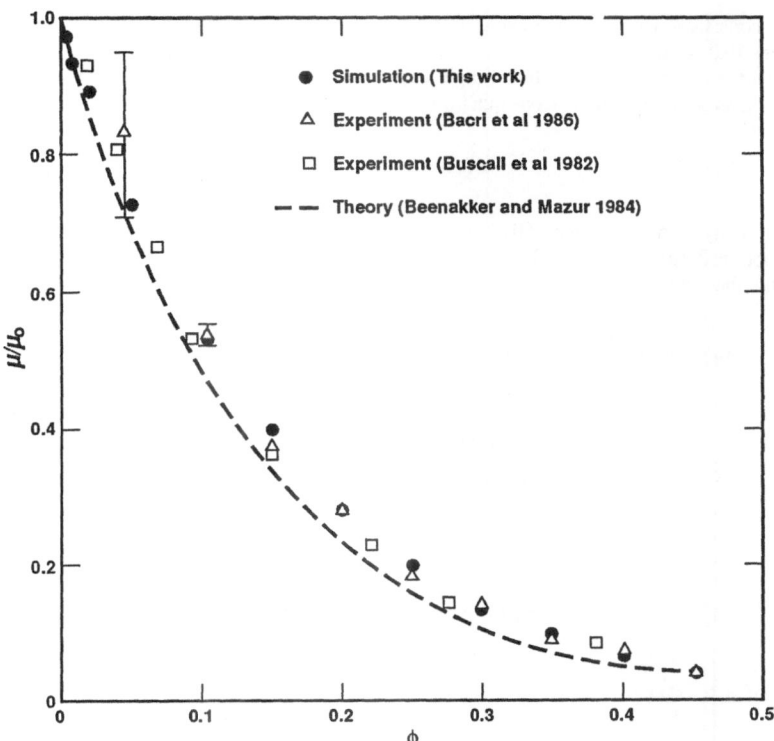

Figure 1. Short-time sedimentation velocity (mobility) of suspended hard spheres. The solid circles are the simulation results; the open symbols are experimental results from Buscall et. al. [20], and Bacri et. al. [21]. The dashed line is the theoretical analysis of Beenakker and Mazur [23].

The dashed lines in Figs. 1 and 2 are theoretical results of Beenakker and Mazur [23,25], based on calculating the hydrodynamic interactions in an effective medium that includes average many-body contributions. The theoretical results are in good agreement overall with the simulations; the worst discrepancy, around $\phi = 0.25$ in the collective mobility, is still only about 20%. The simulation results do not support recent speculation that the collective mobility coefficient of a random dispersion of hard spheres vanishes at a relatively low packing fraction ($\phi \approx 0.15$) [26].

The high-frequency viscosity is shown in Fig. 2. The experimental results of Van der Werff et. al. [27] were derived from oscillating Couette viscometry. The results shown here correspond to a sufficiently high frequency that the solid particles are unaffected by the imposed shear flow, so that the distribution of configurations is just the equilibrium distribution assumed in the simulations. At the lower packing fractions ($\phi < 0.35$) the experimental and simulation results are indistinguishable, but at high packing fractions there is some spread in the experimental data depending on the particle size. The larger size particles tend to have the larger viscosities, which are in better agreement with our simulations than the results for smaller spheres.

An expression for the viscosity has been proposed by Bedeaux [28]

$$\frac{\eta/\eta_0 - 1}{\eta/\eta_0 + 3/2} = \phi[1 + S(\phi)], \tag{3.12}$$

which incorporates a mean-field description of the hydrodynamic interactions when $S(\phi) = 0$. Thus a virial expansion of $S(\phi)$ should lead to a more rapidly converging viscosity than the usual expansion of η/η_0. Van der Werff et. al. [27] used their experimental data to deduce an expansion for S,

$$S(\phi) = (1.41 \pm 0.14)\phi - (1.19 \pm 0.34)\phi^2; \tag{3.13}$$

but this is inconsistent with the theoretical second virial coefficient for hard-sphere dispersions [10,29], which implies that at low packing fractions $S(\phi) \rightarrow 1.00\phi$. Our simulation data is fitted quite well over almost the whole density range by this simple expression for S. However, a best fit to the simulation data includes a small contribution from quadratic and cubic terms,

$$S(\phi) = \phi + \phi^2 - 2.3\phi^{3.} \qquad (3.14)$$

Since the viscosity is a rather insensitive function of S at low packing fraction, the difference between viscosities obtained from Eqs. (3.13) and (3.14) is negligible for $\phi \leq 0.35$. At higher packing fractions Eq. (3.14) fits the simulation data and the 76 nm-sphere experimental data better than Eq. (3.13).

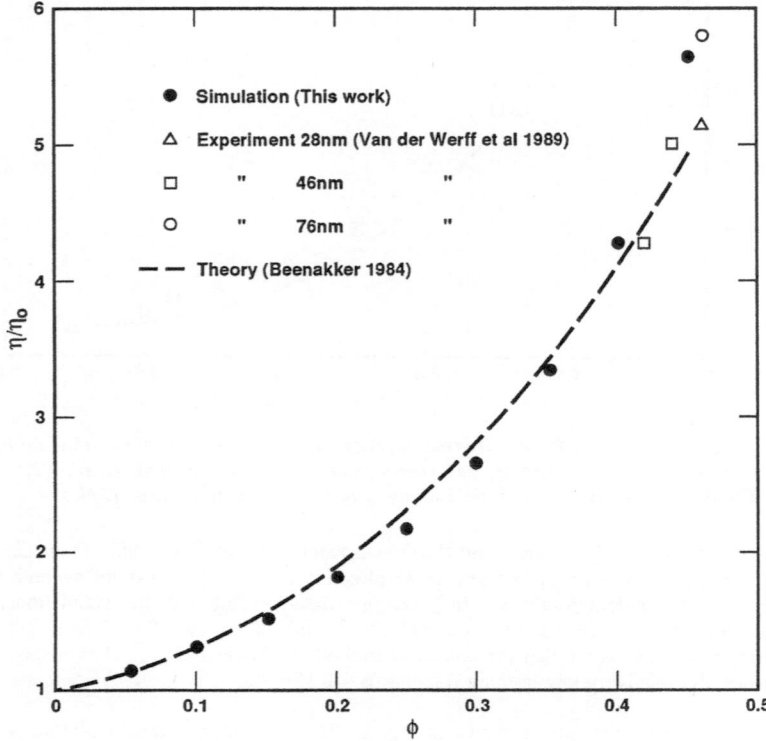

Figure 2. *High-frequency viscosity of suspensions of hard spheres. The solid circles are the simulation results; the open symbols are experimental results with different size spheres from Van der Werff et. al. [27]. The dashed line is Beenakker's theoretical analysis [25].*

4. DISCUSSION

These numerical results demonstrate that it is feasible, with current computer technology, to calculate the hydrodynamic interactions between small numbers of spherical particles precisely. At low packing fractions this means that of the order of 100 particles could be studied; but at high packing fractions, when a large number of force moments are necessary, the number was reduced to less than 20. Nevertheless we have obtained accurate estimates of the transport properties of random dispersions of spheres which are in excellent agreement with experiment.

A significant barrier to further progress is the heavy computational demands of these calculations. At present we are mainly limited by the core memory needed to store the generalized mobility matrix G, which requires $9N^2(p_{max} + 1)^4$ words. However, if we store only the matrix elements that require k-space sums and compute the others as needed, then the storage requirement could be reduced to less than $500N^2$ words for a high-order moment approximation, if all the available symmetries were utilized.

Thus systems of several hundred particles would fit in the core memory of a modern supercomputer. A further constraint is that the triangular decomposition of the $\mathbf{G}$ matrix requires of the order of $(N p_{max}^2)^3$ operations; but by using an iterative method, such as conjugate directions, the simultaneous equations [Eqs. (2.14)] can be solved in a time of order $(N p_{max}^2)^2$, for a specified set of velocities or forces. Thus for high Peclet-number flows, where Brownian motion can be ignored, it should be possible to simulate systems of about one hundred particles at high packing fractions ($p_{max} = 6$ or 7) and several hundred particles at lower packing fractions. However, to include random displacements (or velocities) all the elements of the $6N \times 6N$ mobility (or friction coefficient) matrix must be computed, which inevitably requires of the order of N^3 operations [30].

An interesting alternative to integral-equation methods is the use of lattice-gas cellular automata to model the fluid phase [31]. Lattice-gas models are simplified molecular models in which particles with a discrete set of velocities move from one node to another of a space-filling lattice, undergoing collisions with other particles occupying the same nodes. For a sufficiently large number of particles these models are equivalent to the continuum Navier-Stokes equations [32], with the very significant advantage that thermal fluctuations, which give rise to Brownian motion, are also included [31]. Moreover the computational requirements scale as N instead of N^2 or N^3 as is the case for the integral equation methods. Recent studies of the hydrodynamic interactions between moving spheres [33] have shown that there is quantitative agreement between the results of lattice-gas simulations and lubrication theory down to gaps of the order of a lattice spacing. The combination of particle models of the solid phase and lattice-gas models of the fluid phase can handle the whole range of solid-fluid suspensions, from sub-micron particle sizes where Brownian motion is important, to macroscopic size particles.

In the future both the creeping-flow and lattice-gas models of the fluid phase will be used to study the rheological properties of suspensions by computer simulation. These simulations will incorporate the dynamical effects of the hydrodynamic forces on the structure of the suspension [30], which are crucial to quantitative predictions of low-frequency suspension rheology [34].

ACKNOWLEDGEMENTS

I would like to thank Prof. John Brady (CalTech) for providing a complete list of asymptotic expansions for the lubrication forces. I would also like to to thank Prof. S. Kim (University of Wisconsin) for providing a copy of the computer program, written by himself and Dr. R. Miflin (Princeton University), to calculate hydrodynamic interactions between pairs of particles. This work was supported by the U.S. Department of Energy and Lawrence Livermore National Laboratory under Contract No. W-7405-Eng-48.

REFERENCES

[1] J. Happel and H. Brenner, *Low-Reynolds Number Hydrodynamics*,
     (Martinus Nijhoff, Dordrecht, 1986).
[2] A. Einstein, *Investigations on the Theory of the Brownian Movement*,
     (Dover Publications, New York, 1956).
[3] D.L. Ermak and J.A. McCammon, *J. Chem. Phys.* **69**, 1352 (1978).
[4] W. Van Megen and I. Snook, *J Chem. Soc. Farad. Trans. II.* **80**, 383 (1984).
[5] J.F. Brady and G.Bossis, *J. Fluid Mech.* **155**, 105 (1985).
[6] C.W.J. Beenakker and P. Mazur, *Physica A* **120**, 388 (1983).
[7] L. Durlofsky, J.F. Brady and G. Bossis, *J. Fluid Mech.* **180**, 21 (1987).
[8] R.J. Phillips, J.F. Brady and G. Bossis, *Phys. Fluids* **31**, 3462 (1988).
[9] A.J.C. Ladd, *J. Chem. Phys.* **88**, 5051 (1988).
[10] A.J.C. Ladd, *J. Chem. Phys.* **90**, 1149 (1989).
[11] P. Mazur and W. Van Saarloos, *Physica A* **115**, 21 (1982).
[12] A.J.C. Ladd, *J. Chem. Phys.* submitted (1989).
[13] D.J. Jeffrey and Y. Onishi, *J. Fluid Mech.* **139**, 261 (1984).
[14] S. Kim and R.T. Miflin, *Phys. Fluids* **28**, 2033 (1985).
[15] D.J. Jeffrey and R.M. Corless, *PhysicoChem. Hydrodyn.* **10**, 461 (1988).
[16] B.U. Felderhof, *Physica A* **82**, 611 (1976).
[17] B.U. Felderhof, *Physica A* **159**, 1 (1989).
[18] R.L. Treloar and A.J. Masters, *Mol. Phys.* **67**, 1273 (1989).

[19] H. Hasimoto, *J. Fluid Mech.* **5**, 317 (1959).

[20] R. Buscall, J.W. Goodwin, R.H. Ottewill and T.F. Trados, *J. Colloid Interface Sci.* **85**, 78 (1982).

[21] J.-C. Bacri, C. Frénois, M. Hoyos, R. Perzynski, N. Rakotomalala and D. Salin,
      Europhys. Lett. **2**, 123 (1986).

[22] D.J. Cebula, R.H. Ottewill, J. Ralston and P.N. Pusey, *J. Chem. Soc. Faraday Trans. I* **77**, 2585 (1981).

[23] C.W.J. Beenakker and P. Mazur, *Physica A* **126**, 349 (1984).

[24] J.F. Brady and L.J. Durlofsky, *Phys. Fluids* **31**, 717 (1988).

[25] C.W.J. Beenakker, *Physica A* **128**, 48 (1984).

[26] B. Cichocki and B.U. Felderhof, *Physica A* **154**, 213 (1989).

[27] J.C. van der Werff, C.G. de Kruiff, C. Blom and J. Mellema, *Phys. Rev. A* **39**, 795 (1989).

[28] D. Bedeaux, *J. Colloid Interface Sci.,* **118**, 80 (1987).

[29] B. Cichocki and B.U. Felderhof, *J. Chem. Phys.* **89**, 1049 (1988).

[30] G. Bossis and J.F. Brady, *J. Chem. Phys.* **87**, 5437 (1987).

[31] A.J.C. Ladd, M.E. Colvin and D. Frenkel, *Phys. Rev. Lett.* **60**, 975 (1988).

[32] U. Frisch, D. d'Humières, B. Hasslacher, P. Lallemand, Y. Pomeau and J-P. Rivet,
      Complex Systems **1**, 649 (1987).

[33] A.J.C. Ladd *unpublished work* (1989).

[34] J.C. van der Werff, C.G. de Kruiff and J.K.G. Dhont, *Physica A* **160**, 205 (1989).

MICROSCOPIC SIMULATIONS OF INSTABILITIES

M. Mareschal

CP231, Université Libre de Bruxelles

Boulevard du triomphe, B1050, Brussels, Belgium

1. INTRODUCTION.

In the early stage of the development of Molecular Dynamics (MD) the perspective of dealing with particles in order to solve problems in non-linear high-speed aerodynamics and boundary layers was already present[1]. However most of the results obtained in the last thirty years or so mainly concerned equilibrium properties. In this field, some very interesting, and to some extent unexpected, results were obtained: the velocity correlation function in an atomic dense fluid has been measured by Rahman[2] and later explained on the basis of the motion of a sphere in a continuous compressible fluid[3], with a frequency dependent viscosity. The continuum hydrodynamic equations seem to be a good model for the fluid motion up to very small distances and times at equilibrium[4]. And even when the simple hydrodynamic picture seems to become inadequate, models have been proposed which consist in generalizing hydrodynamics to include a possible dependence of the transport coefficients in the wave vector and frequency[5,6].

The non-equilibrium behavior, on the other hand, has not been as much studied, although the situations encountered are much more diverse and rich than in equilibrium. The main reason for this has been the limitation in computing power available. Early studies by Hoover and Ashurst[7] concerned the simulation of a liquid made of up to a few thousand atoms under a direct shear stress or temperature gradient. The response of the system to the imposed constraint seems to remain linear up to very high values of the forcing, where the departure from linearity could be accounted for by a constraint dependence in the transport coefficients. However, this dependence remains small and is limited to very high values of the forcing. This method has been also used to study color diffusion[8]: here the system made of hard spheres with different colors is in thermal equilibrium and the constraint consists in an imposed color concentration at the simulation cell boundary. The measured transport coefficient is slightly smaller than the Green-Kubo value (the Green Kubo integrand being extrapolated in the long time limit): the difference is due to a cutoff in the wavevectors imposed by the finite size of the simulation cell. Similar subsequent studies have been reported which confirm the finding that MD systems made of a few hundred to a few thousand particles could be sufficient to reproduce hydrodynamical behavior under simple constraints[9], or even in more complex situations like in strong shocks in atomic liquids[10].

This was confirmed in a MD study by Meiburg[11] on the motion of a plate in a moderately dense fluid made of 40,000 atoms. Although this simulation was made in order to compare MD with the direct simulation Monte Carlo (DSMC) for the Boltzmann equation[12], it initiated related studies by MD of, for instance, the flow past a cylinder[13]. It became apparent that with a few hundred thousand particles, one could, at least qualitatively, reproduce the macroscopic behavior.

These last examples were done using a large number of particles, more than 100,000. The question remains however as to what is the minimum number of atoms necessary in order to generate a macroscopic flow. Can one make a quantitative study of the comparison between a MD measurement and the solutions of the corresponding Navier-Stokes-Fourier equations? We have analyzed this question in the framework of the Rayleigh-Bénard problem which has often served as a generic example of non-equilibrium behavior[14]. At a given value of the Rayleigh number-a dimensionless number characteristic of the imposed constraints, namely a vertical temperature gradient and an accelerating force directed in opposite directions- a transition towards a convective state made of rolls takes place. The loss of stability of the reference state and the appearance of a spatial structure are good candidates to test the ability of microscopic systems to reproduce complex behavior. This will be shown in section two of this contribution.

The behavior of chemical systems constrained in non-equilibrium states has also received much attention these last years[15]. In particular, solutions of reaction-diffusion equations have shown properties thought to be sufficient in the understanding of recent experiments displaying periodic oscillations in open chemical reactors. The variety of phenomena due to the non-linear kinetics of chemical systems has been stressed in many studies: simple models can be used to display wave propagation, time periodicity, spatial patterns formation and chaos, most of these having experimental counterparts. The complexity of chemical systems with respect to atomic pure fluid hydrodynamics is responsible for the fact that most models in this area are not meant to be realistic. This is the case of the so-called Brusselator model which involves a trimolecular step. However the inspection of these models has been very useful in the theoretical approach meant to describe non-equilibrium ordering in chemistry.

One should also stress that the precise experimental control of the (non-equilibrium) state is more difficult in chemical systems: the introduction of reactants and the removal of products is usually incompatible with homogeneity; this, in turn, can be responsible for a coupling with the chemistry making many measurements impossible. So, for example, the correlations which appear in the fluctuations of density have been measured experimentally in the case of hydrodynamical systems and not yet in the case of chemical systems. This is an area where the microscopic simulations could be useful, allowing for an "experimental probe" of the theories which compute the effects of the fluctuations at and near instabilities.

We shall report here on preliminary results made with a simulation of the Boltzmann equation of a system made of a reacting gas mixture where the collisions will be allowed to perform chemical changes leading to a kinetic description of the Brusselator model. The model and its implementation with the DSMC method will be discussed in the part three of this paper together with the results showing the emergence of the time periodic solution beyond the bifurcation point.

We end by a few remarks concerning the future developments which can be expected in the next few years from the simulations which will become possible on future largest computing facilities.

2. THE RAYLEIGH-BENARD INSTABILITY

a. The problem

Let us consider a fluid layer of infinite horizontal extent and enclosed between two parallel plates held at temperatures T (bottom) and T+ΔT (top). The flow which will emerge in the system will depend on the Rayleigh number[16]

$$Ra = \frac{\beta \, \Delta T \, g \, L^3}{\nu \, D_T} \tag{1}$$

where β is the isothermal compressibility, g the amplitude of the external force, L the layer thickness, ν the kinematic viscosity and D_T the thermal diffusitivity of the fluid. When this number is larger than a critical value-between 650 and 2000, depending on the boundary conditions (BC) used to describe the fluid-plate boundaries- then a velocity field sets in, forming rolls, which are all parallel pointing to an arbitrary direction. The third dimension is not essential in this problem. If we project the velocity field onto the plane perpendicular to the rolls, the pattern obtained will be characteristic of the three dimensional flow. This is important for the simulation since a two-dimensional system requires less particles than a three dimensional one. Of course the system which can be simulated on a computer is of a finite size (or equivalently with a finite periodicity) and will not be really incompressible. Therefore Ra will only be indicative of the location of the instability but the precise determination of the critical values of the parameters to be imposed will be determined by a more complete stability analysis.

The critical Ra is obtained in laboratory experiments with a temperature difference of a few degrees and a layer thickness of a few centimeters for most fluids. However, the cubic dependance on the thickness is responsible for a dramatic decrease of Ra if we consider a system made of a few thousand atoms having sizes which are a few hundred atomic diameters. Of course, as it has been the case in the temperature gradient direct simulations of Hoover and Ciccotti, the temperature difference can be increased in order to have measurable effects. The external force too can be increased and as a matter of fact, it should be increased since its effect on the density would then balance the effect of a too strong temperature gradient (the fluid at these scales cannot be considered as incompressible). The amplitude of the external acceleration, g, can then be estimated by requiring that the potential energy necessary to overcome the field is balanced by the thermal energy given by the heating, so that $mgL = k_B \Delta T$.

The transport coefficients which are in the denominator of Ra have a known dependance on the density and temperature[17]. For hard spheres, they are proportional to the square root of the temperature and they show a minimum when increasing the number density: this is related to the well understood competition of a decreasing mean free path and an increasing potential part of the transport. The number density which is then optimal in order to have the highest Ra possible is then an intermediate density near this minimum of the transport coefficients, the kinematic viscosity and the thermal diffusitivity. The last parameter, β, is the isothermal compressibility which is inversely proportional to the temperature and has a known density dependance[18].

Introducing all these explicit forms for these coefficients, and putting $n = N/\lambda L^2$, with λ the aspect ratio, the Rayleigh number can be written in the form (valid for a two dimensional system),

$$R_a = \frac{Cst}{\lambda} \cdot \left[\frac{\Delta T}{T}\right]^2 \cdot N$$

(2)

where the constant is a function of the number density alone, having a maximum for values nd^2 between 0.2 and 0.4. The temperature which appears in equation 2 is a mean temperature: indeed, since the system is not homogeneous, a sizeable difference occurs in the transport properties inside the fluid cell. The comparison of the direct measure of the heat conductivity as the ratio of the heat flux to the temperature gradient shows that it is meaningful to interpret the transport coefficient which is measured as the one corresponding to the mean temperature and number density. This expression shows that the only way to reach higher values for Ra is to increase the number of particles, once the optimum choice for the density and temperature gradient has been done. In the case of hard disks, which we have considered, a value of 800 for the Rayleigh number can be obtained with a number of particles of 5000 in a geometry having an aspect ratio equal to 2 and at a number density $nd^2 = 0.2$. This is above the critical value which we have computed for the system, provided we can implement the no-slip boundary conditions.

This also shows how difficult it will be to consider very high values of the Rayleigh number, for which the transition between different turbulent regimes has been observed. Indeed, for every increase of Ra by an order of magnitude, the number of particles to consider is also increased, leading in turn to a higher characteristic time which is also proportional to N. The computational time is then increased at least - that is for the most efficient codes which scales like N- by two order of magnitude. So that even if we extrapolate the increase of the computing power available on large computing configurations, the simulations feasible will be limited to low values of Ra, one or two order of magnitude larger than the critical value. Though, very interesting phenomena take place in this region, in particular with fluid mixtures.

b. The simulations

The fluid is simulated by N hard disks with a unit diameter d and enclosed in a rectangle of sizes L (vertical) by λL (horizontal). The two vertical boundaries are made of specularly reflecting walls, which means that the most natural boundary condition to use in the macroscopic description is a no-slip BC. On the bottom and top boundaries, thermal reservoirs are needed to impose the local temperatures. This is done in a way which has already been described a number of times except for two minor differences: first, when a particle collides with the boundary, it is kept on the boundary and reinjected at a constant frequency in the fluid -this was first done in order to avoid the reflection of thermally exited sound waves at the boundary. The normal velocity component of the reemitted particle is thermalized by sampling from a maxwellian distribution at the local temperature times the velocity component, the equilibrium velocity distribution function of a flux of particles across a surface at equilibrium. Second, the longitudinal component is not changed so that the velocity along the boundary can be different from zero and this can be represented at the macroscopic level by a no-slip BC.

The trajectories inside the rectangle are those of hard disks in an external force field. It is worth mentioning that, even if the effect of the field is to modify the trajectories from straight lines between the collisions to parabolas, the determination of the collision time for any given pair is not changed so that the program written to compute the trajectories is not greatly changed from a usual hard disk code[19].

TABLE 1. Simulation parameters. The parameters listed are given in the system units (d=1, m=1, T= 1/k$_B$). The equation of state used is the Carnaham-Stirling expression at n=0,2. The transport coefficients and the collision frequency are computed from the expressions given in reference 16 for k$_B$T=1 and n=0,2.

equation of state	pV/Nk$_B$T= 1,3
pair correlation function at contact	g$_2$(d)=1,31
mean free path	Λ=1,12
compressibility	β=0,72
collision frequency per particle	ν=1,038
viscosity	η= 0,34
bulk viscosity	ζ= 0,09
thermal diffusivity	D$_T$=3,91
height	L$_z$=112 (=L$_x$/2)
external force	$\mathbf{F}$=-1$_z$ g and g=0,011
number density	n=0,2
upper temperature	k$_B$T$_u$=0,5
lower temperature	k$_B$T$_l$= 1,6

The system is divided in cells so that a spatial coarse graining can be made to measure local values of thermodynamical fields. However, due to the small number of particles present in each cell- six on the average- this spatial average is not sufficient to extract the velocity field out of the thermal noise so that an extra time averaging is necessary. On the graphs shown on figures 1, the time average corresponds to several thousands of collision times (the collision time being defined as the time to have on the average one collision per particle).

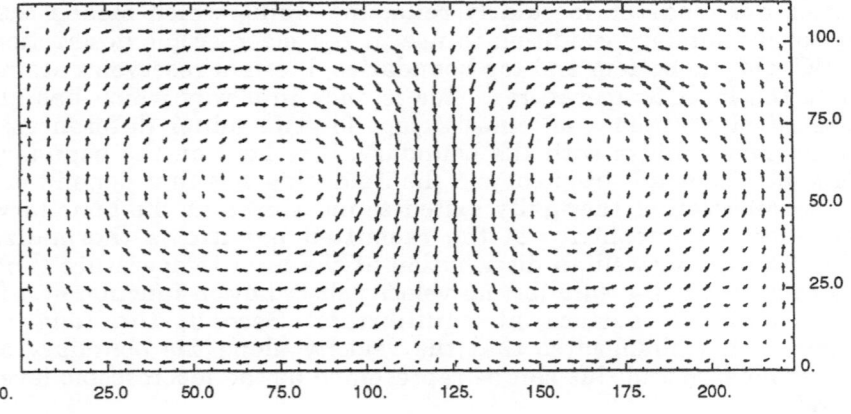

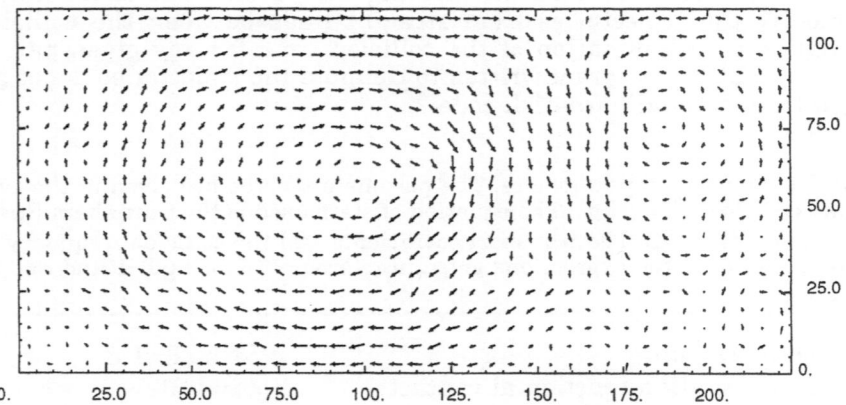

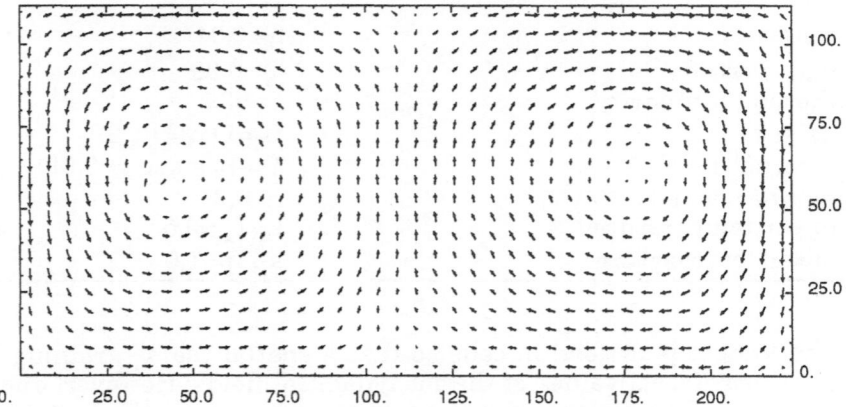

FIGURE 1. Velocity field of a simulation at Ra≈800. The first two graphs result from time averages over 1,000 collisions per particle (cpp), whereas graph 3 is average over 10,000 cpp. They correspond to, respectively, to t= (a) 8,000, (b) 10,500 and 20,000 cpp.

The initial condition is chosen so that a constant number density and a temperature corresponding to a linear profile between the boundaries is present at the start of the runs. The values of the parameters of the fluid are listed in table 1.

c. Results

Several runs have been done. A typical dynamics is shown on figures 1. The first part of the evolution of the system consists of the establishment of a double roll structure, with the fluid going up near the vertical boundaries and, once cooled at the top, coming down in the center of the cell. This structure, although it could be a stable solution, is not maintained in time and after 25 million collisions the size of one of the roll increases and tend to fill entirely the system. Then a second roll starts growing on one side and finally the double roll structure is reestablished but rotating in the opposite direction with respect to the initial pattern. This state seems to remain stable as no change has been measured for times longer than the lateral diffusion time.

This is one of the aspects of non linear evolution, namely that with a given set of parameters there might be more than one final state, depending on the initial conditions or even just on the fluctuations (internal or external).

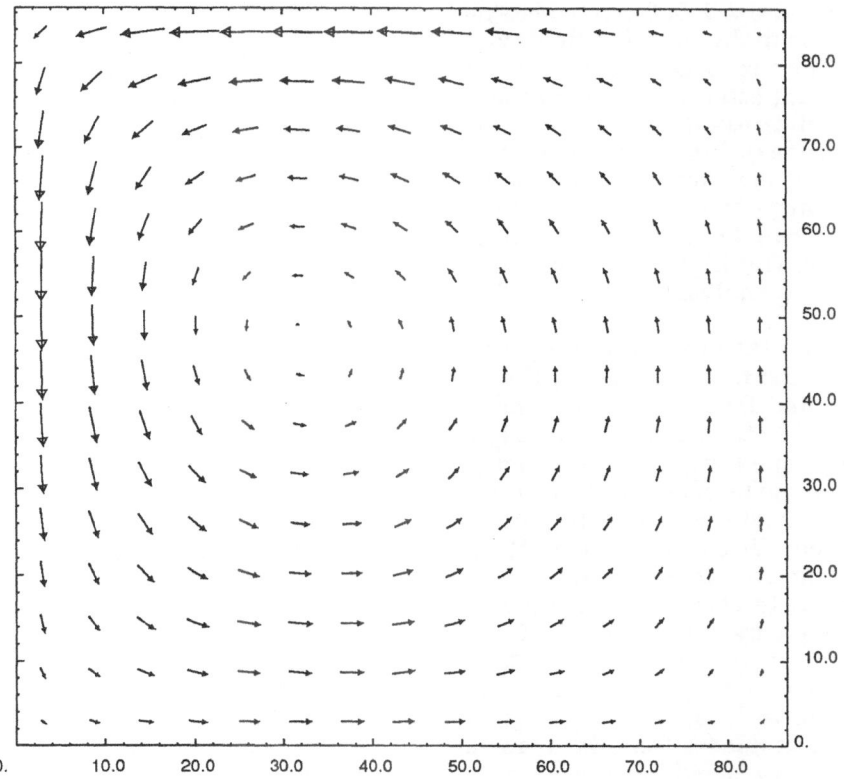

FIGURE 2. Velocity field for a 1,500 particles system, an aspect ratio equal to 1 and an estimated Ra ≈ 800.

One might even have the situation where the system continuously oscillates between two possible states. A difference between this simulation and the simulations of the hydrodynamical equations is precisely that the thermal fluctuations are naturally incorporated in the MD description, whereas they are not in the macroscopic approach. A question to be examined is whether they can be incorporated at the macroscopic level by adding some noise to the deterministic equations[20]. We will come back to this question later.

We have made other simulations changing some of the parameters in order to estimate their influence. First, the aspect ratio chosen is essential to determine what will be the pattern selected by the fluid. If we reduce λ, then instead of a two rolls structure the pattern will evolve towards a one roll structure. The selection by the system is consistent with what has been computed from the numerical solution of the macroscopic equations (see below). It has to be noted that if we decrease λ from 2 to 1.9, then one already has a one roll structure for the flow field. For intermediate values, we have observed the following phenomenon: at $\lambda=1.92$, the final and apparently stable state was different depending on the initial conditions taken, sometimes evolving towards a one roll and sometimes towards a two rolls structure. This could explain why for $\lambda=2$., the system seemingly hesitated between the two configurations before deciding to stay in the two rolls mode.

We have also examined the influence of the size of the system. A 6,000 particles system, with $\lambda=1$, was run under the same conditions as a 1,500 particles and a 1,000 particles system (that is the same Ra). On figure 2 we have shown the velocity field corresponding to a convective state for 1,500 particles system. Obviously, the solution adopted by the microscopic system is the one expected (this, of course can be checked). The values measured are not significantly different for the 1,500 and 6,000 particles systems except for the boundary layer which is more important in the smallest case. As the number of measurement cells is the same, one has four times more particles in the large system to perform the spatial average. And indeed the time average can be reduced by the same factor to keep the same precision. However, the instantaneous values are still much too noisy to permit any direct flow analysis.

Simulations were also run for precritical values of the Rayleigh number (see figure 3). Then, the flow field measured are not random as could be expected from the corresponding macroscopic absence of convection. As a matter of fact, vorticity is already present even for Ra=200-300 but in a much less developed way, a smaller amplitude and much less stability. However on the time scale of a few thousand collision times, patterns of an amplitude approximately one tenth smaller than in the convective case can be measured. Of course such patterns would be smeared out if we were time averaging over longer periods, as other flows would emerge and decay. The transition to convection in this small system is smoothed by the finite size effects and by the presence of fluctuations which permit longer and longer transient vortices to appear in the fluid. As a matter of fact, the finite size of the fluid is also responsible for a smoothing of the transition. We have not been able to localize a Rayleigh number below which no convective motion would take place. Instead, transient structures emerge, the amplitude and the stability of which increase with the Rayleigh number. An interesting problem which remains to be examined is whether these huge fluctuations appearing already well before the critical point are consistent with a fluctuating hydrodynamics (à la Landau-Lifschitz) theory.

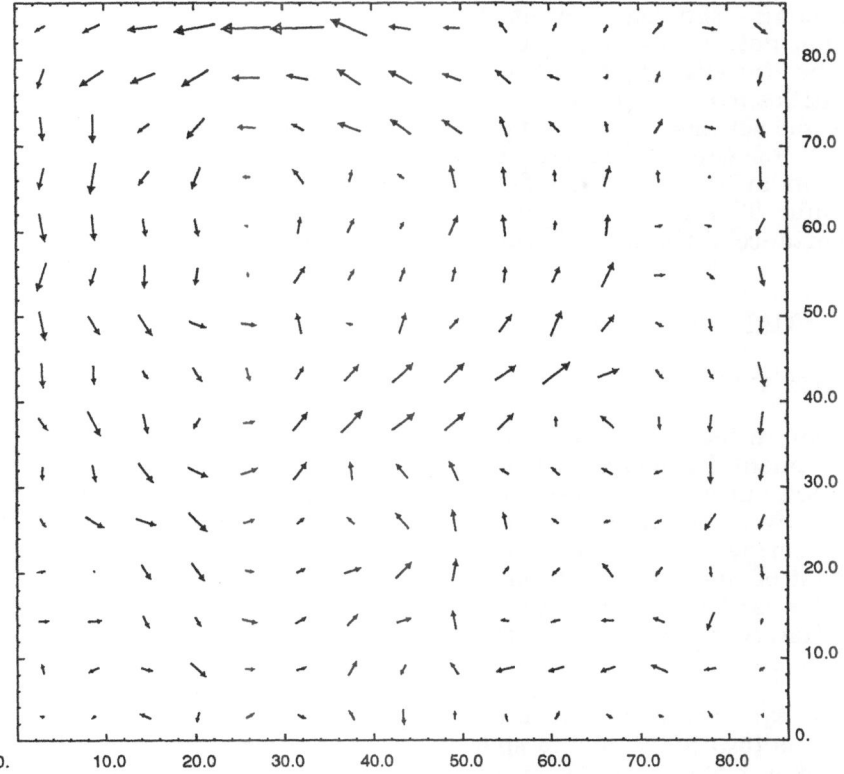

FIGURE 3. Velocity field for the same system as the one corresponding to fig. 2 but a Ra estimated around 200. The time average is over 10,000 cpp, and the amplitude of the velocities measured is one tenth of the previous figure.

d. Macroscopic analysis

The macroscopic analysis done on the problem has been based on the Navier-Stokes equations for a two dimensional hard disk fluid with its known equation of state[18] and transport coefficients[21]. As this comparison has been already published[22], let us simply state the main conclusions of our analysis.

First, the Boussinesq approximation, which consists of treating the fluid as incompressible, is still giving the right order of magnitude for the solutions, a few percent off the exact values for most of the results like the magnitude of the velocity field and so on. This is probably linked to the fact that the density remains quite constant as a function of the height, a consequence of the tuning of the external acceleration with the temperature gradient. Second, the influence of the density and temperature dependance of the transport coefficients is quite small, also a few percent. This also means that the determination of the transport coefficient by a direct measure on a direct non-equilibrium simulation cannot be very precise.

Second, the agreement between microscopic and macroscopic simulation data is striking. Density, velocity and temperature fields agree within the statistical errors of the MD measures, which can be as low as 1% for quantities averaged over one spatial variable. No large deviation from the macroscopic solution has been noticed for the larger systems (5000 and 6000 particles). For the smaller systems, there is a discrepancy at the upper and lower boundaries where the effects due to the wall affect the behavior of the entire fluid. This is not surprising however, as kinetic theory predicts that the importance of the wall effects indeed increases as the size decreases.

3. CHEMICAL OSCILLATIONS

a. The problem

Many interesting phenomena have been studied these last years since the experimental evidence of the existence of periodicity in chemical systems maintained out of equilibrium. The main theoretical effort has been devoted to the study of equations known as reaction-diffusion equations (RD) and which describe the time evolution of reacting mixtures in solutions with small heat of reaction. The application of such models to gas kinetics is less straightforward since the reaction usually involves a coupling with sound modes. However this coupling can be avoided in the model which we shall discuss below.

The systems should have two essential ingredients in order to display bifurcations towards time and spatial periodicity: a non-linear kinetics and a constraint in order to maintain the far from equilibrium conditions. This is usually realized by feeding the reactor with reactants and/or removing products from it. The instabilities so described have important differences with those of hydrodynamical phenomena. One of these is the existence of a characteristic length usually related to the diffusion constant, D, by $L_c=(D/k)^{1/2}$ where k is an effective rate, related to the linearized equations for the chemical species[23]. This quantity is the length over which a fluctuation will diffuse before it decays due to the chemical damping. Its existence is responsible for the fact that density correlations have a range which is not fixed by boundaries as in hydrodynamics (where the correlations extend over the entire system) but which is intrinsic. This fact has been shown by numerical simulations since the experimental evidence has not yet been given[24].

The model which we want to study here is the well known brusselator model which involves a trimolecular step. It has been extensively studied since its introduction in the late sixties as a paradigm of chemical dissipative structures. The chemical equations read

$$A + C \rightarrow A + X$$

$$B + X \rightarrow B + Y$$

$$2X + Y \rightarrow 3X$$

$$C + X \rightarrow C + D \qquad (3)$$

In these equations the concentrations of the species A,B C and D are controlled from the outside while the concentrations of X and Y, denoted by x and y, are evolving according to the equations

$$\frac{dx}{dt} = a - x + x^2y - bx$$
$$\frac{dy}{dt} = -x^2y + bx$$

$$(4)$$

The rate constants for the elementary steps have been incorporated in the definition of the variables a, b and t as is usual in the model. The linear stability analysis is straightforward to perform; introducing the fluctuations dx and dy around the steady state

$$x_s = a$$
$$y_s = \frac{b}{a}$$

$$(5)$$

one readily obtains their time evolution as solution of the linear system

$$\frac{d}{dt}\begin{pmatrix} \delta x \\ \delta y \end{pmatrix} = -\begin{pmatrix} 1-b & -a^2 \\ b & a^2 \end{pmatrix}\begin{pmatrix} \delta x \\ \delta y \end{pmatrix}$$

$$(6)$$

The roots of the characteristic equation are found to have a negative real part for $b<b_c$, at a fixed a value. For $b>b_c$, the stationary state has lost its stability and the system will asymptotically tend towards a limit cycle. The amplitude of the oscillations as well as its frequency should be determined by a non-linear analysis.

b. The simulation

The microscopic simulation is costly since one needs to meet the following requirements: a large system to avoid large fluctuations which might hide the transition and an integration over a long time in order to have a good statistical sampling. Besides, the chemical reactions occur on possibly different time scales so that the total integration time could also be large. All these reasons make it difficult to use a complete dynamical simulation as MD. Instead, we chose the direct simulation method for the Boltzmann equation in order to simulate a dilute gas undergoing chemical changes according to the scheme shown above. This method has already been used for the study of non-equilibrium properties of fluids under either a temperature gradient or a shear. The efficiency of the method in terms of cpu time made it possible to confirm in a quantitative fashion theoretical predictions based on an extension of fluctuating hydrodynamics to stationary non-equilibrium states. A very long integration time was necessary to obtain good statistics for the measurements, which would have been prohibitively costly if one would have used MD.

Of course, it is fair to say that the computation mentioned above cannot be considered as a fully microscopic check of the extension of the phenomenological fluctuation theory to non-equilibrium states as MD would be. The DSMC is not a simulation of exact dynamics. It is a simulation of the Boltzmann equation and more precisely of the fluctuating Boltzmann equation. The results obtained so far, using this simulation, do confirm that it reproduces the correct averages and variances in the equilibrium cases where no doubts exist on the results and that it agrees with the theoretical predictions in situations where no experimental check can be said to be completely satisfactory[25]. It would be hard to imagine that effects linked to the simplifications of the processes described while going from full dynamics to the Boltzmann equation could be responsible for major changes in the measured mean values; as the DSMC technique correctly simulates the fluctuating Boltzmann equation, it constitutes a microscopic test which is nearly as convincing as an exact dynamics.

This last argument is even more convincing if one realizes that the non-equilibrium constraint is purely chemical and that the simulations can be done on systems at thermal equilibrium; indeed, the chemical changes can be modelled as color change[26], with no release of energy so that the system seen with no color or species specification is an equilibrium dilute gas. The constraint consists of keeping some fixed value for the concentrations of A,B,C and D homogeneously in the system. This last requirement of homogeneity in the constraints is quite artificial, since in experimental systems, the constraints are imposed at the boundaries of the reactors. Important effects linked to this imperfect stirring have been discovered and studied. However, the aim of this simulation is not directed towards a realistic study but rather towards the confirmation of theoretical approaches and therefore let us stress again that we are not much concerned with the realism of the conditions for the simulation.

The particles are hard spheres having all the same diameter d and mass m. Their belonging to a given species does not modify their evolution with the exception of possible chemical change at collisions. The collision is decided to be reactive by a call to the number random generator; if we fix the reactive frequency to be n (<1!) times the collision frequency, at every collision between two reactive species particles, the reaction takes place if the random number returned from the routine is smaller than n. We preferred this stochastic procedure to the reactive cross section definition which rely on the relative kinetic energy : indeed this last method combined with the requirement that the number of particles of the A,B.. species are constant can lead to a unequal energy sharing among the species. At every reactive collision which consumes a constrained reactant, a solvent inert particle (S) is turned into a particle of the type which has just been consumed. In the same way, at every production of a particle of a fixed species, one select another one at random and turn it into a solvent particle. Besides, in order to maintain the concentrations of the fixed species homogeneous, at every relaxation time, all the positions of the fixed species are redistributed in the system. Otherwise, as we have experienced it, the composition fluctuations of these species can be responsible for a modification of, for instance, the long range correlations which are present in the system. The A,B.. particles are actually a reservoir of particles which influences the system evolution but is unaffected by it. These last procedures are compatible with the DSMC method and would be difficult to implement in a MD simulation.

The trimolecular step is executed in the following way: at every collision between two X particles, one chooses at random a neighbor of the colliding pair (that any other particle in the same cell) and , if this sphere is a Y particle, it is transformed with a fixed probability into an X. In fact, we model the following two steps reaction

$$X+X->(X+X)*$$

$$(X+X)*+Y->3X \qquad (7)$$

The rate of change of the number N_x of X particles during this step is straightforwardly calculated to be

$$\frac{dN_x}{dt} = k_{xx}\frac{N_y}{N}N_xN_x \qquad (8)$$

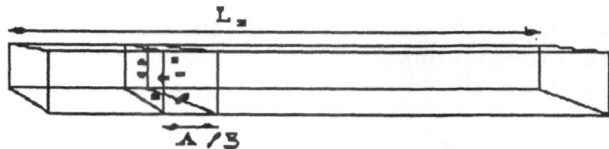

FIGURE 4. Geometry of the system. The number of cells in the x directions ranges from 35 in the smallest system to 630 for the largest system.

TABLE 2. Parameters of the simulations done to simulate the transition to the limit cycle for the Brusselator. The time of a period of oscillation corresponded to 6 million collisions in the 16,000 particles system. Most of these systems were run for more than 2 billion collisions.

N	L_x/L	B	B_c	N_c (in million)
8,000	17	400	400	100
16,000	35	400	800	500
16,000	35	650	800	1,000
16,000	35	700	800	500
16,000	35	750	800	600
16,000	35	800	800	1,000
16,000	35	825	800	500
16,000	35	850	800	700
16,000	35	900	800	500
32,000	70	1,600	1,600	2,000
96,000	210	4,800	4,800	2,000
96,000	210	5,000	4,800	2,000

with k_{xx}, the reactive collision frequency for a binary encounter between two identical particles, the value of which can be easily computed from the Boltzmann collision operator[27]. The second step is very fast compared to the first so that this combination of two steps results in a rate which is the same as if there was only a trimolecular step. Besides, the trimolecular step is just a modelization of two successive collisions occuring on different time scales.

The geometry of the system is that of one dimensional systems: although particles collide in a three dimensional space, the cells defined are aligned along the, say, x direction and only the x coordinate of the atoms need to be followed. The cells are parallelipiped of sides $L_x=L/3$, $L_y=L_z=L$,L being the mean free path. The number density nd^3 is set equal to 0.005, well in the dilute regime. The number of particles in the simulations ranged from 8,000 to 96,000 but most of the simulations were run with 16,000 particles (see table 1). This leads to a number of a bit less than 150 particles per cell.

In table 2, we have listed the parameters of the simulations which were performed on an IBM 4381 computer. As compared to MD, the program runs between 100 to 1000 times more rapidly than would the corresponding MD program. Besides, contrary to what happens in MD, the number of collisions done is independent of the system size. The trajectories were followed during two billion collisions which corresponds to several periods even in the largest system studied.

The essential features which come out of these simulations are the following: first, the macroscopic behavior is reproduced by the DSMC simulations. We have scanned the values of the B parameter so that the compositions in X and Y species have been followed below, at and above the threshold value where the instability should develop a limit cycle. The transition between the precritical behavior and the postcritical one is nicely observed the comparison of the values of the oscillation frequency, the amplitude of the oscillations correspond to those predicted by the macroscopic equations. The location of the threshold value of B itself is in agreement with the Bc value obtained from the linear stability analysis, although the finiteness of the system smooths the transition. Indeed, the figure 3 shows the number of X particles as function of time for values of B equal to 400, 625, 800 (=Bc) and 900. From that figure, it clearly appears that , below the critical B value, the fluctuations in X are increasing with B and that there is a smooth passage to the situation where the oscillations are fully developed.

The oscillations in X and Y, for overcritical values of B, have a characteristic shape: the decrease (respectively increase) in X (Y) is slow and it is followed by a rapid increase (decrease). The trajectory in the plane X and Y is shown on figure 5 for B=400, 800 and 850: the trajectory is run clockwise. The slow part of the motion is in the low left region where the curve concentrate in a small strip, while the rapid motion starts at the top of the graph. This rapid motion is dispersed on a larger region of the phase space. Besides, considering the entire trajectory, most of the points will be in the low left part: a three dimensional density plot in the x-y plane would display a sharp peak corresponding to the slow motion and a small hill spread over the top right region corresponding to the rapid motion. The high dispersion of the curve over the top-right region results from the dispersion of the points where the rapid motion starts: the y values range from 2 to 3 thousand and the corresponding x ranges from 400 to 700.

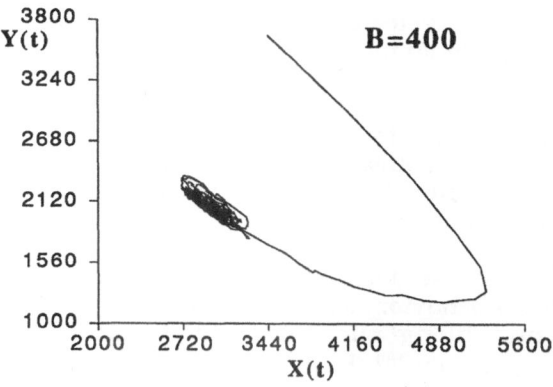

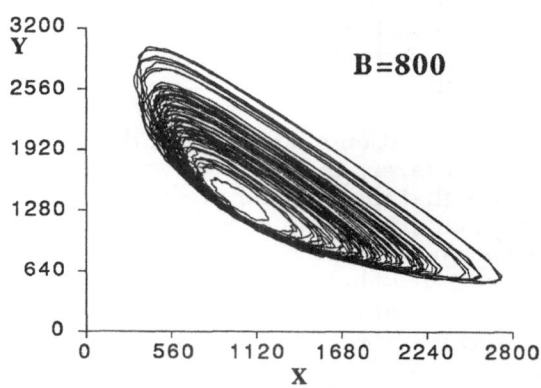

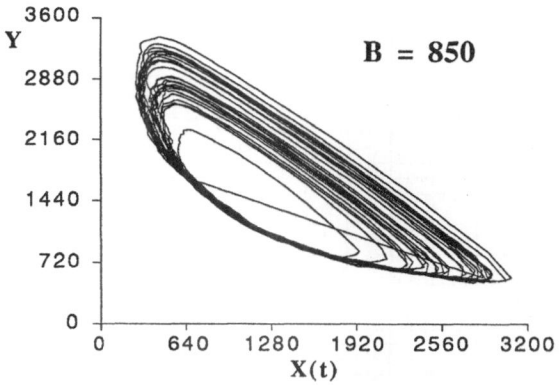

FIGURE 5.Trajectory of the DSMC systems in the x-y phase space for three different
B values, B=400, B=B$_c$=800 and B=850.

The comparison of these results with the corresponding macroscopic
approach is not really related to the same problem as in the hydrodynamic
problem. Indeed, the passage from the Boltzmann equation to the rate
equation is exact in the homogeneous case (since there is no heat of reaction

or activation energy, the velocity distribution remains maxwellian). The problems addressed by the simulations therefore concern basically two other questions:

(1) the simulation is a (stochastic) dynamics of a (finite) set of particles showing an evolution of the mean values and of their fluctuations. Can we easily extract the signal out of the noise and understand it in terms of the macroscopic rate equations? (how good is the simulation?)

(2) Concerning the analysis of the fluctuations, does the simulation reproduce the expected behavior near of equilibrium and can it be understood far from equilibrium along the lines developed in usual theoretical approaches of these questions[28]? (how good are the theories?)

The first question can be answered positively from our results. Indeed, the comparison between the mean values measured on the DSMC systems fits the macroscopic values with less than 1 % of deviation. Both, (well) before the transition and (well) after, the microscopic and macroscopic description match each other very well.

Concerning the distribution of states now, the quantitative comparison is still under study. However, the qualitative features which have been observed tend to indicate that the behavior predicted by the master equation or Fokker Planck approach for this model is indeed observed: before the transition the distribution of states is centered around the stationary stable state. For $B>B_c$, a sharp distribution in the slow motion region and a large dispersion in the rapid motion region for $B>B_c$. The transition from precritical to overcritical B's is smoothed also by the finite size of this system. As can be seen on figure 6, the amplitude of the oscillations continuously increase to their overcritical value while increasing B.

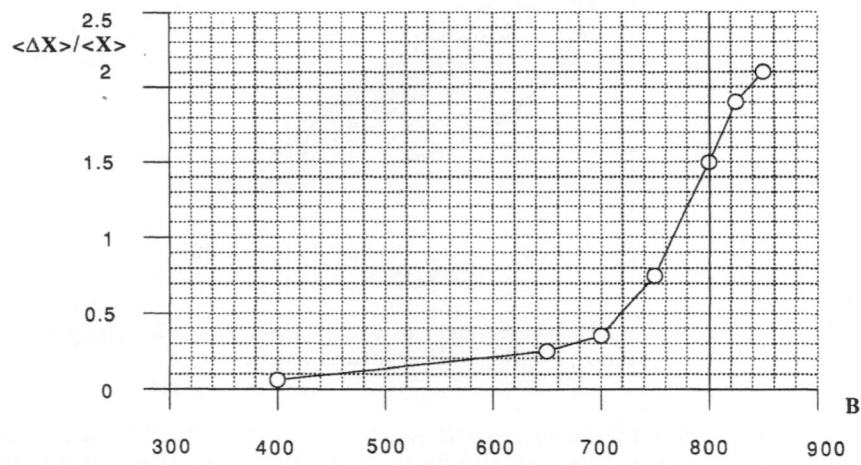

FIGURE 6. Amplitude of the oscillations as a function of the number of particles of the B species; the critical value of B obtained from a linear stability analysis is 800.

The states generated by the time evolution are indeed distributed around the stable stationary state for undercritical values of B. On figure 5, the effect of increasing B can be visualized. One has the formation of a crater like distribution for B greater than B_c. The intensity of the peaks is not constant along the trajectory: indeed in the upper right part of figures 5(c) or figure 7, the motion is very rapid (the increase in X and the decrease in Y particles) and very few states are located there. On the contrary, in the lower left part, one has a huge peak in the distribution of states since the motion is very slow there. The number of X particles is minimum and the number in Y particles still increases before the system jumps quickly. These are indeed the distribution found by other studies based on stochastic approaches of these phenomena in the Brusselator. Figure 7 is more explicit : it is an enlargement of a trajectory obtained at twice the critical B value. One can see that the motion of the system is slow and oscillates a lot around the deterministic trajectory; the location of the beginning of the fast motion (nearly straight lines) is scattered on a relatively large area. The same is found by inspection of a Langevin simulation as has been reported by Schraner et al. in reference 28.

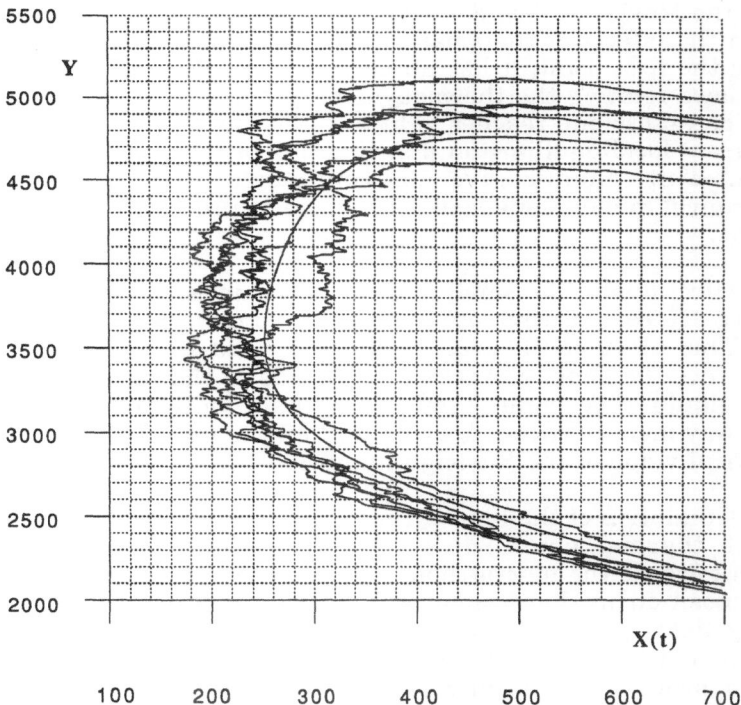

FIGURE 7.This trajectory has been obtained for a value of (B-Bc)/Bc≈2, well beyond the critical point. The graph shows part of the dsmc trajectory oscillating around the deterministic limit cycle which is also represented on the graph (it can easily be distinguished from the others since it is smooth). The trajectory is run clockwise.

One should stress that these simulations will provide an interesting experimental probe of the statistical mechanics of this far from equilibrium system.

Another important aspect which can be observed in these simulations is linked to the spatial dispersion. Indeed, the largest system is 200 mean free paths long: a particle would cross it by diffusion after 40,000 collisions. This corresponds to a time which is much longer than the oscillation time (in this case about 1,000 collisions per particle). One can then expect to observe a desynchronization between the left and right part of the system. This can indeed be seen and, again, a more precise comparison with a theoretical approach is under way.

4. CONCLUSIONS

The two examples discussed here were typical of non-equilibrium structures. The amazing result which comes out of these simulations is that the non-equilibrium structures can be simulated directly from a molecular modelling. It is remarkable that the macroscopic description remains valid to such small time and space scales in non-equilibrium cases in the same manner as it came as a surprise that hydrodynamics can be used to interpret equilibrium correlation functions.

The point we have made in the hydrodynamic case is related to the validity of the Navier Stokes description in the presence of large constraints and also to the effects of the small number of particles for the simulations. Concerning the first point, it has already been observed that deviations from the linear laws relating the thermodynamical fluxes and forces are small for atomic fluids unless one reaches values of the forcing which would induce a deviation from local equilibrium distributions of more than 5 % on a mean free path. This should give us confidence that many interesting hydrodynamical phenomena can indeed be reproduced from microscopic simulations.

The effect due to the small number of particles, on the other hand, has not been really seen except on the smaller simulations made with 1,500 particles. The effect of the wall inside the fluid is not much important otherwise since the density chosen corresponds to that of a dense gas rather than of a liquid. The same simulations made with a liquid would have required many more particles to avoid wall effects into the fluid and to attain a supercritical value of the Rayleigh number.

The simulation of the limit cycle of the Brusselator, one the other hand, is done at thermal equilibrium. The transition from the microscopic description towards the macroscopic require an average over the fluctuations of the composition in the system. The dsmc simulation provides us with a fluctuating description of the reaction diffusion system. The non-linear macroscopic behavior is well recovered and more precise comparison is needed in order to make any statement concerning the adequacy of a Langevin, Fokker-Planck or master equation description. Spatially inhomogeneous systems should also be possibly studied using this approach, at least in 1 or 2 dimensions.

Now, the question could also be asked as to what are the limits of our ability to simulate microscopically, how will it evolve in the future and also what kind of problems will this help to solve?

The simulations reported recently in the literature have often dealt with systems as large as 100,000 particles. And even if that requires long computation times on sophisticated supercomputers, one can easily predict that, within the coming years, an hydrodynamical simulation of a microscopic system will become more and more accessible to laboratories around the world. Simulations of up to a million particles can become routine work with the access to the several gigaflops machines. One has of course to be cautious just because the cpu time scales linearly with the physical time which often scales the square of the number of particles used.

Now, regarding the question of what kind of problems will be studied using these methods, three types of question could be distinguished.

First, the microscopic simulation could provide an alternative to the finite difference or spectral methods in solving the Navier-Stokes equations when these methods are not valid. This has been the main reason of the interest developed recently for the methods which use cellular automaton techniques to integrate lattice gases[29]. Indeed, in this case, the traditionnal MD technique is far too costly to be ever able to compete with these codes where one does simplify the dynamics to the minimum required for the simulation of hydrodynamics. The main accelerator in these simulations is the use of bit arithmetic instead of floating points. The price to pay is a loss of the microscopic character of the model due to the absence of a real microscopic dimension. In many circonstances, it is believed that these methods will be more efficient than usual numerical integrators of the Navier-Stokes equations[30]. The MD approach, on the other hand, is of a more fundamental nature: it answers questions such as what is the minimum size required to have an hydrodynamic behavior.

Another interesting recent approach has been the development of stochastic methods to simulate dilute gases[31]. These methods are two to three orders of magnitude faster than MD to produce flow patterns or shock profiles. They are mostly used in the case of an extremely low concentrations where the Navier Stokes description itself become questionable. This is the second category of problems where microscopic simulations could become useful: to use a molecular modelling in order to investigate the validity of a macroscopic description in some extreme or difficult to investigate situations. The nature of the boundary condition to be used in a two phase flow in a pipe for instance has been looked at by a direct simulation using 4,000 particles[32]. An interesting question could also be the use of direct simulation of non-equilibrium states to test the validity of the Burnett corrections to the linear laws.

The third category of problems is the study of the critical phenomena and the possibility of using MD to answer questions on the nature of a transition or on the importance of fluctuations in the critical behavior. The size effects in this domain are of course very important and the possibility of using large systems will be decisive in the feasibility of these studies as this is the case for the chemical problem reported here.

5. ACKNOWLEDGEMENTS

The work reported here has benefitted from discussions with and encouragements from Giovanni Ciccotti, Eddy Kestemont, Malek Mansour, Andreas Puhl, Grégoire Nicolis and Ilya Prigogine.

The access to computers has been possible through the grants given by the Fonds National de la Recherche Scientifique (Belgium). I acknowledge also the financial support of the Actions de Recherche Concertées of the Belgian Government.

REFERENCES

1 B. J. Alder, T. Wainwright, in Transport Processes in Statistical Mechanics, I. Prigogine (editor), J. Wiley-Interscience, New York (1958).

2 A. Rahman, Phys. Rev. **136**, 405 (1964)

3 R. Zwanzig and M. Bixon, Phys. Rev. **A2**, 2005 (1970)

4 W. E. Alley, B. J. Alder, S. Yip, Phys. Rev. **A27**,3174 (1983)

5 W. E. Alley and B. J. Alder, Phys. Rev. **A27**, 3158 (1983).

6 J. P. Boon, S. Yip, Molecular Hydrodynamics, Mac Graw-Hill, New York (1980)

7 W. G. Hoover and W. T. Ashurst, in Theoretical Chemistry, vol 1: Advances and perspectives (Academic Press, New York, 1975)

8 W. W. Wood in Fundamental Problems in Statistical Mechanics III, E. G. D. Cohen editor, North-Holland, (1975)

9 A. Tenenbaum, G. Ciccotti, R. Gallico, Phys. Rev. **A25**, 2778 (1982); C. Trozzi and G. Ciccotti, Phys. Rev. **A29**, 916 (1984)

10 B. L. Holian, W. G. Hoover, B. Moran, G. K. Straub, Phys. Rev. **A22**, 2798 (1980)

11 E. Meiburg, Phys. Fluids, **29**, 3107 (1986)

12 G. A. Bird, Molecular Gas Dynamics, Clarendon Press, Oxford (1976)

13 D. C. Rapaport, E. Clementi, Phys. Rev. Lett. **57**, 695 (1987)

14 S. Chandrasekhar, <u>Hydrodynamic and Hydromagnetic Stability,</u> Clarendon Press, Oxford (1961).

15 G. Nicolis and I. Prigogine, <u>Self-Organization in Nonequilibrium Systems,</u> J. Wiley, New York (1977)

16 L. D. Landau, E. M. Lifshitz, <u>Fluid Mechanics,</u> Pergamon Press, London (1958)

17 P. Résibois, M. Deleener, <u>Classical Kinetic Theory of fluids,</u> J. Wiley-Interscience, New York (1977)

18 J. A. Barker, D. E. Henderson, Rev. Mod. Phys. **48**, 587 (1976)

19 J. J. Erpenbeck, W. W. Wood, in <u>Modern Theoretical Chemistry,</u> edited by B. J. Berne, vol.6B, Van Nostrand-Reinhold, New York (1983)

20 A. L. Garcia, M. Malek Mansour, G. C. Lie, E. Clementi, J. Stat. Phys. **47**,209 (1987)

21 D. M. Gass, J. Chem. Phys. **54**, 1898 (1971).

22 M. Mareschal, M. Malek Mansour, A. Puhl, E. Kestemont, Phys. Rev. Lett. **61**, 2550 (1988);A. Puhl, M. Malek Mansour and M. Mareschal, Phys. Rev. **A40**, 1999 (1989)

23 G. Nicolis AND M. Malek Mansour, Phys. Rev. **A29**, 2845 (1984)

24 G. Nicolis, A. Amellal, G. Dupont, M. Mareschal, J. Mol. Liq. 41,5 (1989)

25 A. L. Garcia, M. Malek Mansour, G. C. Lie, E. Clementi, J. Stat. Phys. **47**,209 (1987); see also the contribution by A. Garcia in this volume.

26 J. Boissonade, Phys. Lett. **74A**, 285 (1979); ibid., Physica **113A**, 607 (1982); P. Ortoleva and S. Yip, J. Chem. Phys., **65**, 2045 (1976)

27 A. A. Frost, R. G. Pearson, <u>Kinetics and Mechanism,</u> J. Wiley and sons, New York (1961)

28 see for instance G. Nicolis, in <u>Order and Fluctuations in Equilibrium and Non-equilibrium Statistical Mechanics,</u> G. Nicolis, G. Dewel and

J. W. Turner, editors, J. Wiley, New York (1981); see also P. H. Richter, I. Procaccia and J. Ross, Chemical Instabilities, Adv. Chem. Phys. 217-268 (1980) and R. Schramer, S. Grossmann, P. H. Richter, Z. Phys. B35,363-381 (1978).

[29] R. Monaco, (editor) Discrete Kinetic Theory, Lattice Gas Dynamics and the Foundations of Hydrodynamics, World Scientific, Singapore (1989).

[30] see for instance the contribution of S. Zaleski in the reference 24.

[31] G. A. Bird in this volume.

[32] J. Koplik, J. R. Banavar, J. F. Willemsen, Phys. Fluids. **A1**, 781 (1989)

SHOCK WAVES, FRAGMENTATION, AND HYPERVELOCITY IMPACTS BY MOLECULAR DYNAMICS

Brad Lee Holian

Theoretical Division, Los Alamos National Laboratory
Los Alamos, NM 87545, USA

ABSTRACT

A wide variety of nonequilibrium processes can be investigated from the atomistic viewpoint by the method of molecular dynamics (MD), where the equations of motion of thousands of interacting atoms are solved on the computer. MD simulations of shock waves in three-dimensional fluids have shown conclusively that shear-stress relaxation is achieved through the atomic rearrangement of transverse viscous flow, for which Navier-Stokes hydrodynamics has been shown to be accurate. MD simulations of homogeneous adiabatic expansion have provided significant insight into the process of fragmentation for hot dense fluids, in that the fragment distribution is exponential in fragment mass; moreover, the average mass can be reasonably estimated by a simple model based on energy balance. Finally, both compression and expansion are involved when a high-velocity sphere impacts and penetrates a thin wall; atomistic (MD) simulations of this highly nonequilibrium flow resemble in many ways the continuum (hydrodynamic) simulations, especially when the number of atoms in the simulation is sufficiently large.

1. INTRODUCTION

The method known as molecular dynamics (MD) has proved itself to be the most successful tool yet developed for studying the collective motion of large numbers of atoms engaged in nonequilibrium flows[1]. Supercomputers have made it possible to study a wide range of complex (nonlinear and nonlaminar) flow problems by MD, three of which I will discuss in this paper: shock waves[2], fragmentation of fluids[3], and high-velocity impacts[4]. These MD simulations involve as many as 10,000 atoms interacting with central, pairwise-additive, short-range forces. The principal limitation to these calculations is that the macroscopic physical phenomena that we wish to simulate by MD must have maximum length-scales that can be contained within the MD computational cell. Also, the time-scales must be short enough for practical MD simulations. For example, the characteristic distances that must be considered in MD shockwave simulations include the thickness of the shock wave, the distance-of-run required to achieve a steady-wave profile, and sufficient cross-sectional area to model the transverse flow (dissipation) that makes the shock wave steady.

In the case of fluids, MD calculations of shock waves, showing viscous rearrangement of atoms in the immediate vicinity of the shock front, compare surprisingly well with solutions to the Navier-Stokes (NS) equations of hydrodynamics, even for strong shock waves[2,5]. The NS calculations require as input the equation of state (EOS) for the fluid, which was obtained from equilibrium atomistic simulations, as well as linear hydrodynamic transport coefficients, which were obtained from nonequilibrium MD simulations (these simulations were carried out much earlier, independent of the shockwave simulations). In the case of solids, MD shockwave calculations have been carried out, but progress beyond identifying the microscopic process of plastic flow has been very slow. For weak shock waves in solids, the initial presence of a significant number of dislocations, the extended defects that are believed to facilitate plastic flow, means that a very large number of atoms are required for a believable simulation[2]. In contrast to solids, where defects are almost nowhere (and often large-scale, such as dislocations and grain boundaries), fluids have fine-grained defects (density fluctuations) that are almost everywhere. For this reason, we will concentrate in Section 2 of this paper on MD simulations of fluid shock waves.

In Section 3, we will discuss MD simulations of fragmentation in fluids[3]. At sufficiently high rates of expansion, a fluid can be expected to break apart along low-density surfaces originating in thermal density fluctuations. As pointed out in the case of shock waves, the defect structure of fluids is fine-grained, relative to solids, so that it is reasonable to expect that fragmentation in fluids will yield smaller pieces at a given strain-rate, that is, better statistics for the distribution of fragments. We also can increase the effective sample size by considering a two-dimensional fluid. From the atomistic simulations, we can answer two fundamental questions in modeling fragmentation: 1) What is the nature of the distribution of fragment mass? 2) How does the average fragment mass vary with the initial expansion energy, or strain-rate? As we will show, the number of fragments greater than a given mass decays exponentially with mass. The average fragment mass can be reasonably estimated from an energy-balance model, where the expansion energy balances the energy it takes to create the free surface area of the fragments[3].

Finally, in Section 4 we conclude with MD simulations[4] that combine both the physics of rapid shock compression to a high-density and temperature state, and the rapid expansion and fragmentation from that state, namely, the high-velocity impact onto and penetration through a thin plate by a sphere. We present results of a three-dimensional (3D) simulation, where the sphere contains several hundred atoms, as well as a 2D simulation for several thousand ball atoms. The formation of a cloud of debris (ball and wall material) is strikingly similar to continuum (hydrodynamics) simulations[6] especially for large numbers of atoms. All three of these examples of simulations of complex nonequilibrium flows make use of strong driving conditions (high strain-rates) as well as the small scale of fluid defects (density fluctuations) in order to connect the atomistic picture to the continuum. Faster computers with larger memories, including machines with parallel architecture, will expand the microscopic (atomistic) horizons of MD simulations even further toward the macroscopic (continuum) realm of observable complex flows.

2. SHOCK WAVES IN FLUIDS

The technique for generating shock waves in either solids or fluids via MD mimics a symmetric planar impact experiment, with the effects of free surfaces minimized by utilizing periodic boundary conditions[2,5]. A planar shock wave is initiated at time t = 0 by shrinking the x-direction periodic length L_x according to $L_x(t) = L_x(0) - 2u_p t$, while keeping the cross-sectional lengths L_y and L_z fixed; u_p is the piston, or particle velocity. The periodic boundary then behaves very much like the interface of a planar impact experiment. Before initiating the shock wave at t = 0, a filamental rectangular parallelepiped of atoms is equilibrated at an initial relatively-low density ρ_0 and temperature T_0. The length of the parallelepiped in the direction of shock propagation is typically 50-100 face-centered cubic (fcc) unit cells long, or 100-200 planes of atoms in the x-, or <100>-direction. The transverse (y and z) dimensions are usually 3-4 unit cells, or 18-32 atoms per plane. For the fluid shocks, of course, the equilibration process includes time for the initial fcc crystal to melt. Feedback methods have been developed to achieve a desired temperature for the equilibrated initial state[7].

A schematic of shock waves generated by shrinking periodic boundary conditions is shown in Fig. 1. A pair of oppositely-running shock waves move out with shock-front speed u_s from the boundaries toward the middle of the computational cell, leaving behind shocked material at a higher density ρ_1 and temperature T_1, moving along with the interfacial "pistons" at speed u_p (reshocking from an initially-shocked state can be achieved by letting the shock waves collide; results discussed here are for the singly-shocked state).

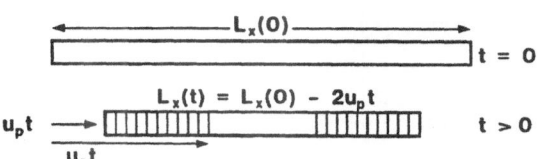

FIGURE 1. Schematic of shockwave generation by shrinking periodic boundary conditions; $L_x(t)$ is the periodic length in the shock-propagation direction. (Whenever an atom crosses a periodic boundary, $\pm 2u_p$ must be added to that atom's velocity, with the sign given by the direction of crossing.) Shocked material at time t > 0 is shaded.

The shockwave profiles for density, velocity, pressure tensor, temperature, internal energy, and heat flux vector, are obtained by lumping particles and their individual kinetic and potential contributions into bins, in order of their x coordinates, in multiples of the initial number of atoms in a yz cross-sectional plane. These planar Lagrangian mass elements, particularly in the case of solids, lead to smoother profiles than fixed Eulerian boxes. For steady shock waves, time averages of profiles can be gathered for both waves simultaneously by riding along with the shock fronts, as shown in Fig. 2. To the left of the stationary front at x = 0, unshocked material (ρ_0, T_0) flows at velocity u_s toward a piston at the far right, where material is compressed to ρ_1 and heated to temperature T_1, and moves with the receding piston at velocity u_s - u_p. By the mass conservation equation

$$\frac{\partial \rho}{\partial t} + \nabla.(\rho \mathbf{u}) = 0$$

we see that a steady ($\partial \rho / \partial t = 0$), planar (function of x only) shock wave leads to a constant mass flux ρu throughout the profile:

$$\rho(x)u(x) = \rho_0 u_s = \rho_1(u_s - u_p)$$

Thus the total volumetric strain in the shock is $\varepsilon_x = \rho_0/\rho_1 - 1 = -u_p/u_s$, and the total strain rate at the shock front (x = 0) is $\dot{\varepsilon}_x = \varepsilon_x u_s/\lambda = -u_p/\lambda$, where λ is the shockwave thickness.

In the Navier-Stokes, or linear hydrodynamics view of fluid flow, a steady shock wave is formed when the process of longitudinal compression (leading to steepening of the wave) competes with the dissipative process of viscous flow (spreading of the wave). The NS transport coefficients, the thermal conductivity κ and the longitudinal viscosity, $\eta_L = \eta_V + 4/3\eta_s$, where η_V is the volume (bulk) viscosity and η_s is the shear viscosity, as well as the EOS [pressure $P(\rho,E)$ as a function of density ρ and internal energy E] can be obtained from independent MD calculations. The EOS[8] and the Green-Kubo transport coefficients, which can be related to equilibrium fluctuations of the pressure tensor and heat flux vector for η_L and κ, respectively, can be obtained from equilibrium MD; the nonlinear transport coefficients for finite strain rates and temperature gradients via nonequilibrium MD (NEMD) can be extrapolated to zero rates to obtain the NS transport coefficients[9]. The

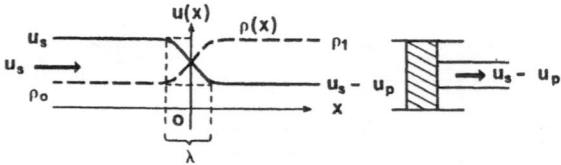

FIGURE 2. Schematic of density $\rho(x)$ (dashed line) and fluid velocity u(x) profiles in the frame of reference for a steady shock wave, whose front of thickness λ is located at x = 0, separating unshocked material on the left (ρ_0, T_0) moving at velocity u_s, from shocked material at the right (ρ_1, T_1) moving at u_s - u_p.

indirect Green-Kubo calculations are time-consuming because of the statistical errors inherent in fluctuations, while the direct NEMD calculations require extrapolation of several computer experiments[10]. Most of the results to date for transport coefficients are from NEMD.

In Fig. 3 are shown shockwave profiles generated by NEMD[5], with NS comparisons sketched in as dashed curves. The atoms in this simulation interacted via a Lennard-Jones pair potential (repulsive inverse-12th power of the separation plus attractive inverse-6th power). Using energy and distance parameters in this potential appropriate for argon, the shock strength was such that the final temperature achieved was about 12,000 K, or near the point of ionization. Even so, NS provides a good approximation to the MD results, underestimating the MD viscosity (shock thickness) by only 30-40%. This is the worst case; much better agreement is obtained for weaker shockwaves[2,5], where NS and NEMD profiles agree within 4%.

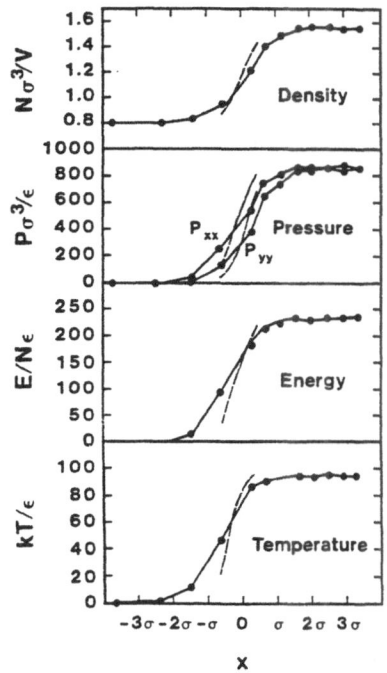

FIGURE 3. Density, pressure, internal energy, and temperature profiles for a strong steady shock in a Lennard-Jones fluid calculated by non equilibrium molecular dynamics[5]. Corresponding Navier-Stokes results are shown as dashed lines. (The units for a Lennard-Jones system are: the atomic mass m, the crossing point of the pair potential σ, and the depth of the potential minimum ε.) The initial conditions for this strong shock are: $\rho_0\sigma^3/m = 0.8442$, $kT_0/\varepsilon = 0.722$, $u_p(m/\varepsilon)^{1/2} = 22.4$.

In subsequent work, it was discovered that even better agreement can be obtained[2] between NS and MD if the temperature component in the direction of propagation of the shock wave, T_{xx}, is used to compute $\eta_L(\rho,T)$, rather than the mean temperature at the shock front, $T = (T_{xx} + T_{yy} + T_{zz})/3$, which can be as much as a factor of two lower than T_{xx}. This nonlinear (non-NS) correction serves to narrow the NS shock thickness at low shock strengths; that is, at lower temperatures, using T_{xx} reduces the apparent viscosity towards the MD value. At the other extreme (strong shock), using T_{xx} rather than T at the shock front enhances the viscosity, as in the dilute gas limit, increasing the thickness to within 10% of the MD result. Attempts to relate frequency and wavelength corrections to the viscosity, in the framework of generalized hydrodynamics, are not so successful. Even less successful is the attempt to include the nonlinear effecton the shear viscosity due to shear-thinning with increased strain-rate. No satisfactory explanation of these observations has yet been proposed.

These fluid shockwave results have recently revealed an interesting and potentially useful correlation[2]. The peak shear pressure, $1/2(P_{xx}- P_{yy})$, which occurs near the shock front, appears to be a constant fraction (~10%) of the Hugoniot jump in pressure, $P_1 - P_0$. Hence, for a viscous fluid (the shear pressure for the constant-volume process of compression in the x-direction and expansion in the y- and z-directions is $-\eta_s \dot{\epsilon}_x$, the ratio

$$\frac{(P_1-P_0)}{\frac{1}{2}(P_{xx}-P_{yy})} = \frac{-\rho_0 u_s u_p}{\eta_s \dot{\epsilon}_x} = \frac{\rho_0 u_s \lambda}{\eta_s}$$

is a "shock Reynolds' number" of ~10, essentially independent of shock strength. It is interesting that approximately the same ratio holds for plastic shockwave deformation in MD solids, though of course a viscous-fluid concept like a Reynolds' number is inappropriate. Thus, for strong fluid or solid shock waves, narrow enough to be calculated by MD, it appears that there exists a roughly constant ratio for the balance between compressional steepening of a shock wave and dissipative spreading due to atomic rearrangement[2].

3. FRAGMENTATION OF FLUIDS

The random segmentation of a one-dimensional (1D) infinite line provides an example of a geometrical statistical theory whose fragment distribution is in reasonable accord with the results of 1D dynamic fragmentation experiments. Unfortunately, in more dimensions, numerous geometric constructions are possible, leading to widely differing distributions[11]. Furthermore, because of inherent test complexity, laboratory fragmentation experiments are not able to differentiate unambiguously among the various competing geometric statistical theories. MD computer experiments on fragmentation do not suffer from the practical limitation of laboratory testing and allow us to select the microscopically correct statistical theory. In these MD simulations[3], condensed matter undergoes homogeneous adiabatic expansion and fragmentation, using a method of moving periodic boundaries closely related to the inhomogeneous method discussed in the previous section on fluid shock waves. The expansion technique is reminiscent of the classical, Euclidian-space picture of the "Big Bang," where the infinite sample Universe is modeled as an expanding checkerboard of periodically repeated units, each containing N atoms. Using MD, we have been able to establish that the homogeneous distribution of fragment or cluster masses is exponential, and that the average cluster mass can be simply explained by an energy balance between the kinetic energy of expansion and the potential energy of broken surface bonds. The microscopic origin of continuum fragmentation models[12], as discovered in these novel MD simulations, lends credence to their application to a wide range of problems, from the breakup of oil shale, to the destruction of armor, and even to the distribution of galaxies in our Universe.

As in the case of inhomogeneous shock waves, the homogeneous expansion method we have invented for the study of fluid fragmentation by MD minimizes the effects of free surfaces by imposing periodic boundaries. Within each of the periodic units there are N particles whose initial coordinates and momenta correspond to a specified equilbrium state of a fluid (though the method can equally well be applied to a solid). At time zero, the side lengths of the periodic unit are made to expand at a constant isotropic rate: for example, in the x direction, $L_x(t) = L_x(0)(1 + \dot{\varepsilon}_x)$. In contrast to the method for shock wave simulations, however, a constant homogeneous velocity gradient is applied to all the particles within the unit: for example, the x velocity of particle i (coordinate x_i) becomes $u_i(0+)$ = $u_i(0) + \dot{\varepsilon}_x x_i(0)$, where $\dot{\varepsilon}_x$ is the initial linear expansion strain rate (Hubble constant in cosmological terms) and $u_i(0)$ is the initial random <u>thermal</u> velocity. From time zero onward, the expansion is adiabatic (no more energy is added to the system), and the particles obey Newton's (Hamilton's) equations of motion, with expanding periodic boundary conditions - that is, if an atom with coordinate x_i and velocity u_i leaves, for example, the left-hand boundary of the periodic cell, it is replaced by an image particle that enters the right-hand boundary with $x_i{}^* = x_i + L_x$ and $u_i{}^* = u_i + \dot{L}_x$.

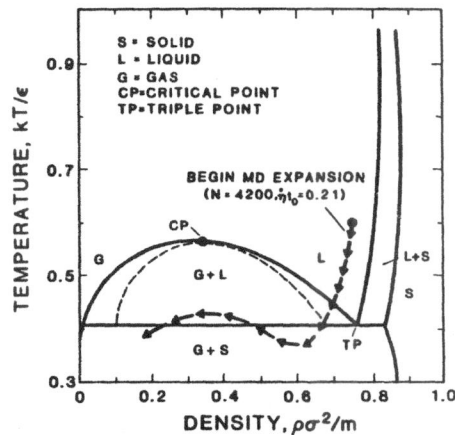

FIGURE 4. Phase diagram for the 2D Lennard-Jones system[13]. The MD expansion path for N = 4200 atoms and volumetric strain rate $\dot{\eta}t_0$ = 0.21 is shown by arrows; the spinodal line is shown by a dashed curve.

The MD simulations of expansion and fragmentation were for a 2D Lennard-Jones fluid, in order to study a statistically significant number of fragments for a "realistic" material. By that, we mean that the Lennard-Jones pair potential has an attractive well and therefore has a cohesive mechanism for realistic clustering. The 2D Lennard-Jones phase diagram[13] has been determined from atomistic simulations, and has a liquid-vapor coexistence region, as shown in Fig. 4. The expanding fluid in the MD fragmentation experiments follows a track through the density-temperature phase diagram that resembles a van der Waals loop (approximating an isentrope); the example shown in Fig. 4 is for the initial state equilibrated at a density $\rho\sigma^2/m$ = 0.75, approximately that of the triple point, and a temperature

$k_BT/\varepsilon = 0.6$, slightly higher than that of the critical point. The initial volumetric strain rate is $\dot{\eta}t_0 = 0.21$. (In D dimensions, the volumetric strain rate is related to the linear strain rate by $\dot{\eta} = D\dot{\varepsilon}_x$. Unfortunately, in the text of Ref. 3, linear strain rate is used, while the figures are labelled by volumetric strain rate; here we use different symbols to avoid confusion.) The unit of time is $t_0 = \sigma(m/\varepsilon)^{1/2}$; Lennard-Jones units are defined in the caption to Fig. 3. The bulk temperature along the expansion path, <u>assuming</u> local thermodynamic equilibrium, is computed using the velocity fluctuation of particles about their local expansion velocity.

Fragmentation of the bulk liquid begins somewhere near the spinodal line, where incipient cavities form due to thermal fluctuations, as seen in Fig. 5. The termination of the expansion experiment is shown in Fig. 6, clearly a snapshot of the nonequilibrium state, since many of the droplets are far from spherical; also, according to the kinetic temperature, only 20% of the mass should be in the condensed (nonvapor) phase at equilibrium.

Cluster statistics for the experiment at the volumetric expansion rate $\dot{\eta}t_0 = 0.21$, using a cluster bond length of $r_c = 1.24\sigma$, are presented in Fig. 7, where the logarithm of the cumulative number of fragments of mass M or greater is plotted against M, in units of the atomic mass m. The distribution is well represented by a sum of two exponentials, a monomeric peak ($\mu = 1$) and a broad shoulder with average fragment mass $\mu = 20.6$. As pointed out by Grady and Kipp[11], 1D fragmentation of an infinite line can be derived in a manner analogous to maximizing the entropy in order to obtain the Boltzmann distribution in statistical mechanics. However, in 2D and 3D, it is

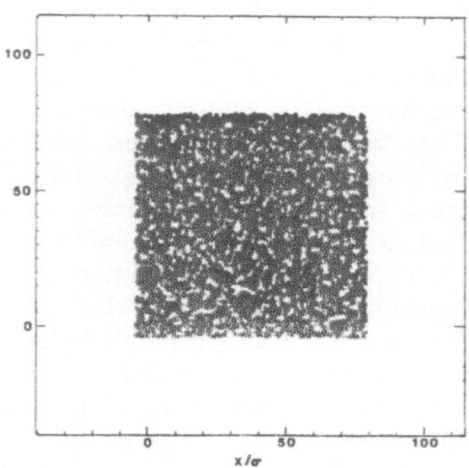

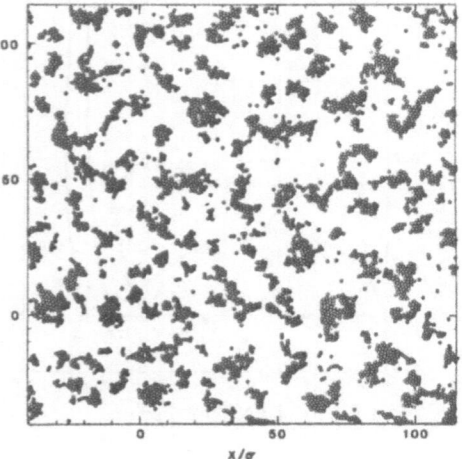

FIGURE 5. Snapshot of particle positions early in the 2D MD expansion ($\rho\sigma^2/m = 0.612$, $k_BT/\varepsilon = 0.37$, N = 4200, $\dot{\eta}t_0 = 0.21$, $t/t_0 = 1.0$).

FIGURE 6. Snapshot of the final particle positions for the 2D MD expansion experiment ($\rho\sigma^2/m = 0.175$, $k_BT/\varepsilon = 0.39$, N = 4200, $\dot{\eta}t_0 = 0.21$, $t/t_0 = 10.0$).

possible to imagine a variety of geometric constructions to model fragmentation <u>mathematically</u> (rather than physically), so as to obtain fragment distributions that are exponential in fragment diameter (square root of mass in 2D) or that are a power-times-exponential of fragment mass (as in the Voronoi model of fragmentation). These models are not supported by the MD simulations. Laboratory observations (including, interestingly enough, the observed distribution of galaxies produced by the expansion of the <u>real</u> Big Bang), are also best described by exponential distributions, though often the conclusions from experiments are clouded by difficulties in controlling the experimental homogeneity of expansion[3]. The MD simulations are, in fact, very homogeneous, as can be seen in Fig. 6, and as can be deduced from local temperature distributions.

Finally, the mean or median fragment mass as a function of the expansion strain rate can be predicted from a continuum model due to Grady[12], which postulates that the energy of expansion, including both kinetic and elastic potential energy, must balance the energy required to form the free surface of a spherical (circular, in 2D) droplet of mass μ. As the fluid expands and cools, it goes into tension (negative pressure), storing energy as elastic potential energy. Thermal fluctuations in density give rise to low-density regions that are candidates for formation of droplet free surfaces. The time-scale for density fluctuations across the droplet is set by the sound speed c_0 in the fluid, so that the expansion strain at the moment of fragmentation is related to the strain rate by $\varepsilon_x = \dot{\varepsilon}_x t = R\dot{\varepsilon}_x/c_0$, where R is the radius of the droplet. Thus, the elastic energy per unit mass is $1/2\, R^2\dot{\eta}^2$. The free surface potential energy per unit mass is $1/2\, D(r_c/R)E_{coh}$, where E_{coh} is the cohesive energy per unit mass and r_c is the thickness of the shell of broken surface bonds. Since the kinetic energy is small compared to the elastic, equating the elastic and surface energies gives the curve in Fig. 8 of

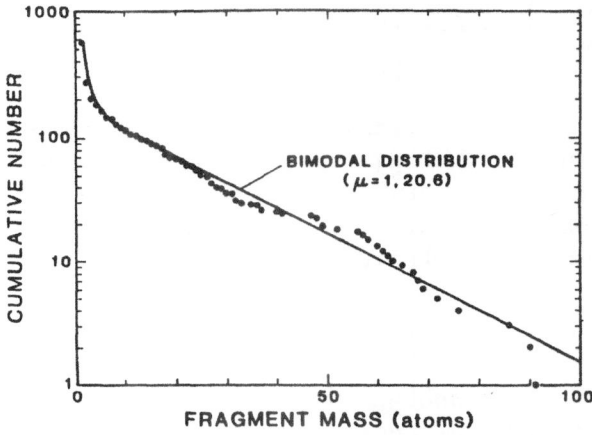

FIGURE 7. Cluster statistics for the final state of the 2D MD expansion experiment (see Fig. 6): the logarithm of the cumulative number of fragments of mass M or greater (dots) versus M, along with fitted bimodal [Σexp(-M/μ)] distribution (solid curve).

average fragment mass μ as a function of the volumetric strain rate $\dot{\eta}$, compared to the MD simulations. For the 2D Lennard-Jones system, the cohesive energy per unit mass is approximately $1/2\, n_D$, where n_D is the number of nearest neighbors (2, 6, and 12 for D = 1, 2, and 3, respectively).

In Fig. 8, both the mean cluster mass, as determined from the slope of the broad shoulder in the log-linear plots of cumulative number versus mass (that is, excluding the monomeric peak), and the median cluster mass (half the mass of the sample is in clusters bigger than the median) are shown. For a single exponential distribution, the median is 1.68 times the mean; for a bimodal distribution, the ratio is a little less. Nevertheless, this does not account for the factor of 3 by which the energy-balance theory overestimates the average cluster mass. Part of the difficulty is the fractal nature of the connectivity of clusters, as shown in Fig. 6: by a change of the cluster bond length from 1.24 to 1.33, the median cluster mass increases as much as 50%. As higher strain rates are achieved in the MD experiments, the clusters become smaller and more spherical, and hence in better agreement with the simple model.

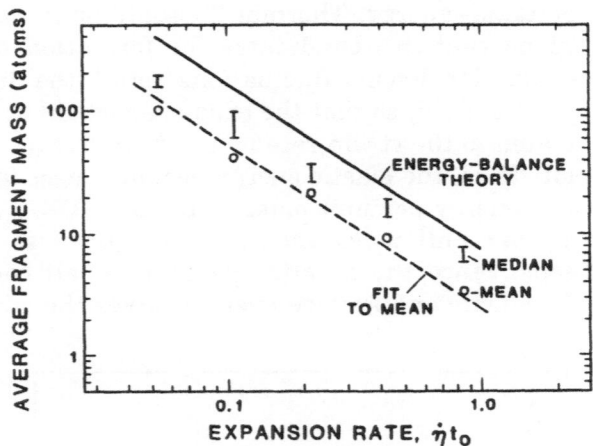

FIGURE 8. Average fragment mass versus initial volumetric strain rate (2D MD experiments, N = 4200 atoms). Both arithmetic mean and median cluster size are shown (see text). Grady's energy-balance theory (solid curve) and a fit to the mean cluster size (dashed curve) both have slope -4/3.

4. HYPERVELOCITY IMPACTS

Experimental x-ray photographs have been taken of the debris cloud produced by a lead ball striking and penetrating a lead wall at a velocity sufficient to vaporize the debris. (Lower velocities lead to solid fragmentation under conditions that are highly inhomogeneous and therefore inherently more complex than the homogeneous, single-strain-rate fragmentation discussed in the previous section; at velocities high enough that the debris temperature is above melting, the absence of solid strength and fracture effects simplify the continuum treatment). Recent continuum hydrodynamics (hydrocode) simulations[6] of these hypervelocity impacts have shown that one important aspect of the distribution of mass in the cloud is particularly sensitive to the degree of accuracy with which the solution is obtained numerically; namely, while the experimental cloud is hollowed out at its center, a hydrocode will predict a cloud that is <u>densest</u> at the center, if a low-order numerical method is used (as is typical practice). In order to investigate the degree to which continuum behavior persists down to the atomistic level, I used MD to simulate a hypervelocity impact of a 3D sphere of 683 Lennard-Jones particles hitting a plate of 8000 particles[4]. The most interesting question to be addressed by this simulation is whether a calculation involving this number of atoms exhibits behavior similar to the experiment, including the hollow balloon of debris, using the same velocity and ratio of ball diameter to wall thickness, scaled down to the atomistic level.

In order to relate the lead-on-lead experiments to time and distance scales appropriate to MD calculations, two parameters are of importance: 1) the ratio of the sphere diameter d to the thickness of the plate w, and 2) the ratio of the kinetic energy of the impacting sphere, $1/2\, Nmu_p^2$, to its binding, or cohesive energy, NmE_{coh}. (The mass of each of the N atoms in the sphere is m). In the lead-on-lead experiments, the ratio $d/w = 2.4$ (e.g., d = 1.5 cm and w = 0.635 cm), while $1/2\, u_p^2/E_{coh} = 25$ (i. e., u_p = 6.6 km/s). For the MD calculations, the diameter of the ball (N = 683 atoms) was about 10 σ (~40Å), where σ is the crossing point of the Lennard-Jones pair potential. The wall was composed of 5 close-packed fcc planes in the x-direction (a thickness of about 4.5 σ), with a cross-sectional area of 400 σ^2. The forward edge of the ball and the back edge of the wall were separated initially by 1.8 σ, just beyond the range of the potential. Both the ball and wall were equilibrated at an initial density of $\rho\sigma^3/m = 1$ and temperature, arbitrarily chosen to be about one-tenth the melting point, $kT/\varepsilon = 0.1$, where ε is the well depth of the Lennard-Jones pair potential. For the Lennard-Jones potential, the cohesive energy per atom in 3D is about 6 ε; hence, the value of $20(\varepsilon/m)^{1/2}$ was chosen for u_p. A convenient scale for time is d/u_p, which for the hydrodynamic problem is 2.3 μs, while for the molecular dynamics problem it is 0.5 $\sigma(m/\varepsilon)^{1/2}$ (consequently, 30 μs in the hydrodynamics simulation, when the x-ray photograph was taken experimentally, is equivalent to a time of ~7 in the MD simulation).

In Fig. 9, we show the initial configuration of the ball and wall for the MD calculations. In Fig. 10, the expanding debris cloud is displayed for MD time t = 7; the projection onto the xy plane is shown, that is, the side view of the debris cloud as viewed through binoculars from a long distance away. (At t = 1.5, the shock wave emanating from the hole into the plate reached the yz periodic boundaries, and from that time onward, the edges of the wall were made into free boundaries; this does not affect the debris-cloud evolution.) At t = 7, the debris cloud has developed a characteristic mushroom shape, with the beginnings of clusters near the front edges. On the surface of the debris bubble are atoms that have been punched out from the plate. The atoms from the sphere are in the forward interior of the cloud. We note that the debris cloud is not particularly hollow, similar to the low-order hydrocode calculations. Also, atoms from the rear of the ball have been blown off in the opposite direction from the original projectile velocity, forming a comet-like tail that extends back through the hole and to the farthest reaches rearward in the backsplash. By tagging projectile particles, we find that the clusters farthest radially from the incident projectile axis are coagulating <u>wall</u> particles. The condensation (clustering) as the debris cloud expands and cools is an interesting phenomenon deserving of further investigation.

In general, the MD results mirror those of the hydrocodes, even though the physical scale of these atoms is many orders of magnitude smaller. That hydrodynamic (continuum) behavior extends downward to such small time and distance scales is, by now, a well-known feature of atomistic simulations[1]. When the mass per unit area of the MD debris cloud is displayed, we see that hollowing out of the debris cloud is occurring just forward of the plate. Like the first-order hydrocode calculations[6], the MD cloud is densest at the middle, and in this important respect, does not resemble the experimental results. The reason that the hydrocode fails to reproduce the experiment is that there are advection errors in the Eulerian solution, which are known to smear out features such as a shell of material; a smaller computational mesh size, or higher order treatment of the advection tends to solve this difficulty[6]. For atomistic as opposed to

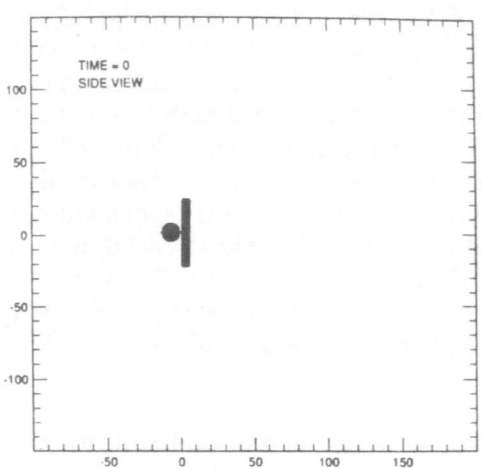

FIGURE 9. 3D hypervelocity impact: side view (xy-plane) of the initial configuration of the 683 ball atoms and 8000 wall atoms. The ball diameter is 10 σ and the wall thickness is 4.5 σ. (Coordinates on axes in Figs. 9-11 are in units of the crossing point of the pair potential, σ.)

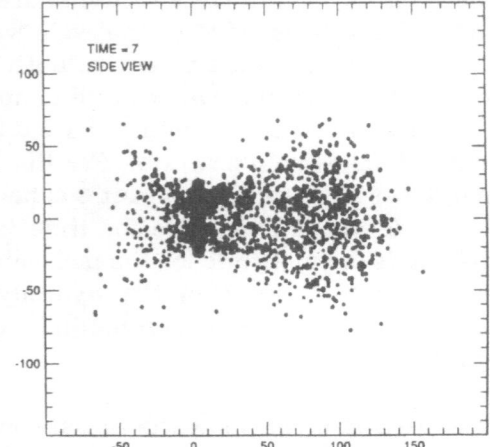

FIGURE 10. 3D hypervelocity impact: side view after t = 7 [the unit of time is $\sigma(m/\varepsilon)^{1/2}$], corresponding to 30 μs in the lead-on-lead experiment. The initial velocity of the ball was to the right at 20 $(\varepsilon/m)^{1/2}$.

continuum simulations, however, it is important to remember that there is an extra distance scale, namely, the range of the pairwise interaction potential (σ). Until an order of magnitude more ball atoms are simulated in 3D, we can expect MD calculations to appear inherently more "gluey" than an accurate continuum treatment, that is, compared to blobs of continuum, the atoms in these MD calculations tend to cluster more easily in the final stages of the debris-cloud expansion, as well as failing to fragment as easily in the early stages of the bubble formation.

We have tested this hypothesis by studying two-dimensional systems. If N atoms are required to define a macroscopic sample in 3D, $N^{2/3}$ atoms are necessary in 2D. In Fig. 11, we show a snapshot from a calculation of a debris cloud in 2D for 4698 ball atoms (equivalent to one-third of a million particles in 3D!) striking a wall of 8968 atoms at velocity $u_p = 12(\epsilon/m)^{1/2}$. In contrast to the 3D case in Fig. 10, the 2D debris cloud in Fig. 11 is hollow. We have seen similar results in 2D for 1558 ball particles - but <u>not</u> for a ball of 522 atoms - suggesting that there is a threshold in the number of atoms required

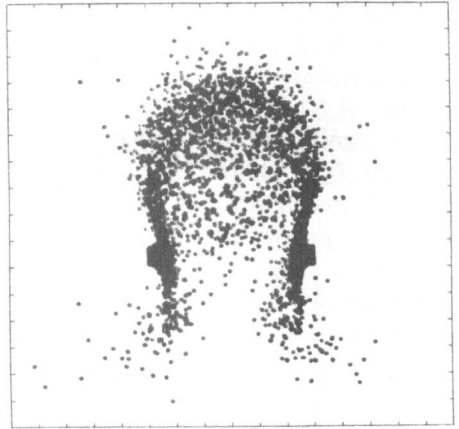

FIGURE 11. 2D hypervelocity impact after t = 30 [the unit of time is $\sigma(m/\epsilon)^{1/2}$], corresponding to 10 μs in the lead-on-lead experiment. The ball, composed of 4698 atoms (diameter 80 σ), traveling to the right at a velocity of 12 $(\epsilon/m)^{1/2}$, struck the wall (8968 atoms, width 33 σ) at t = 0 . (The scale of the window is 800 σ x 800 σ.)

in an MD simulation before a particular feature such as this looks "macroscopic." In other words, if the size of the simulated system is too small, the thickness of the wall of the debris bubble is on the order of its radius; for larger systems, the thickness of the bubble wall can be resolved. Even larger MD simulations would be required in order to resolve the details of more fine-grained features, such as the appearance of a possible double-wall structure in the experimental photograph of the lead-on-lead debris cloud[6]. Preliminary work underway suggests that many-body, local density-dependent interactions, which can describe defects more realistically than simple pairwise-additive potentials, enhance deformation in hypervelocity impacts as compared to the equivalent pair potential. (The cohesive energy, bulk modulus, and normal density can be made the same for both pair- and many-body potentials, but the vacancy-formation energy and shear moduli will be different.) The result is to make the shape of the debris cloud less spherical and, in general, to make the impact "splashier."

In conclusion, these three examples of highly complex nonequilibrium flows (shock waves, rapid expansion, and hypervelocity impacts and penetration) can be studied at the atomistic level by molecular dynamics simulations. The continuum effects make themselves apparent, even at the level of 10,000 atoms. Supercomputers have made these calculations feasible, and in the future, will enable us to study even larger systems, including plastic flow in solids.

REFERENCES

1. See for example, contributions in *Molecular-Dynamics Simulation of Statistical-Mechanical Systems*, edited by G. Ciccotti and W. G. Hoover (North-Holland, Amsterdam, 1986).
2. B. L. Holian, Phys. Rev. A 37, 2562 (1988).
3. B. L. Holian and D. E. Grady, Phys. Rev. Letters 60, 1355 (1988).
4. B. L. Holian, Phys. Rev. A 36, 3943 (1987). An x-ray photograph of a lead-on-lead hypervelocity-impact debris cloud is reproduced in Fig. 12 of this reference, as obtained from experiments performed by D. J. Liquornik at GM Delco (Santa Barbara, California) and reported by G. W. Pomykal, Lawrence Livermore Laboratory Report No. DDV-86-0010 (1986, unpublished).
5. B. L. Holian, W. G. Hoover, B. Moran, and G. K. Straub, Phys. Rev. A 22 p 2798 (1980); see also W. G. Hoover, Phys. Rev. Letters 42, 1531 (1979).
6. K. S. Holian and B. L. Holian, Int. J. Impact Engng. 8, 115 (1989).
7. D. J. Evans and B. L. Holian, J. Chem. Phys. 83, 4069 (1985); see also B. L. Holian, in Ref. 1, p. 241.
8. Monte Carlo and molecular dynamics EOS data are summarized in F. H. Ree, J. Chem. Phys. 73, 5401 (1980).
9. W. G. Hoover and W. T. Ashurst, "Nonequilibrium Molecular Dynamics" in *Theoretical Chemistry*, vol. 1 (Academic Press, New York, 1975), 1; W. G. Hoover, A. J. C. Ladd, R. B. Hickman, and B. L. Holian, Phys. Rev. A21, 1756 (1980).
10. B. L. Holian and D. J. Evans, J. Chem. Phys. 78, 5147 (1983).
11. D. E. Grady and M. E. Kipp, J. Appl. Phys. 58, 1210 (1985).
12. D. E. Grady, J. Appl. Phys. 53, 322 (1982).
13. J. A. Barker, D. Henderson, and F. F. Abraham, Physica (Utrecht) 106A, 226 (1981).

HYDRODYNAMIC FLUCTUATIONS AND THE DIRECT SIMULATION MONTE CARLO METHOD

Alejandro L. Garcia

Dept. of Physics, San Jose State Univ.

San Jose, CA 95192-0106

ABSTRACT: The use of particle simulations in the study of hydrodynamic fluctuations in nonequilibrium systems is reviewed. Some results for Rayleigh-Bénard convection measured by a Direct Simulation Monte Carlo program are presented.

I. INTRODUCTION

One of the early problems to which electronic computers were applied was the measurement of the statistical properties of fluids.[1] Computer simulations of particle dynamics are attractive since microscopic details, such as correlation functions, are available. The first molecular dynamics (MD) programs dealt with only equilibrium systems but the combination of new algorithms and advanced computer technology has expanded the field to include nonequilibrium systems. This work ranges from simple systems (constant shear or heat flux) to the more recent work in complex flows; these proceedings present a good sampling of this spectrum: Prof. Hoover shows us how to work with nonequilibrium systems of no more than three particles; on the other hand, there are papers describing van Karman vortex shedding behind an obstacle and Rayleigh-Bénard convection.

Microscopic fluctuations are often studied using Molecular dynamics (MD) simulations; the characteristic length scale for their correlations is a few atomic diameters. In experiments, these microscopic fluctuations are measured by neutron scattering. At larger length scales one enters the hydrodynamic regime where the fluctuations are observable by light scattering.[2] At equilibrium, the Landau-Lifshitz theory accurately predicts the experimentally observed spectrum. A few years ago it was realized that in a highly nonequilibrium system the hydrodynamic correlation functions would be slightly modified from their equilibrium form. Specifically, it was predicted [3] (and later observed [4]) that the Brillouin peaks are asymmetric when the fluid is subjected to a strong temperature gradient. This effect is caused by the fact that the static density-velocity correlation function, $\langle \delta\rho(r)\delta v(r') \rangle$ is nonzero in the presence of the temperature gradient. Several good reviews of this work have appeared [5,6].

Microscopic Simulations of Complex Flows
Edited by M. Mareschal, Plenum Press, New York, 1990

This paper is divided into two parts. In the first part, I review the use of particle simulations in the study of hydrodynamic fluctuations in simple nonequilibrium systems. The latter half of the paper discusses the more recent work on complex flows, specifically Rayleigh-Bénard convection. An important branch of simulation work is excluded here: the coupling of hydrodynamic and chemical fluctuations. This exciting and rapidly advancing field is discussed, at least partly, in the contributions by Michel Mareschal and Florence Baras in this volume.

Given the informal atmosphere of the meeting I decided to organize the review part of the paper around a theme: the hunt for the elusive $\langle \delta T(x,t) \delta T(x',t) \rangle$ correlation. Since this static correlation is not readily accessible experimentally, it has been primarily studied by computer simulation. I have purposely made this a personal account; putting in some background behind the work and including details not found in the original papers. It has been my privilege and pleasure to know many of the people who have worked on these computer simulations; I only hope that the reader finds the style of the presentation more interesting than distracting.

II THE HUNT FOR THE ELUSIVE $\langle \Delta T(X) \Delta T(X') \rangle$ CORRELATION

a) Master Equation models

In the early 80's, Prigogine's group began studying thermo-chemical problems (such as combustion) using the Master Equation formalism. Gregoir Nicolis and Malek Mansour introduced a simple way of deriving a Master equation for the one-dimensional thermal conduction problem.[7] The corresponding Langevin equation is derived using only the properties that a) in the deterministic limit it reduces to the Fourier law and b) that the transition rate between states obeys detailed balance at equilibrium. They obtained the following interesting result; for a fluid contained between thermal plates at x=0 and x=L, the static correlation of temperature fluctuations has the form

$$\langle \delta T(x) \, \delta T(x') \rangle = \frac{k_B T_0^2}{C_V} \delta(x-x') + \{ \delta T(x) \, \delta T(x') \} \tag{1}$$

where

$$\{ \delta T(x) \, \delta T(x') \} \equiv \frac{k_B \gamma^2}{C_V \, L} x \, (L-x') \tag{2}$$

and $x \leq x'$; C_V is the specific heat per unit length, γ is the imposed temperature gradient ($\gamma \equiv (\Delta T/L)$) and k_B is Boltzmann's constant. The first term on the r.h.s. is the equilibrium contribution to the temperature fluctuations modified by the fact that the average temperature, $T_0(x)$, is a function of location. The term $\{ \delta T(x) \, \delta T(x') \}$ is the nonequilibrium contribution to the correlation function; this term is illustrated in Figure 1. Note that this nonequilibrium contribution is long-ranged (linear) and proportional to the square of the imposed temperature gradient $\gamma \equiv \Delta T/L$.

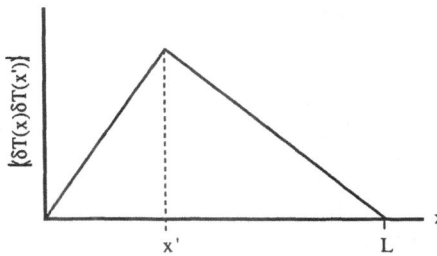

FIGURE 1. Sketch of $\langle \delta T(x) \delta T(x') \rangle$ as defined in equation (2).

I was still a graduate student at the University of Texas when Prof. Nicolis showed me a preprint of that paper. He asked me if it would be possible to measure $\langle \delta T(x) \delta T(x') \rangle$ using a Monte Carlo simulation of the Master equation presented in the paper.[8] Unfortunately, as they point out in the paper, the transition rate in their model has some unphysical properties. The difficulty arises from the approximation that the thermal diffusivity is strictly constant; only if one includes the (weak) dependance it has on temperature is a proper transition rate possible.

I was, however, already familiar with another model which did not have this problem. Nicolis, Baras and Malek-Mansour [9] had derived the Master equation for a dilute gas when the transport is Knudsen flow between cells. (Knudsen flow occurs when two containers are connected by an aperture which is only a few mean free paths in diameter) They derived an expression for the fluctuations in a single cell connected to two reservoirs at different temperatures; I had already confirmed this result by a simple Monte Carlo simulation. Using a similar program with a chain of cells connected by Knudsen apertures, I observed the long-range, linear correlation of temperature fluctuations predicted by equation (1).[10] While this result was encouraging, the Knudsen system was only a curious but unrealistic model.

b) Molecular Dynamics

About the same time that I was getting these results on the Knudsen system, Michel Mareschal and Eddie Kestemont were visiting Texas for a few months. They came to study the same problem by a different approach, using a molecular dynamics simulation of 3000 hard disks under a strong temperature gradient. Working first on our VAX and then on the Cyber, they made runs of some 2 million collisions each for various temperature gradients. The system was only some 220 molecular diameters in length and the temperature gradients were very large ($g \approx 10^8$ K/cm). Measurements of the moments of the local velocity distribution, however, were in very good agreement with a Gaussian distribution; this shows that local thermal equilibrium is maintained even under such extreme nonequilibrium conditions [11].

Mareschal and Kestemont measured $\langle \delta T(x) \, \delta T(x') \rangle$ and found, in agreement with (1), that the nonequilibrium temperature fluctuations were proportional to the square of the imposed temperature gradient [12]. Unfortunately, they did not have enough statistics to accurately measure the spatial dependence of the correlations; i.e. they did not reproduce Figure 1. The problem is complicated by the slow relaxation of the lowest order modes in the system. One thing was clear: observation of these subtle effects required long run times. In my Knudsen flow model, I needed to run for over 10^8 events to get the correlation function to about 10% error. A few years later, Lar Hannon would run another MD program and attempt to measure $\langle \delta T(x) \, \delta T(x') \rangle$ only to find that even the supercomputer resources at IBM Kingston were insufficient [13].

c) Direct Simulation Monte Carlo

While finishing my dissertation, I was trying to find a realistic system for which I could hope to observe $\langle \delta T(x) | \delta T(x') \rangle$. By chance, my thesis advisor, Jack Turner, was serving as a consultant on an aerospace project involving the evaporation of a solid into vacuum. The problem was being studied using an algorithm called Direct Simulation Monte Carlo (DSMC) method. This simulation was introduced by G.A. Bird in the early 70's and it is widely used in rarefied gas dynamics [14]. Turner showed me the problem; lent me a copy of Bird's book and asked me to look it over.

After learning the algorithm, I realized that the evaluation of collisions is very similar to the Master Equation formulation of Kac for a homogeneous gas [15]. However, the motion of the particles is computed deterministically from their positions and velocities. The two processes are combined by "splitting": at each time step the particles are moved and a few undergo collisions. If the timestep is sufficiently small the DSMC correctly models a dilute gas (see Prof. Bird's contribution in these proceedings). One of the main advantages of DSMC is that it runs over 100 times faster than comparable MD codes. When I came to Brussels in early 1985, I suggested to Malek Mansour and Michel Mareschal that we try using Bird's method to measure the nonequilibrium temperature fluctuations.

The initial results from the simulation were very encouraging.[16] The DSMC reproduced the equilibrium fluctuations perfectly, including the finite size corrections. In the nonequilibrium system, the peak of the measured density-velocity static correlation function was linearly proportional to the temperature gradient, in agreement with theory and light scattering experiment. [5] Finally, the measured $\langle \delta T(x) | \delta T(x') \rangle$ was in good agreement with Figure 1 although the error bars were still unsatisfactorily large.

d) Landau-Lifshitz theory

While the Cyber labored away, we began working on the theory using fluctuating hydrodynamics. The fluctuating Fourier equation is obtained from the Landau-Lifshitz theory when one assumes that the temperature fluctuations are uncoupled from the density and velocity fluctuations. The solution of this equation for the temperature gradient problem also gives (1) [17,18]. For a dilute gas, however, the equations for density, velocity and temperature are coupled making the problem far more complicated. However, we expected that the results for $\{\delta T(x)\,|\delta T(x')\}$ would be qualitatively similar.

While much theoretical work had been done on the temperature gradient problem, most was motivated by light scattering experiments in liquids. Our computer simulation differed significantly from these experiments in several ways. (1) The system was extremely small (10 mean free paths) so finite size effects were very important. (2) To get an observable effect we used an extremely large temperature gradient. Since our medium was a dilute gas the density variation was also large. (3) We could measure all hydrodynamic quantities while light scattering only probed the density-density time correlation function.

After various attempts to analytically solve the Landau-Lifshitz equations for a dilute gas, Malek Mansour hit on solving them numerically. Our first effort was to numerically integrate the fluctuating Fourier equation since we knew the exact solution. Discretizing in space, the partial differential equation reduces to a set of ordinary stochastic differential equations.[19] I remember that we had the simulation of this Langevin equation running in a few days and good results after about a week. Euphorically confident, we then tried to write a similar Langevin simulation for the dilute gas equations and quickly hit an impasse. It was not so easy to properly handle the boundary conditions; specifically, there could be no boundary condition on the density.

The second attempt at numerically solving the fluctuating hydrodynamic equations involved using the static correlation equations. For a Langevin equation of the form,

$$\frac{dc_i}{dt} = f_i(c_1,...,c_n) + F_i(t) \tag{3}$$

where $F_i(t)$ is a white noise with variance,

$$<F_i(t)\,F_j(t')> = Q_{ij}\,d(t - t') \tag{4}$$

then

$$\frac{d}{dt} <c_i(t)\,c_j(t)> = <f_i(c_1,...,c_n)\,c_j(t)> + \frac{1}{2}\,Q_{ij} \tag{5}$$

Applying this identity to the linearized fluctuating hydrodynamic equations yields a coupled set of equations for the static correlations.[19,20] Malek noticed that the equation for $\{\delta T(x)\,|\delta T(x')\}$ is closed without having to specify any boundary conditions for the density. After discretizing in space, the problem reduces to solving a set of simultaneous linear equations.

In early 86, Malek and I went to upstate New York as invited scientists. It was an exciting time to be in IBM Kingston; Lar Hannon had recently found vortex formation in the flow behind an obstacle and Dennis Rapaport was beginning to work with him on this problem [21]. Using the lCAP supercomputer, we could make new DSMC simulations using much larger systems (50 mean free paths between the thermal plates).

As mentioned above, our data from smaller systems (10 mean free paths) gave temperature correlations as in Figure 1. The larger system showed richer behavior; the correlations took the form shown in Figure 2. The agreement with the numerical solution of the correlation equations was excellent.[20] In fact, we first obtained Figure 2 from the correlation equations and it was so unexpected we spent a long time trying to find the bug in the program. Only later, when the DSMC simulation gave the same result did we realize that we were really observing multimodal behavior.

While in Kingston, Malek resolved the problem of how to numerically solve the full equations without specifying boundary conditions for the density. This difficulty with the boundary conditions is overcome by using a half-grid formulation. The density is specified on grid points which lie between the grid points for the velocity and temperature. The density grid contains only interior points (no points on the boundary) so no boundary conditions are needed for density.

The hydrodynamic correlation functions in the temperature gradient problem are now well known. The Couette flow problem (constant shear) has received similar attention and again, the results from DSMC simulations agree very well with fluctuating hydrodynamic theory.[22] A recent application of these results has been the testing and validation of Cellular Automata (CA) simulations. Chopard and Droz developed a two-speed CA model and measured the hydrodynamic fluctuations in the temperature gradient problem. Unfortunately, their preliminary results are inconclusive.[23]

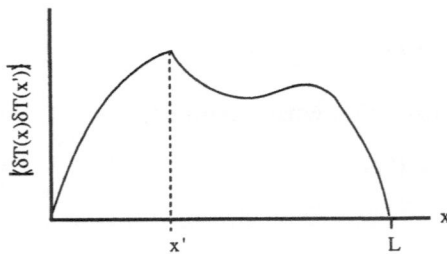

FIGURE 2. Schematic picture of $\langle \delta T(x) \delta T(x') \rangle$ as observed in larger systems.

III. RAYLEIGH-BENARD CONVECTION

Recent work has shown that particle simulations may be used to study complex flow problems. Rayleigh-Bénard convection is a paradigm instability; at a critical Rayleigh number there is a bifurcation between the states of purely conductive heat flow and buoyancy-driven convection. [24] The nature of the hydrodynamic fluctuations near this transition point has been studied theoretically by a variety of methods.[25] Ahler's group has performed several careful experiments and have measured the variation in the heat flux near the onset of convection.[26] Quantitative comparison, however, between fluctuating hydrodynamics calculations and laboratory experiments reveal significant, unaccountable discrepancies.[27] Furthermore, many theoretical predictions for the fluctuations near the convective threshold remain untested due to experimental difficulties.

Mareschal and Kestemont showed that it was possible to observe the Rayleigh-Bénard instability using Molecular Dynamics [28]; this work has been duplicated by other groups [29,30]. The observed density, velocity and temperature fields agree very well with those predicted by the standard Navier-Stokes equations.[31,32]

The DSMC method may also be used to study this problem but only at the expense of working with large systems. The dimensionless number characterizing the instability is the Rayleigh number,

$$R = \frac{\alpha \gamma g L^4}{\nu \kappa} \tag{6}$$

where g is the gravitational field, L is the depth of the system, $\gamma \equiv \Delta T / \Delta z$ is the uniform temperature gradient, $\alpha \equiv 1/\rho \, [\partial \rho / \partial T]_P$, is the coefficient of volume expansion, κ is the thermometric conductivity and ν is the kinematic viscosity.

The critical Rayleigh number depends on the boundary conditions at the walls; in the limit of large aspect ratio the critical Rayleigh number is 1708 for no-slip boundaries and 658 for slip boundaries. In the simulation described below, I used slip boundaries but the aspect ratio was unity raising the critical Rayleigh number to about 780.

For a dilute gas, the density profile is a function of the temperature profile as, $\rho \propto T^{-a}$, where $a = 1 - mg/k_B\gamma$, m is the particle mass. Taking the value of the gravitational field as $g = k_B\gamma/m$ the density is approximately constant throughout the system. The thermometric expansivity, thermal conductivity and kinematic viscosity may be written as

$$\alpha = 1/T \quad ; \quad \nu = 2/5 \, \kappa \tag{7}$$

$$\nu = \frac{10}{32} l \sqrt{2\pi k_B T/m} \tag{8}$$

where λ is the mean free path. From the above

$$R = \frac{256}{125\pi} (\Delta T/T)^2 \, (L/\lambda)^2 \cong 0.652 \, (\Delta T/T)^2 \, (L/\lambda)^2 \tag{9}$$

Even with an extremely strong temperature gradient $\Delta T/T$ will be of order one; to achieve the critical Rayleigh number one needs a system with a length of about 35 λ (for slip boundaries and an aspect ratio of one). Because the DSMC method needs about 10 particles per computational cell and that the cells be no larger than about a mean free path, we need to use over 12,000 particles. This is to be compared with Molecular Dynamics where convection may be observed in a system as small as 1500 particles.[32]

I ran a large DSMC simulation with 50,000 particles in a square box $50\lambda \times$ $50\lambda \times 1\lambda$ in size (i.e. the aspect ratio equals one). The sidewalls are slip, insulating walls; a particle striking them rebounds elastically. The top and bottom walls are semi-slip, thermal walls; a particle striking them is thermalized in the directions perpendicular to the convective flow (y and z directions) while its velocity in the x-direction is unchanged. Similar boundary conditions were used by Mareschal and Kestemont in their MD simulations of Rayleigh-Bénard.[28]

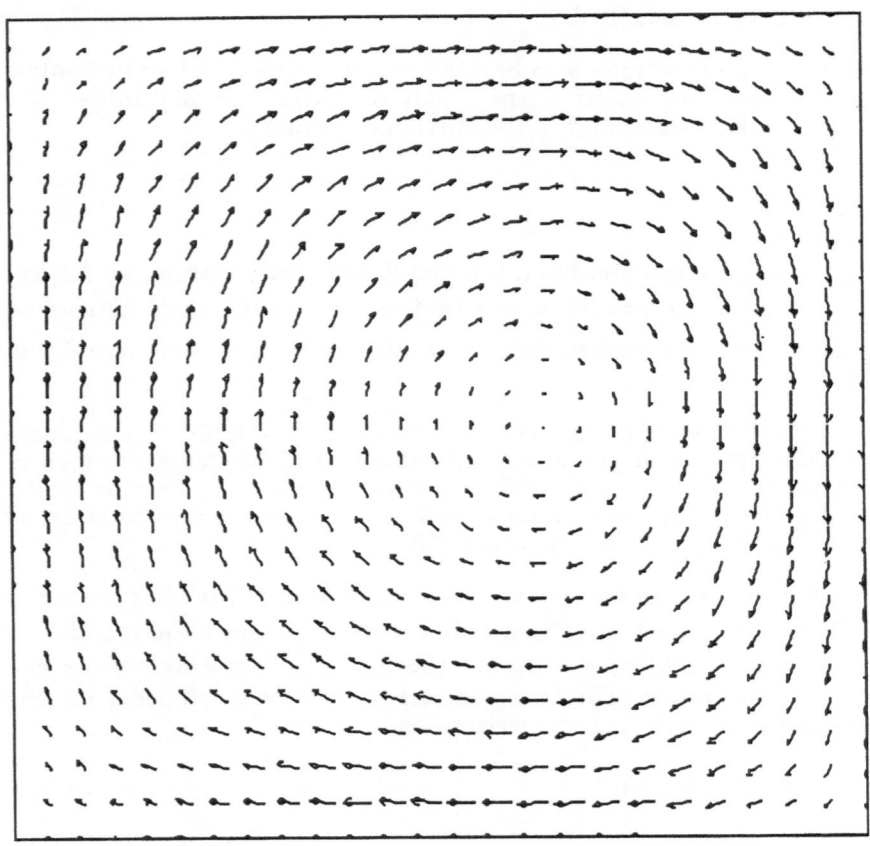

FIGURE 3. Velocity field from the DSMC simulation of the Rayleigh-Bénard problem. See the text for the parameters used in the simulation.

The top and bottom temperatures are 0.5 and 2.0, respectively [33]. Because of the temperature jump at the boundary, the effective boundary temperatures are .636 and 1.874. The gravitational field is g = 0.1; as mentioned above, the imposed gravitational field is chosen to maintain the density approximately constant. The Rayleigh number is approximately 1300, almost twice the critical Rayleigh number. The system was run for about 200 million collisions; a noticeable roll developed after about 40 million collisions. After about 150 million collisions the system reached a steady state and statistics were accumulated over the last 40 million collisions. On a SUN 4/260 the program processed about 2.2 million collisions per CPU hour.

The observed average flow field is illustrated in Figure 3. Note that the roll is not symmetric since the density is lower in the hot, rising fluid and higher in the cold, falling fluid. Conservation of mass requires that the hot stream be wider than the cold stream. [34] The fluid is highly non-Bousinessq; this can be seen from the fact that the contours of constant density (Figure 4) look very different from the contours of constant temperature (Figure 5).

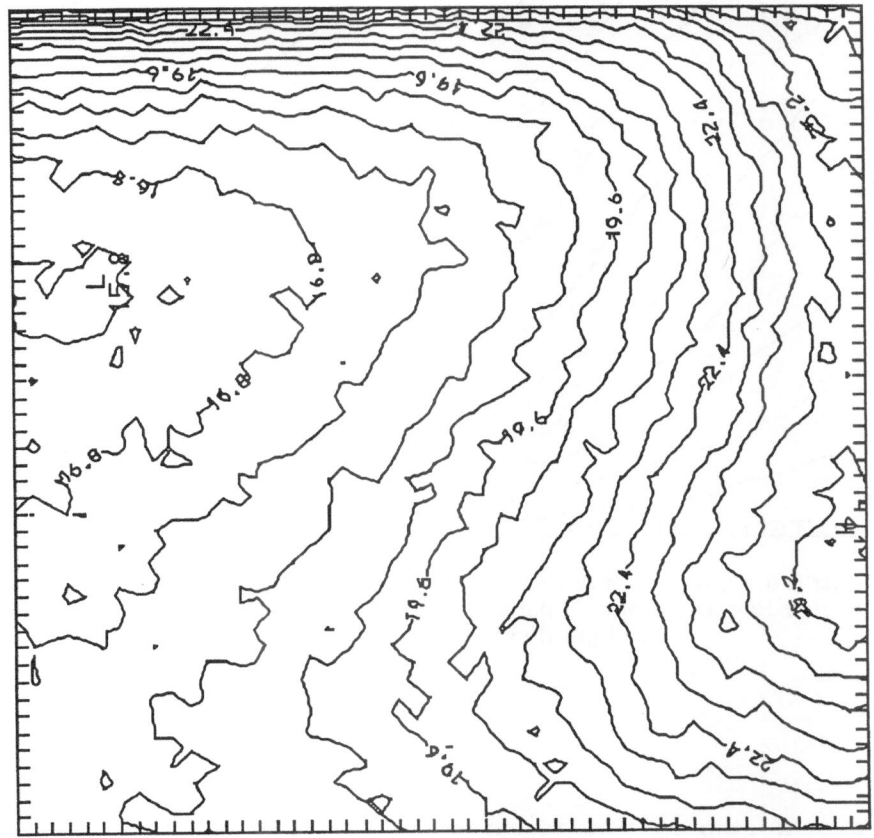

FIGURE 4. Contour plot of the density field from the DSMC simulation of the Rayleigh-Bénard problem. Compare with Figure 5; note that the isotherms are not parallel to the isopycnal lines.

The full Navier-Stokes equations for a dilute gas were solved numerically and the resulting solutions agree closely with the average density, velocity and temperature fields measured in the DSMC simulation. My earlier DSMC runs showed only mediocre agreement with the Navier-Stokes integrator because: 1) I was using half as many particles (only 10 per cubic mean free path) and 2) I was only thermalizing the velocity in the direction perpendicular to the wall. This led to a considerable temperature jump at the wall. A complete discussion of these results will appear elsewhere.

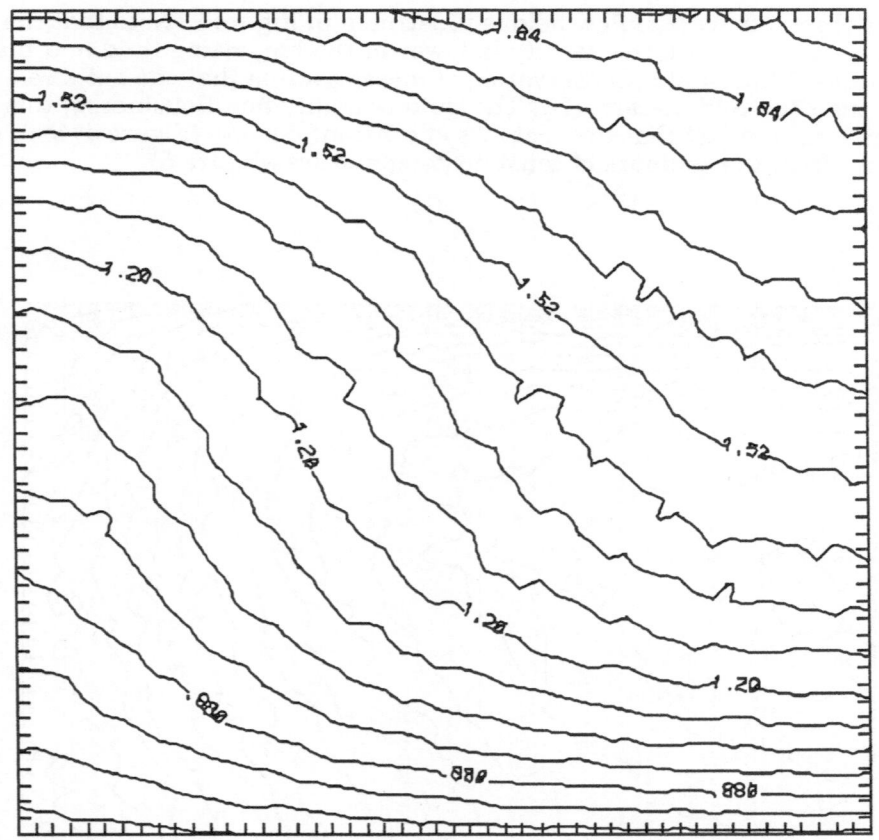

FIGURE 5. Contour plot of the temperature field from the DSMC simulation of the Rayleigh-Bénard problem. Compare with Figure 4; note that the isotherms are not parallel to the isopycnal lines.

The next step in this research is the study of the hydrodynamic fluctuations in the Rayleigh-Bénard problem. Due to the considerable computational effort involved in the simulations it is difficult to get reliable results to compare with theoretical predictions. Towards this end, Cecile Penland and I are beginning to use some of the sophisticated data analysis methods developed in climatology.

ACKNOWLEDGEMENTS

It was my pleasure to collaborate with Michel Mareschal, Malek Mansour, Florence Baras, John William Turner, Andreas Puhl and Eddie Kestemont on the work present in this paper. I wish to thank Prof. Prigogine for his encouragement and insightful suggestions. Special thanks to Malek Mansour for the use of his Navier-Stokes integrator and for his unsolicited remarks whenever my programs wouldn't run.

REFERENCES

1. Molecular Dynamics simulation of Statistical Mechanical Systems, eds. G. Ciccotti and W.G. Hoover, Enrico Fermi Summer School of Physics Series, Vol. 97 (1987).

2. B.J. BERNE and R. PECORA, Dynamic Light Scattering, Wiley, New York (1976).

3. I. PROCACCIA, D. RONIS and I. OPPENHEIM, Phys. Rev. Lett. **42** 287 (1979); T. R. KIRKPATRICK, E. G. D. COHEN and J. R. DORFMAN, Phys. Rev. Lett. **44** 472 (1980).

4. D. BEYSENS, Y. GARRABOS and G. ZALCZER, Phys. Rev. Lett. **45** 403 (1980).

5. A. M. TREMBLAY, in Recent Developments in Nonequilibrium Thermodynamics, J. Casas-Vasquez, D. Jou and G. Lebon Eds., Springer Verlag, Berlin (1984).

6. R. SCHMIDT, Phys. Reports **171** #1 (1988).

7. G. NICOLIS and M. MALEK MANSOUR, Phys. Rev. A **29** 2845 (1984)

8. In this paper the term Monte Carlo does not refer to the Metropolis Monte Carlo algorithm used to study equilibrium ensembles. The term is used for any simulation with a stochastic element in its dynamics.

9. G. NICOLIS, F. BARAS and M. MALEK MANSOUR, in Nonlinear Phenomena in Chemical Dynamics, A. PACAULT and C. VIDAL Eds., Spinger-Verlag, Berlin (1981).

10. A. GARCIA, Thesis, The Univ. of Texas at Austin (1984); A. GARCIA, Phys. Lett. **119** 379 (1987).

11. A. TENEBAUM, G. CICCOTI and R. GALLIO, Phys. Rev. A **25** 2778 (1982).

12. M. MARESCHAL and E. KESTEMONT, Phys. Rev. A **30** 1158 (1984).

13. L.Hannon, private communication.

14. G.A. BIRD, Molecular Gas Dynamics, Claredon Press, Oxford (1976).

15. M. KAC, in Probability Theory and Related Topics in Physical Science, Wiley-Interscience, New York (1959).

16. A. GARCIA, Phys. Rev. A **34** 1454 (1986).

17. M. MARESCHAL, in Nonlinear Phenomena in Chemical Dynamics, A. PACAULT and C. VIDAL Eds., Spinger-Verlag, Berlin (1981).

18. A. DIAZ-GUILERA and J. M. RUBI, Phys. Rev. A **34** 462 (1986).

19. A. GARCIA, M. MALEK MANSOUR, G. C. LIE, and E. CLEMENTI, J. Stat. Phys. **47** 209 (1987).

20. M. MALEK MANSOUR, A. GARCIA, G. LIE and E. CLEMENTI, Phys. Rev. Lett., **58** 874 (1987).

21. D. C. RAPAPORT and E. CLEMENTI, Phys. Rev. Lett. **57** 695 (1987); L. HANNON, G. C. LIE and E. CLEMENTI, J. Stat. Phys. **51** 965 (1988).

22. A. GARCIA, M. MALEK MANSOUR, G. C. LIE, M. MARESCHAL and E. CLEMENTI, Phys. Rev. A **36** 4348 (1987).

23. B. CHOPARD and M. DROZ, Helv. Phys. Acta, **61** 893 (1988).

24. S. CHANDRASEKHAR, Hydrodynamic and Hydromagnetic Stability, Dover Press, New York (1981).

25. V. M. ZAITSEV and M. I. SHLIOMIS, Sov. Phys. JETP **32** 866 (1971); H. N. W. LEKKERKERKER and J. P. BOON, Phys. Rev. A **10** 1355 (1974); T. R. KIRKPATRICK and E. G. D. COHEN, J. Stat. Phys. **33** 639 (1983); R. SCHMITZ and E. G. D. COHEN, J. Stat. Phys. **38** 285 (1985).

26. R. P. BEHRINGER and G. AHLERS, J. Fluid Mech. **125** 219 (1982); G. AHLERS, M. C. CROSS, P. C. HOHENBERG and S. SAFRAN, J. Fluid Mech. **110** 297 (1981); C. W MEYER, G. AHLERS and D. S. CANNELL, Phys. Rev. Lett. **59** 1577 (1987); G. AHLERS, C. MEYER and D. CANNELL, J. Stat. Phys. **54** 1121 (1989).

27. H. Van BEIJEREN and E. G. D. COHEN, Phys. Rev. Lett. **60** 1208 (1988); ibid, J. Stat. Phys. **53** 77 (1988).

28. M. MARESCHAL and E. KESTEMONT, Nature **323** 427 (1986); ibid, J. Stat. Phys. **48** 1187 (1987) and their contribution in these proceedings.

29. D. C. RAPAPORT, Phys. Rev. Lett. **60** 2480 (1988) and his contribution in these proceedings.

30. J. A. GIVEN and E. CLEMENTI, J. Chem. Phys. **90** 7376 (1989).

31. M. MARESCHAL, M. MALEK-MANSOUR, A. PUHL and E. KESTEMONT, Phys. Rev. Lett. **61** 2550 (1988).

32. A. PUHL, M. MALEK-MANSOUR and M. MARESCHAL, Phys. Rev. A **40** 1999 (1989).

33. The temperature is scaled such that at temperature T=1, the most probable molecular speed equals one. See reference 14 for more details.

34. I wish to thank Malek Mansour for this comment.

A SIMPLE MODEL OF HYDRODYNAMIC INSTABILITY

M. Malek Mansour

Faculté des Sciences, Université Libre de Bruxelles

Campus Plaine, C. P. 231, 1050 Brussels / Belgium

1   INTRODUCTION

For the theoretical physicist, molecular dynamics (MD) is a sort of ideal laboratory for testing a theory, the validity of some approximation scheme, etc... A nice aspect of this procedure is its generality: By setting up a microscopic model, with given interaction rules, and by specifying the nature of the boundary conditions and external forces, an MD experiment can be performed which, without corresponding closely to a real world system, captures nevertheless some essential aspects of the phenomenon of interest. For instance, hard sphere or hard disk fluids have been used since the early development of MD[1]. Another example is the so called "Lorentz gas", where the trajectory of a single molecule around randomly distributed scatterers is studied through MD[2].

With the ever growing power of super computers, much attention has been focused recently on non-equilibrium systems featuring complex behavior, such as shock waves[3], flow past an obstacle[4], and hydrodynamic instabilities[5,6,7]. The main purpose here is to shed some light on the onset of such complex behavior, for which laboratory experiments are often difficult to perform. One example is the behavior of fluctuations and correlation functions near an instability threshold. In this respect, recent laboratory experiments for the case of Bénard instability have pointed out a discrepancy of four orders of magnitude between the data and the existing theories based on fluctuating hydrodynamics[8,9]. The Bénard problem is one of the prototypes of hydrodynamic instabilities and, since the early work of Zaitsev and Shliomis[10] in 1971, many theoretical works have been devoted to this problem[11,12,13]. All these theories are based on approximation procedures which, although intuitively sound, are difficult to justify from an experimental point of view. For instance, in the case of Bénard problem reported in ref. (8), it is difficult to assert whether it is fluctuating hydrodynamics that fails near the convective instability, or the inevitable approximation schemes underlying the computations.

The main difficulty with the Bénard problem comes from the coupling of the viscous mode to the heat mode. This coupling is crucial for the onset of convective instability and can in no way be neglected. On the other hand, in the well known Taylor problem[14] the above coupling does not interfere with the onset of instability and can indeed be neglected. Here, a fluid is confined between two rotating cylinders. As the angular velocity of the inner cylinder increases, the pressure gradient toward the outer cylinder becomes more and more important, as a consequence of the centrifugal force. If this angular velocity exceeds some well defined threshold, the fluid undergoes an instability and regular cellular flow patterns between cylinders emerge. Unfortunately, owing to the cylindrical geometry, the mathematical treatment of this system is even more involved than the Bénard problem. It would therefore seem that there exists no real system leading to a hydrodynamical instability for which the mathematical treatment remains simple. Now, in the spirit of MD, one can always design a working-model which can be realized through computer experiments. The purpose of this short communication is to present one such simple model. In section 2, we describe its main features and sketch some simplifying hypotheses. The linear stability analysis is presented in section 3 and some remarks concerning the behavior of the fluctuations are given in section 4.

2. THE MODEL

As we already pointed out, the main difficulty with the Taylor problem is its cylindrical geometry. This difficulty can be avoided by considering a Cartesian geometry and adding a fictitious force mimicking the centrifugal force in the original Taylor problem.

Consider a fluid confined in a three-dimensional rectangular box, as depicted in fig. 1. The plate located at the plane ZOX is assumed to move with a given velocity, V_0, along the X axis while the plate parallel to it and located at $Y = L_y$, is held fixed. Assume now that an external force is acting on each particle whose magnitude is a given increasing function of the X component of the particle velocity and is oriented in Y direction. Clearly, this model mimics the physics of the Taylor problem. Indeed, as we increase V_0, the fictitious force increases as well and imposes a larger pressure gradient to the fluid. It is then expected that if this pressure gradient exceeds some threshold, the fluid undergoes an instability and regular flow patterns (convective rolls) between the plates perpendicular to the Y direction emerge. We show below that this is indeed the case, by performing a quantitative analysis.

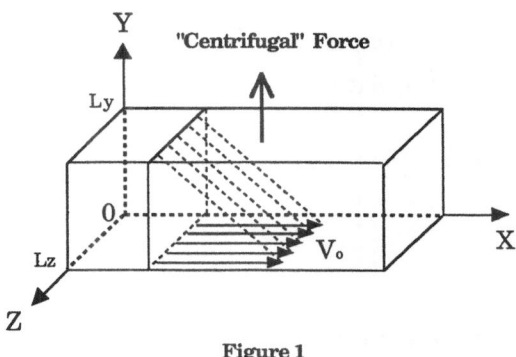

Figure 1

The hydrodynamic equations for our model read:

$$\partial_t \rho = - \vec{\nabla}. \rho \vec{V} \tag{1}$$

$$\rho \, \partial_t \vec{V} = - \rho \, (\vec{V}.\vec{\nabla}) \, \vec{V} - \vec{\nabla} P + \eta \nabla^2 \vec{V}$$

$$+ (\zeta + \frac{\eta}{3}) \, \vec{\nabla}(\vec{\nabla}.\vec{V}) + \rho \, f(V_x^2) \, \vec{1}_y \tag{2}$$

where ρ is the mass density, P the pressure, ζ and η the bulk and shear viscosity respectively and $f(V_x^2)$ the acceleration field corresponding to the imposed external force. As already mentioned, the coupling with the heat equation is of no importance for the kind of instability we are dealing here. We therefore choose the following equation of state :

$$\vec{\nabla} P = C^2 \vec{\nabla} \rho \tag{3}$$

where C is the (isothermal) sound velocity. The boundary conditions are chosen as follows : The system is periodic in X direction. The box is assumed to be rigid in Y and Z directions (constant volume), i.e., the components of the velocity field perpendicular to theses walls are set equal to zero:

$$V_y(y=0,L_y) = V_z(z=0,L_z) = 0$$

(4.a)

The planes perpendicular to the Z direction are perfectly slip :

$$\frac{\partial}{\partial z}V_x(z=0,L_z) = \frac{\partial}{\partial z}V_y(z=0,L_z) = 0$$

(4.b)

The planes perpendicular to the Y direction are perfectly stick in the X direction, and perfectly slip in the Z direction :

$$V_x(y=0) = V_0 \ ; \ V_x(y=L_y) = 0$$

(5.a)

$$\frac{\partial}{\partial y}V_z(y=0,L_y) = 0$$

(5.b)

Note that the latter boundary conditions cannot be realized in laboratory experiments, but can be easily set up in a molecular dynamics experiment. As we shall see, they greatly simplify the mathematical analysis of the problem.

The above boundary conditions, supplemented by an initial condition for the density and the velocity field, entirely determine the solution of the problem. We note that no boundary condition can be imposed on the density as its values near the walls are defined by the internal dynamic of the system.

3.   LINEAR STABILITY ANALYSIS

We first consider the stationary solution of the equations (1) and (2). Taking into account the symmetry of the problem, one easily finds:

$$\vec{V}_s = U_0(y) \ \vec{1}_x$$

(6)

$$\rho_s = \rho_s(y)$$

(7)

with

$$U_0(y) = V_0 \left[1 - \frac{y}{Ly} \right]$$

(8)

and

$$\frac{1}{\rho_s}\frac{\partial \rho_s}{\partial y} = \frac{1}{C^2} \ f(V_x^2) \approx O\left(\frac{V_0^2}{C^2}\right).$$

(9)

Next, we consider the linearised hydrodynamical equations about the above stationary state. Setting

$$\rho = \rho_s \left(1 + \delta\rho\right) \tag{10}$$

$$\vec{V} = \vec{V}_s + \vec{\delta V} , \tag{11}$$

and using eq. (9), one easily finds :

$$\partial_t \, \delta\rho = - \, \vec{\nabla} . \, \vec{\delta V} - \frac{\delta V_y}{\rho_s} \frac{\partial \rho_s}{\partial y} \tag{12}$$

$$\partial_t \, \vec{\delta V} = \gamma \, \delta V_y \, \vec{1}_x - U_o(y) \frac{\partial}{\partial x} \, \vec{\delta V} - C^2 \vec{\nabla} \, \delta\rho + \frac{\eta}{\rho_s} \, \nabla^2 \, \vec{\delta V}$$

$$+ \frac{(\zeta + \eta/3)}{\rho_s} \vec{\nabla}(\vec{\nabla}.\vec{\delta V}) + \left(\frac{\partial f}{\partial V_x}\right)_s \, \delta V_x \, \vec{1}_y \tag{13}$$

where γ represents the absolute value of the velocity gradient :

$$\gamma = - \frac{\partial}{\partial y} U_o(y) \geq 0 \tag{14}$$

The analysis of these equations in their general form is quite difficult, mainly because the coefficients are space dependent. To simplify the problem, we first take the space average over the X direction, i.e. we consider only "reduced" hydrodynamical variables defined as :

$$\delta h(y,z) = \frac{1}{L_x} \int_0^{L_x} dx \; \delta h(x,y,z) \tag{15}$$

where δh stands for $\delta\rho$ or $\vec{\delta V}$. Note that this is equivalent to considering axisymmetric perturbations in the original Taylor problem[14]. Furthermore, we limit ourselves to values of $V_0 \ll C$, so that the space variation of the stationary state density can be neglected (see eq. 9) :

$$\rho_s = \rho_0 = \text{Constant} \tag{16}$$

Finally, we choose the following form for $f(V_x^2)$:

$$f(V_x^2) = \frac{b}{2} \frac{V_x^2}{U_o(y)} \, \gamma \tag{17}$$

where b is a dimensionless constant which can be used as a control parameter. The main advantage of this choice comes from the fact that it leads to a constant term in the linearised equations, i.e.

$$\left(\frac{\partial f}{\partial V_x}\right)_s = b\,\gamma \tag{18}$$

Under the above considerations, it can be easily checked that the hydrodynamical variables can be expanded in the following basis:

$$\begin{pmatrix} \delta V_x \\ \delta V_y \end{pmatrix} = \sum_{k_y=0}^{\infty} \sum_{k_z=0}^{\infty} \sin\left(\frac{k_y\,\pi}{L_y}y\right) \cos\left(\frac{k_z\,\pi}{L_z}z\right) \begin{pmatrix} \delta u_{\mathbf{q}} \\ \delta v_{\mathbf{q}} \end{pmatrix}$$

$$\delta V_z = \sum_{k_y=0}^{\infty} \sum_{k_z=0}^{\infty} \cos\left(\frac{k_y\,\pi}{L_y}y\right) \sin\left(\frac{k_z\,\pi}{L_z}z\right) \delta w_{\mathbf{q}}$$

$$\delta\rho = \sum_{k_y=0}^{\infty} \sum_{k_z=0}^{\infty} \cos\left(\frac{k_y\,\pi}{L_y}y\right) \cos\left(\frac{k_z\,\pi}{L_z}z\right) \delta\rho_{\mathbf{q}}$$

$$\tag{19}$$

with

$$\frac{\partial}{\partial t}\begin{pmatrix} \delta\rho_{\mathbf{q}} \\ \delta u_{\mathbf{q}} \\ \delta v_{\mathbf{q}} \\ \delta w_{\mathbf{q}} \end{pmatrix} = \begin{pmatrix} 0 & 0 & -q_y & -q_z \\ 0 & -\nu_1 q^2 & \gamma & 0 \\ C^2 q_y & \gamma & -\nu_1 q^2 - \nu_2 q_y^2 & -\nu_2 q_z q_y \\ C^2 q_z & 0 & -\nu_2 q_z q_y & -\nu_1 q^2 - \nu_2 q_z^2 \end{pmatrix} \begin{pmatrix} \delta\rho_{\mathbf{q}} \\ \delta u_{\mathbf{q}} \\ \delta v_{\mathbf{q}} \\ \delta w_{\mathbf{q}} \end{pmatrix}$$

$$\tag{20}$$

where we have set

$$\nu_1 = \frac{\eta}{\rho_0}\ , \quad \nu_2 = \frac{(\zeta + \eta/3)}{\rho_0}\ , \quad b = 1 \tag{21}$$

and

$$q_y = \frac{k_y \pi}{L_y} \quad , \quad q_z = \frac{k_z \pi}{L_z} \quad , \quad q^2 = q_y^2 + q_z^2 .$$

(22)

To solve the perturbation equation (20), we have to compute the eigenvalues of a 4x4 matrix. Although in principle this can be done exactly, the results appear to be quite awkward. Nevertheless, one can easily check that there exists a particular value of the velocity gradient, γ_q, for which the real part of one of the eigenvalues vanishes. This condition is given by

$$\gamma_q = \nu_1 \frac{q^3}{q_z}$$

(23)

which is reminiscent of the stability condition for the Bénard problem. If $\gamma > \gamma_q$, then the mode corresponding to the wave-number q becomes unstable. As the velocity gradient increases, the first instability occurs for $q = q_c$, which corresponds to a pair of integers (k_y^c, k_z^c) closest to the minimum of γ_q. The marginal stability values of the velocity gradient are obtained for $k_y^c = 1$ and depend crucially on the aspect ration $A_r = Lz / Ly$. They are given by :

$$\gamma(k_z) = \nu_1 \left(\frac{\pi}{L_y}\right)^2 \frac{\left[(k_z/A_r)^2 + 1\right]^{3/2}}{k_z/A_r}$$

(24)

It can be easily checked that for A_r equal 1 or 2, the first instability occurs for $k_z^c = 1$ (one convective roll), whereas for $A_r = 3$, the mode $k_z^c = 2$ sets in first (two convective rolls). These two modes become simultaneously unstable for a particular value of the aspect ratio given by

$$A_r = 2^{1/3} \left(4^{1/3} + 1\right)^{1/2} \approx 2.067$$

(25)

The study of the relative stability of these two modes requires a non-linear stability analysis, similar to the one discussed by Knobloch et al. and Busse et al.[15] for Bénard problem in an incompressible fluid (Bousinesq approximation), except that here we do not need to impose the incompressibility assumption[16]. Work on this direction is in progress and will be presented elsewhere.

In order to compute the complete solution of eq. (20), we notice that both computer experiments and neutron scattering experiments have shown that the hydrodynamic regime can only be granted if the dimensionless parameters $\nu_1 q / C$ and $\nu_2 q / C$ remain small[17] (smaller than 0.01 for example). Using this fact as well as our main hypothesis, i.e. $V_0 \ll C$, it is easy to show that to the dominant order the eigenvalues of the equation (20) take the following form :

$$\lambda_{\pm}^{v} = -v_1 \, q^2 \pm \frac{q_z \gamma}{q} \tag{26}$$

$$\lambda_{\pm}^{c} = \pm i \, C \, q - \Gamma_s \, q^2 \tag{27}$$

where we have introduce the sound damping coefficient

$$\Gamma_s = \frac{(\zeta + 4\eta/3)}{2\,\rho_0} \tag{28}$$

As can be seen, the sound modes λ^c are not affected by the non-equilibrium constraints. This is not the case of shear modes λ^v. In particular, there exists a critical value of the velocity gradient for which one the shear modes crosses the real axis (cf. eq. 24)

4. CONCLUDING REMARKS

In this short communication we have presented a simple model which can exhibit an instability, similar to the one occurring in Taylor problem[14]. The main advantage of our model resides in two facts. First, it can be easily simulated through molecular dynamics computer experiments since it avoids the cylindrical geometry. Next, it is designed so that a complete bifurcation analysis can be carried out without requiring the incompressibility assumption. This last point is crucial if we wish to study the behavior of the fluctuations. Indeed, as first pointed out by Zaitsev and Shliomis[10], the incompressibility assumption does not allow the computation of the static (equal time) correlation functions, since it imposes fictitious correlations between the velocity components. In our case, the analysis carried out in section 3 can be easily generalized to deal with fluctuations. For example, using the Landau Lifshitz fluctuating hydrodynamics[18], one can easily show that the static density-density correlation function is given by

$$<\delta\rho_q \, \delta\rho_{q'}> \; - \; \frac{4 \, k_B \, T_0}{\rho_0 \, V \, C^2} \, \delta_{q,q'}^{Kr.} = \frac{4 \, k_B \, T_0}{\rho_0 \, V \, C^4} \, \gamma^2 \left\{ \frac{(v_1 + \Gamma_s) \, q_y^2}{\Gamma_s \, q^4} \; + \right.$$

$$\left. + \; \frac{v_1^2 \, q_y^2 \, q^2}{v_1^2 \, q^6 - q_z^2 \gamma^2} \right\} \delta_{q,q'}^{Kr.} \tag{29}$$

where V is the volume of the system. The second term in the left hand side represents the equilibrium contribution. As can be seen, the non-equilibrium contribution is proportional to the square of the external constraint γ. The first term in the right hand side of eq. (29) gives rise to a long range correlation function extending over the entire volume of the system and is present independent of any instability[19]. The second term expresses directly the influence of instability as it diverges when γ approaches to $\gamma(k_z^c)$. On the basis of a linear theory, it is thus not possible to associate the existence of long range correlations solely to the presence of a bifurcation point, as it is usually the case in equilibrium critical phenomena. A nonlinear analysis is necessary in order to clarify this point. In this respect, an interesting question is whether a Landau-Ginzburg theory can be derived form our simple model and, if so, what would be the influence of the compressibility (sound modes) on the nonlinear terms. We believe that a more advanced mathematical analysis of our model conducted in parallel with a molecular dynamics computer experiment will bring some light on the problem of hydrodynamical instability and, perhaps, help to clarify the experimental results presented in ref. 8.

ACKNOWLEDGEMENTS

I am very grateful to professors G. Nicolis, J. W. Turner and P. Borckmans for helpful comments.

REFERENCES

1. B. J. Alder and T. E. Wainwright, J. Chem. Phys., 27, 1208 (1957)

2. W. W. Wood and F. Lado, J. Comp. Phys., **7**, 528 (1971);
    C. Bruin, Physica, **72**, 261 (1974);

3. B. L. Holian, Phys. Rev. **A37**, 2562 (1988); W. G. Hoover, Phys. Rev. Lett. **42**, 1531 (1979)

4. D. C. Rapaport, E. Clementi, Phys. Rev. Lett. **57**, 695 (1987);
    L. Hannon, G. Lie and E. Clementi, J. Sc. Comp.**1**, 145 (1986);
    D. C. Rapaport, Phys. Rev. **A 36**, 3288 (1987);
    E. Meiburg, Phys.Fluids, **29**, 3107 (1986)

5. M. Mareschal, E. Kestemont, Nature, **323**, 427 (1987);
    M. Mareschal, E. Kestemont, J. Stat. Phys. **48**, 1187 (1987);
    D. C. Rapaport, Phys. Rev. Lett. **60**, 2480 (1988)

6. M. Mareschal, M. Malek Mansour, A. Puhl, E. Kestemont, Phys. Rev. Lett. **61**, 2550 (1988);
    A. Puhl, M. Malek Mansour and M. Mareschal, Phys. Rev. **A 40**, 1999 (1989)

7. A nearly up to date account of the field is given in "*Molecular Dynamics Simulation of Statistical Mechanical Systems*", G. Ciccotti and W. G. Hoover, Eds., North Holland (1986)

8. G. Ahlers, C. Meyer and D. Cannel, J. Stat. Phys. **54**, 1121 (1989)

9. H. Van Beijeren and E. D. G. Cohen, Phys. Rev. Lett. **60**, 1208 (1988); H. Van Beijeren and E. D. G. Cohen, J. Stat. Phys. **53**, 77 (1988)

10. V. M. Zaitsev and M. I. Shliomis, Sov. Phys. JETP, **32**, 866 (1971)

11. H. Lekkerkeker and J. P. Boon, Phys. Rev. **A 10**, 1355 (1974)

12. R. Graham, Phys. Rev. **A 10**, 1762 (1974); R. Graham and H. Pleiner, Phys. Fluids, **18**, 130 (1975); J. Swift and P. C. Hohenberg, Phys. Rev. **A 15**, 319 (1977); T. R. Kirkpatrick and E. D. G. Cohen, J. Stat. Phys. **33**, 639 (1983)

13. A systematic derivation of the results of ref. 10 is given in : R. Schmitz and E. G. D. Cohen, J. Stat. Phys., **38**, 285 (1985); R. Schmitz and E. G. D. Cohen, J. Stat. Phys., **40**, 431 (1985)

14. S. Chandrasekhar, *Hydrodynamic and Hydromagnetic Stability*, Clarendon Press, Oxford (1961)

15. E. Knobloch and J. Guckenheimer, Phys. Rev. **A27**, 408 (1983); F. H. Busse and A. C. Or, ZAMP **37**, 608 (1986)

16. A molecular dynamic study of the compressible Bénard problem is reported in ref. 6.

17. See for instance W. E. Alley and B. J. Alder, Phys. Rev. **A27**, 3158 (1983) ; W. E. Alley, B. J. Alder and S. Yip, Phys. Rev. **A27**, 3174 (1983)

18. L. D. Landau, E. M. Lifshitz, Fluid Mechanics, Pergamon Press, London (1958)

19. T. R. Kirkpatrick and E. D. G. Cohen, Phys. Rev. **A 26**, 995 (1982); M. Malek Mansour, J. W. Turner and A. Garcia, J. Stat. Phys. **48**, 1157 (1987)

PART III

APPLICATIONS (I): NON EQUILIBRIUM SYSTEMS

IRREVERSIBLE PROCESSES FROM REVERSIBLE MECHANICS*

William G. Hoover,[1,2] Carol G. Hoover,[3] Will J. Evans,[2]
Bill Moran,[4] Joanne A. Levatin,[4] and Errol A. Craig[1]

1. Department of Applied Science; University of California at
   Davis/Livermore
2. Department of Physics; Lawrence Livermore National
   Laboratory
3. National Magnetic Fusion Energy Computer Center
4. Earth Sciences Department; Lawrence Livermore National
   Laboratory

[All at Livermore, California, 94550, United States of America.]

ABSTRACT

We describe and illustrate methods for treating many-body irreversible
processes using time-reversible deterministic Nosé-Hoover thermostats. In
phase space, Lyapunov-unstable multifractal strange attractors are the
common feature representing any of these nonequilibrium flows, be they
steady, periodic, or transient. This generic behavior is illustrated here for
three prototypical one-body problems: steady field-driven <u>diffusive</u> flow in a
Galton Board, time-periodic boundary-driven <u>viscous</u> flow of a Lorentz gas,
and transient, but time-periodic, <u>compressible</u> flow characterizing a one-
dimensional free expansion followed by compression and thermalization.

1. INTRODUCTION

"Real" macroscopic processes are thermodynamically "irreversible".
They generate entropy. The microscopic reason underlying this
irreversibility is "chaos", or a sensitivity to initial conditions. The discovery
and elucidation of chaos is beautifully chronicled in Gleick's book[1] "_Chaos_".
Despite its aesthetic appeal, chaos is not always welcome. In his recent
California lecture series Yorke[2] termed chaos the "AIDS of Dynamics".

Since the Second World War engineers have used ever-faster
computers to simulate the macroscopic continuum mechanics of ever-more-
complex flows of gases, liquids, and solids. More recently, particularly in the
last five years, atomistic computer experiments have been successfully
modeling these same flows, with the motion equations of microscopic
atomistic mechanics[3].

Microscopic Simulations of Complex Flows
Edited by M. Mareschal, Plenum Press, New York, 1990

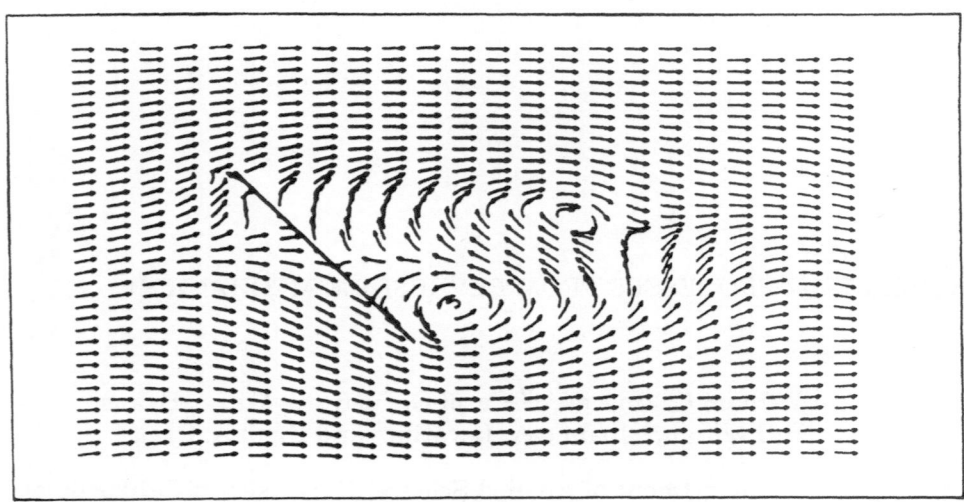

FIGURE 1. Hard-sphere streamlines in a three-dimensional flow past an inclined splitter plate. Each arrow indicates the instantaneous direction of velocity of approximately 50 hard spheres.

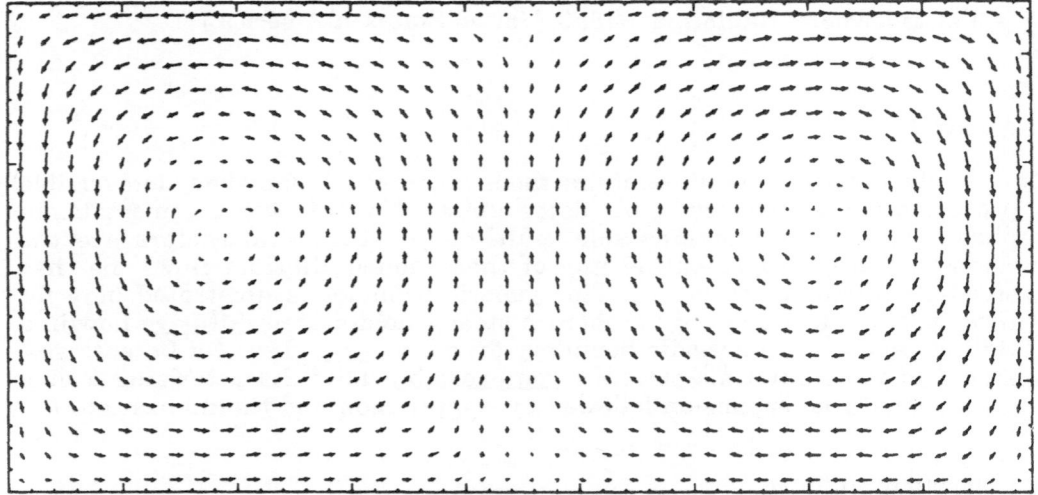

FIGURE 2. Hard-disk velocities for a time-averaged Rayleigh-Bénard flow. The two convective rolls shown are driven by the lower (hot) and upper (cold) boundaries in the presence of a vertical gravitational field.

Meiburg[4] established that microscopic atomistic mechanics can simulate macroscopic hydrodynamic flows. He simulated the flow of a dense viscous fluid of hard spheres past a "splitter plate" by following the motion of 50,000 atoms in three space dimensions. His spatially-averaged streamlines are shown in Figure 1.

A more complicated flow, with stationary hot and cold boundaries, is the heat-conducting Rayleigh-Bénard flow[5] shown in Figure 2. In the Rayleigh-Bénard system, a sufficiently strong gravitational field, combined with two heat reservoirs, hot at the bottom and cold at the top, can drive a compressible fluid into rotational motion. For sufficiently strong driving, buoyant forces, fed by thermal expansion, generate eddy currents which then dissipate vortical energy through viscosity. In this Rayleigh-Bénard case the fixed-temperature boundary conditions are stationary but the responding currents typically fluctuate in time. More complicated time-dependent problems can involve boundary conditions with either periodic or transient time dependences.

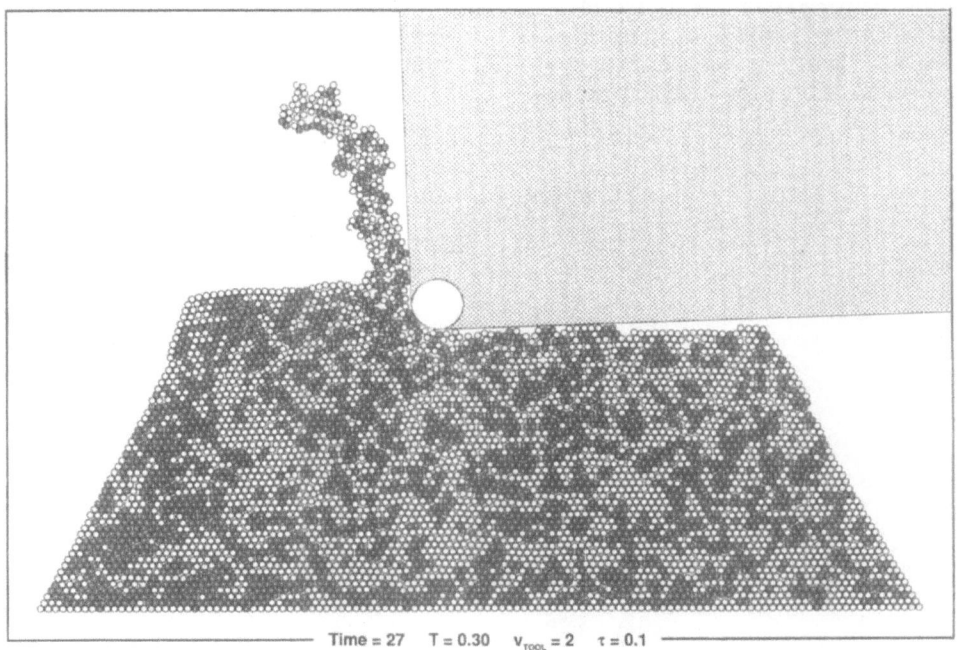

Time = 27   T = 0.30 v_{TOOL} = 2 τ = 0.1

FIGURE 3. Two-dimensional embedded-atom simulation of the diamond turning of a metal workpiece. The "tool", representing a diamond chip with a nanometer radius of curvature, is driven into the crystal at about one-tenth the speed of sound. The crystal is thermostatted at about two-thirds the melting temperature by the two rows of Nosé-Hoover particles adjacent to the base of the crystal. The atoms with higher-than-average potential energy have been shaded.

Both the macroscopic hydrodynamic and the microscopic atomistic approaches have been applied to the convective Rayleigh-Bénard problem. In Lorenz' seminal paper[6] describing the impact of chaos on weather, "Deterministic Nonperiodic Flow", a classic macroscopic caricature of this flow is described. This was the paper that initiated widespread interest in chaotic dynamics.

Even the nonchaotic completely-static small-Rayleigh-number case has some interest as an application for irreversible thermodynamics, the study of processes which generate entropy. In a heat-conducting fluid, even in the absence of gravity, the external entropy loss, at the bottom, $-(dQ/dt)/T_{HOT}$, is less than the external entropy gain, at the top, $+(dQ/dt)/T_{COLD}$. The net result is the entropy increase guaranteed by the Second Law of Thermodynamics.

More complicated solid-state flows, such as the nanometer-scale machining simulation[3] illustrated in Figure 3, involve sources or sinks for plastic or viscoelastic work as well as for heat. Here the microscopic approach is absolutely necessary because the characteristic lengths involved, a few atomic spacings, are too small for the continuum approximations to plastic flow and heat flow to apply.

Meiburg's splitter-plate simulation[4] involved 50,000 particles far from equilibrium. Even earlier the purely-equilibrium Monte Carlo method had been applied at similarly large scales. Beginning with Farid Abraham's announcement[7] of his 161,604-atom computer experiments simulating rare-gas atoms adsorbed on graphite, it became widely appreciated that we can simulate truly macroscopic systems, with sizes corresponding to those studied in laboratory experiments.

FIGURE 4. DeGroot's SPRINT Multiprocessor. The 64 $500 processors can simulate the cutting process shown in FIGURE 3 at the same clock speed as does a $10,000,000 CRAY computer.

To simulate ever-larger sizes, current research in both microscopic and macroscopic simulations is concentrating on using "multiprocessors". These machines increase computer capabilities about one thousandfold. A recent tabletop model, de Groot and Parker's "SPRINT" machine[8], with 64 parallel processors, is shown in Figure 4. Scaling up such machines should lead soon to a widespread capability for simulating microseconds of real time dynamics for systems with millions of degrees of freedom. Such simulations require hardware outlays of millions of dollars, and programming efforts measured in man-years. Nevertheless, the longterm trend toward parallelism is already clearly established. The major problem with present and projected large-scale calculations is displaying the results in order to "understand" them. This display step is still hard, and time consuming, in three dimensions. Large data files typically require *days* to transport and process.

On the other hand, much of the fundamental physics can be understood by studying much smaller two-dimensional systems for which detailed averages over *millions* of bins for *billions* of timesteps can be accumulated on current supercomputers. Analogs of many-body macroscopic diffusive, viscous, and heat-conducting dissipative flows can be found on an even much smaller microscopic scale, with just a few degrees of freedom, using the same many-body time-reversible atomistic equations of motion. We return to this idea, with three examples, in Section 5.

We have adopted Boltzmann's interest: relating macroscopic irreversibility to microscopic mechanics. It has only recently been recognized that Second-Law irreversibility can arise naturally even in small systems with two or more degrees of freedom. This recognition was slow in coming because traditional textbook mechanics (Newtonian or Lagrangian or Hamiltonian) predates thermodynamics, and accordingly there was no traditional means for describing *thermo*dynamic properties such as temperature and heat flow. The link between mechanics and thermodynamics is basic and straightforward nevertheless. It can be established by adopting the ideal-gas definition of temperature and the microscopic mechanical definitions of momentum and energy fluxes as the analogs of the macroscopic pressure tensor and heat flux vector.

In an atomistic Boltzmann-Gibbs statistical description the microscopic mechanical state of a classical system includes the complete list of coordinates and momenta for all the degrees of freedom. This information is pictured as specifying a point in many-dimensional "phase space". The equations of motion then generate a "flow" of probability, traditionally written in terms of a time-dependent probability density $f(q,p,t)$ in that space. The probability density f is pictured as representing the "ensemble" flow of many similar discrete systems in space, all obeying the same motion equations, but not interacting with each other. The integral of f, over the whole $\{q,p\}$ space, is constant, representing the total number of systems in the "ensemble" of systems undergoing investigation.

For conservative Hamiltonian mechanics the generalized (phase-space) flow "velocity" v is composed of both coordinate and momentum contributions, $v = (dq/dt, dp/dt)$, and follows Hamilton's equations of motion:

$$dq/dt = + (\partial H/\partial p) \qquad dp/dt = - (\partial H/\partial q)$$

The divergence of the probability current fv, computed locally as the difference of flux contributions gives the local (fixed q and p) time-rate-of-change of f:

$$(\partial f/\partial t)_{\text{HAMILTON}} = - \nabla \cdot (fv) = - (\partial/\partial q)(fdq/dt) - (\partial/\partial p)(fdp/dt)$$

Then, using the velocity and acceleration equations from Hamilton's equations of motion, we find that $(\partial/\partial q)(dq/dt) = (\partial^2 H/\partial q \partial p)$ and $(\partial/\partial p)(dp/dt) = - (\partial^2 H/\partial p \partial q)$ sum to zero, leading to the usual comoving (or "Lagrangian") form of Liouville's theorem:

$$df/dt = (\partial f/\partial t) + (dq/dt)(\partial f/\partial q) + (dp/dt)(\partial f/\partial p) = 0$$

Conservation of probability, $f\otimes$, where $\otimes$ is a comoving phase-space hypervolume element, then establishes the relations:

$$dln(f\otimes)/dt = dlnf/dt + dln\otimes/dt = dln\otimes/dt = 0$$

It is not so well known that all these relations hold even in the case that the forces F are explicit functions of time, as, for instance, in a driven oscillator system.

These deceptively simple Liouville Theorem results, that the comoving probability density f and phase-space hypervolume $\otimes$ don't change with time, disguise the incredibly complex deformation and rotations generated by nonlinear chaotic dynamics. The basic phase-space deformation mechanism, the Smale horseshoe, is indicated in Figure 5. Successive passes of a comoving hypervolume through those parts of phase space where bending and folding occur produce a multilayered $(2 \rightarrow 4 \rightarrow 8 \rightarrow 16 \rightarrow ...)$ "fractal" structure with a complexity that increases exponentially fast with time.

Smale's horseshoe deformation is steady. The effects of the process, a repeated bending and folding, steadily penetrate to smaller and smaller scales. The classic space-filling Peano curve, also shown in the same Figure 5, is different. The Peano construction is nonsteady, requiring successivly more-refined deformations, at smaller and smaller scales, as "time" proceeds.

The characteristic time for Smale's exponential bending and folding is the "Lyapunov time". It is of the same order as the collision time and leads to phase-space structures of incredible complexity and beauty for systems with only a few degrees of freedom. In many-body systems the *rotation* of phase-space trajectories occurs on an even faster scale than does the bending and stretching. The resulting geometric complexity lies at present well beyond our descriptive abilities.

The determinism underlying the elegant Liouville description and Smale's phase-space folding would no longer apply if we were to follow the traditional approach to "understanding", or at least simulating, irreversible processes. This well-entrenched recipe proceeds by introducing both friction and "noise" (or "randomness") through Langevin modifications of the motion equations. This approach is obsolete. It suffers from two defects: the stochastic dynamics it generates is neither time-reversible nor deterministic. As a consequence, a Langevin dynamics phase-space flow no longer follows Liouville's Theorem. The twin defects of the Langevin approach complicate analysis and limit reproducibility of results. In this paper we will outline the progress made during the past 15 years in treating irreversible processes with nonequilibrium molecular dynamics. This approach follows Gauss and Lagrange by incorporating (thermal) constraints directly into deterministic and reversible equations of motion. The new equations yield simple analogs of the Liouville Theorem, giving an exact description of the deformation of comoving volume elements in phase space.

The new methods resemble the classical ones in that the fundamental flows in phase space are simple, but the new methods are not Hamiltonian. The starting point is still the same, the phase-space continuity equation satisfied by *any* kind of mechanics with differentiable equations of motion:

$$(\partial f/\partial t)_{\text{ARBITRARY}} = - \nabla\cdot(fv)$$

Here v is again a generalized phase-space velocity, a vector made up of all the time-rates-of-change of the phase-space variables describing the problem. The generalized gradient operator ∇ is similarly a vector sum of phase-space derivatives with respect to these same variables. We illustrate the generalized flow in Figure 6 for a space spanned by coordinates {q}, momenta {p}, and friction coefficients {z}. We will discuss friction coefficients at length in what follows. In any direction in such a space the flow of probability through an area dA during a time interval dt is the product of (i) the probability density f, (ii) the flowvelocity v multiplied by the perpendicular element of area dA, and (iii) the time interval dt. In the example shown in Figure 6, during the time interval dt the flow out of the volume element and parallel to the p direction, through the face dA = dqdζ is f(dp/dt)dAdt. By summing *pairs* of such contributions, from opposite faces, the continuity equation can be derived. For equilibrium Hamiltonian systems this result is called "Liouville's Theorem".

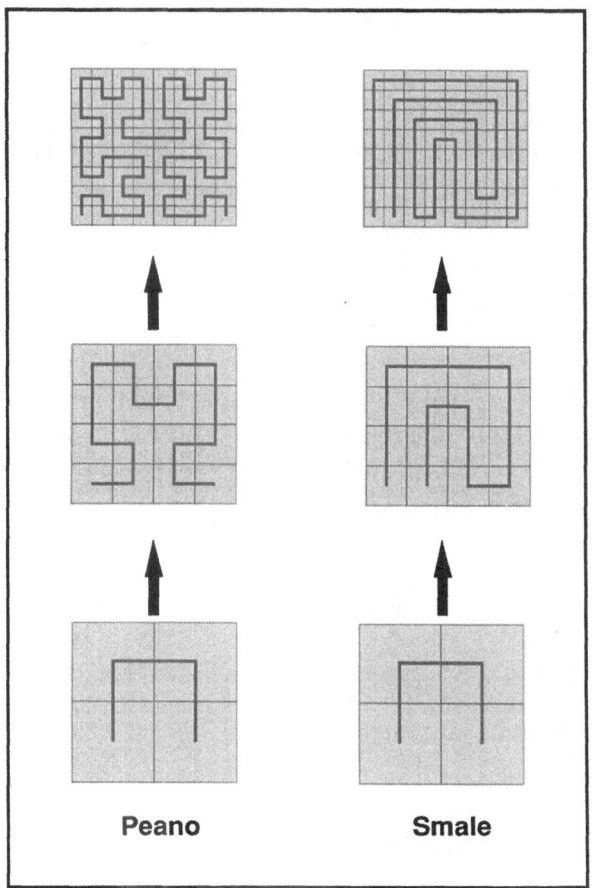

Peano   **Smale**

FIGURE 5. The Peano curve and the Smale Horseshoe deformation. Though both constructions have ergodic limits that come arbitrarily close to every point the Peano curve requires successively smaller and smaller scales of deformation while the Smale construction relies on a repeated large-scale bending and stretching. The Smale Horseshoe is a faithful caricature of real phase-space deformation.

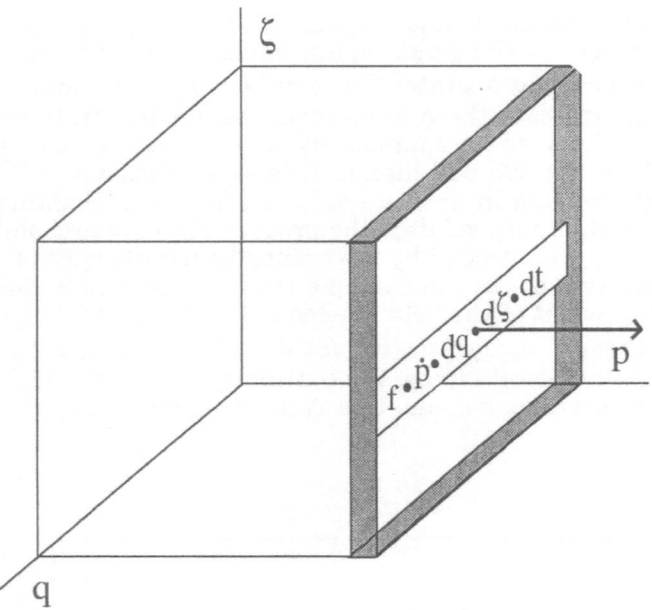

FIGURE 6. The flow from a phase-space hypervolume element is the product of a density f, a perpendicular velocity, dp/dt for the face shown here, a cross-sectional area, here dqdζ, and the time interval dt.

Liouville's Theorem establishes that the motion given by Hamilton's equations, even with time-dependent forces, takes place at *constant* probability density and *constant* hypervolume in a comoving frame. Heat flow gives new results. A *transfer* of heat, ΔQ, from the surroundings into the system, *changes* not only the phase-space probability density, as $e^{-\Delta Q/kT}$, but also the comoving hypervolume, as $e^{+\Delta Q/kT}$, so that df/dt and d⊗/dt are both *nonzero*. Because the time-averaged kinetic energy, $<\Sigma p^2/2m>$, is the direct mechanical measure of thermodynamic temperature, thermal constraints are implemented by modifying the time evolution of the momenta. The simplest possible modification of the equations of motion describing heat transfer incorporates frictional forces,

$$F_{\text{FRICTIONAL}} = -\zeta p$$

where ζ is a "friction coefficient". Such forces have proved very useful, both at and away from equilibrium. We describe the corresponding "Nosé-Hoover" version of this development, which began with Nosé's 1984 work, in the next section.

2. NOSE-HOOVER MECHANICS

A "new" approach related to an "old" idea (Gauss' 1829 Principle of Least Constraint) was developed by Shuichi Nosé[9] in his post-doctoral work with Mike Klein, in Canada. Nosé discovered that he could reproduce Gibbs' canonical and isothermal-isobaric distributions, with fluctuating energy and volume, by modifying traditional mechanics.

Let us consider here the simplest of Nosé's modified dynamics, corresponding to Gibbs' equilibrium canonical distribution,

$$f(q,p)_{CANONICAL} \propto e^{-H(q,p)/kT}$$

Nosé reproduced this distribution by using purely deterministic and time-reversible equations of motion based on a clever modification of Hamiltonian mechanics. It is unfortunate that the approach Nosé pioneered involved a wholly unnecessary "time-scaling" as well as an extraneous distinction between "real" and "virtual" variables. These twin distractions have contributed to misunderstandings of his work's significance[10]. At equilibrium, the end result of Nosé's approach is simply to add a time-varying frictional force to the usual Hamiltonian equations of motion. The result is called the "Nosé-Hoover" equation of motion to indicate the absence of time-scaling and virtual variables:

$$(dp/dt)_{HAMILTON} = F(q)$$

$$(dp/dt)_{NOSÉ-HOOVER} = F(q) - \zeta_{NH}p$$

In this form the friction coefficient ζ_{NH} itself obeys a first-order ordinary differential "integral-feedback" equation linking its time-history to the temperature T imposed on the "thermostatted" degrees of freedom:

$$d\zeta_{NH}/dt = [(p^2/mkT) - 1]/\tau^2.$$

If *several* degrees of freedom are to be constrained in this way, with a common friction coefficient, then the righthand side is to be averaged over *all* these degrees of freedom. The thermostat relaxation time t is "phenomenological" (that is, a free parameter). We follow Maxwell, Boltzmann, and Gibbs in identifying the ideal-gas temperature scale, $<p^2/mk> \equiv T$, as the fundamental (mechanical) definition of temperature. In any stationary or time-periodic process $<d\zeta/dt>$ vanishes and the "definition" of temperature just described is also a dynamical identity.

For ergodic systems, the motion equations just given generate the canonical distribution in phase space. The equations are novel. They include both macroscopic and microscopic variables side by side. But just as in traditional microscopic mechanics, the Nosé-Hoover equations are time-reversible, with both z and p changing sign along with time on a reversed trajectory. Straightforward generalizations of them generate the isothermal-isobaric phase-space distribution. The thermostat forces $\{-\zeta p\}$ operate by taking up or giving off heat according to the past history of the instantaneous ideal-gas temperature, $T(t) \equiv p^2/mk$. If p^2/mk exceeds the desired temperature T for a particular degree of freedom the friction increases. If p^2/mk is less than T then the friction tends to decrease, and, for negative values, *adds* energy to the system. This time-varying friction makes additional random or stochastic forces of the Langevin type unnecessary.

Brad Holian suggested a constructive derivation of these "Nosé-Hoover" non-Hamiltonian equations of motion which is much simpler than Nosé's Hamiltonian approach. Begin by *assuming* a friction-coefficient equation of motion with a friction coefficient $\zeta(q,p)$ depending only on q and p. Then observe that the *only* motion equation for $d\zeta/dt$ consistent with the canonical distribution is Nosé's:

$$d\zeta/dt \equiv [(p^2/mkT) - 1]/\tau^2$$

As a fringe benefit, the equilibrium distribution for ζ (that is, the distribution in an isolated system) turns out to be a simple Gaussian, centered on zero, with the fluctuating frictional effects vanishing [typically as $N^{-1/2}$ or N^{-1} for N degrees of freedom] in the large-system [large N] "thermodynamic limit".

It is interesting that exactly the *same* friction-coefficient form of the equations of motion follows also from Gauss' <u>Principle of Least Constraint</u>, if that Principle is used to impose a constraint force keeping the *kinetic energy* constant.

$$(dp/dt)_{GAUSS} = F(q) - \zeta_{GAUSS}p$$

The recipe for Gauss' friction coefficient looks *different* from Nosé's:

$$\zeta_{GAUSS} = -(d\Phi/dt)/2K$$

where $\Phi(q)$ and $K(p)$ are the potential and kinetic energies. But, in the limit that Nosé's relaxation time t vanishes the coordinate-space trajectories following the Nosé-Hoover equations given above reduce to those generated using Gauss' equations of motion.

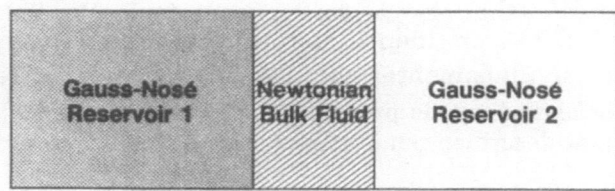

FIGURE 7. Schematic construction of a nonequilibrium steady state. The two boundary regions, stabilized by Gaussian or Nosé-Hoover reservoirs, drive the Newtonian particles sandwiched between them.

More than 30 years after Fermi's pioneering work at Los Alamos, atomistic equilibrium calculations are a well-established and routine undertaking but nonequilibrium techniques are still being developed. And Nosé's ideas, a natural extension of Gauss' "Principle of Least Constraint", are a fertile source of new methods. To study a simple two-temperature heat flow, with hot and cold boundaries, the sandwich construction shown in Figure 7 can be used. In that Figure Newtonian bulk fluid is shown interacting with two "Gauss-Nosé" reservoir regions. Suppose that the lefthand region is hot and the righthand region cold. In the lefthand "hot" region *some* degrees of freedom satisfy thermostatted Nosé-Hoover equations with a "hot" friction coefficient ζ_{HOT}:

$$(dp/dt)_{\text{NOSÉ-HOOVER}} = F(q) - \zeta_{\text{HOT}}p$$

$$d\zeta_{\text{HOT}}/dt = [(p^2/mkT_{\text{HOT}}) - 1]/\tau_{\text{HOT}}^2$$

In the righthand "cold" boundary region at least some degrees of freedom satisfy similar thermostatted equations of motion incorporating a "cold" friction coefficient ζ_{COLD}:

$$(dp/dt)_{\text{NOSÉ-HOOVER}} = F(q) - \zeta_{\text{COLD}}p$$

$$d\zeta_{\text{COLD}}/dt = [(p^2/mkT_{\text{COLD}}) - 1]/\tau_{\text{COLD}}^2$$

Though this may appear to be nothing more than an *"ad hoc"* way of imposing hot and cold temperatures on selected degrees of freedom, this approach has two real advantages over the old Langevin approach:

First, Nosé-Hoover mechanics makes it possible to predict the *direction* of energy flow consistent with the Second Law of Thermodynamics[11].

Second, the new mechanics also establishes a *quantitative* connection between the macroscopic entropy increase and the mechanical Lyapunov spectrum of the underlying microscopic mechanics[12].

We can establish both results by considering a Nosé-Hoover phase-space flow. Such a flow can have *many* friction coefficients, $\{\zeta\}$, with each coupled to its own particular set of degrees of freedom. The instantaneous values of the friction coefficients fluctuate, but the sum of the time averages, $\langle\Sigma\zeta\rangle$, is constrained geometrically. It is easy to see, for any bounded region of phase space away from equilibrium (implying nonvanishing heat transfer, $\langle\Sigma\zeta\rangle \neq 0$), that the *analog* of Liouville's Theorem,

$$df/dt = -f(\partial/\partial p)(dp/dt) = f\Sigma\zeta$$

makes sense *only* if the friction coefficient sum is positive. If instead df/dt were negative, the steady-state f would have to decay to zero and the occupied phase-space volume $\otimes$ would then necessarily diverge. Thus df/dt and $\langle\Sigma\zeta\rangle$ must both be postive. Accordingly, away from equilibrium the phase volume $\otimes$ must go to zero. This vanishing of the phase volume leads to strange-attractor phase-space structure of the type we illustrate in Section 5.

Nosé's equations of motions have another interesting consequence. They imply directly that the friction coefficients are not correlated with their corresponding temperatures. To see this, multiply the equation of motion for $d\zeta_{\text{NH}}/dt$ by $\tau^2\zeta_{\text{NH}}$ and time average:

$$\tau^2\langle\zeta_{\text{NH}}d\zeta_{\text{NH}}/dt\rangle \equiv \langle\zeta_{\text{NH}}[(p^2/mkT) - 1]\rangle = 0$$

Because ζ_{NH}^2 is bounded, the time average vanishes, leading to the conclusion:

$$\langle\zeta_{\text{NH}}[(p^2/mkT)]\rangle = \langle\zeta_{\text{NH}}\rangle$$

This lack of correlation between the instantaneous temperature $p^2/(mk)$ and the friction coefficient ζ_{NH} gives a simple relation between the external surroundings <u>entropy</u> rate-of-change associated with a reversible heat transfer, $(1/T)(dQ/dt)$, and the friction coefficient ζ_{NH}:

$$-(1/T)<dQ/dt> = (1/T)<\zeta_{NH}p^2/m> = <\zeta_{NH}>(1/T)<p^2/m> = k<\zeta_{NH}>$$

Here, positive ΔQ (or negative friction) corresponds to flow into the system, through the frictional forces, from the "surroundings". Because the requirement that the phase-space volume be bounded implies that $<\zeta_{NH}>$ is <u>positive</u> we conclude that the summed heat inputs, divided by the corresponding temperatures, must be <u>negative</u>. In the steady state this means that the corresponding sum for the <u>external</u> heat reservoirs, which extract the heat ΔQ stabilizing the motion, must be positive:

$$dS_{TOTAL}/dt = -\Sigma(dQ/dt)/T > 0$$

Thus Nosé-Hoover mechanics establishes that *the flow of heat must be in the direction consistent with the Second Law*, so that

<u>Work can *only* be converted into Heat</u>

not the other way around. This link between Nosé-Hoover mechanics and the Second Law of Thermodynamics is outlined in Figure 8. In the upper half of that Figure the conservation of probability $f\otimes$, where $\otimes$ is a phase-space volume element, is indicated. Nosé's equations of motion link the change in probability density with time to the friction coefficients $\{\zeta\}$ and simple geometry links the change in volume with time to the Lyapunov exponents $\{\lambda\}$. We discuss these links between the microscopic Lyapunov spectrum, phase-space volume, and fractal strange attractors in the next two sections.

In the bottom section of Figure 8 it is indicated that finite phase-space volume implies that the friction coefficient sum, $\Sigma\zeta$, must be positive, and that this in turn establishes that the production of entropy is positive, the Second Law of Thermodynamics.

Application of Nosé's ideas to irreversible macroscopic problems is straightforward. In a companion paper we describe simulations of high-speed machining operations in which reversible Nosé-Hoover thermostats are used to control the macroscopic temperature of the workpiece. In the present paper we instead develop analogs for macroscopic systems which are chosen for simplicity, to illustrate the qualitative features and logical connections without the complexity of high-dimensionality state-of-the-art molecular dynamics simulations.

The traditional textbook way to study nonequilibrium states is to imagine that they arose as equilibrium fluctuations. For states more than a little different from equilibrium states this is an odd idea. It is conceptually much simpler to focus on driven systems--systems forced away from equilibrium into nonequilibrium steady states. In steady-state nonequilibrium processes chemical potential, velocity, or temperature differences are maintained in the face of diffusive, viscous, and thermal dissipation. Shock or detonation waves combine *all* of these processes. Figure 9 shows the simple prototypical nonequilibrium system illustrating these steady-states. Black and white are used in the Figure to represent differences in species, velocity, or energy, with the lefthand reservoir a source of black particles and the righthand reservoir a source of white ones. Mixing, described by Fick's Laws of diffusion, Newtonian viscosity, or Fourier's heat conduction, occurs in the middle Newtonian region. Mass, momentum, and energy flow steadily through the system. As a consequence of the nonequilibrium flow heat is continuously generated and flows out of the system into the boundary regions.

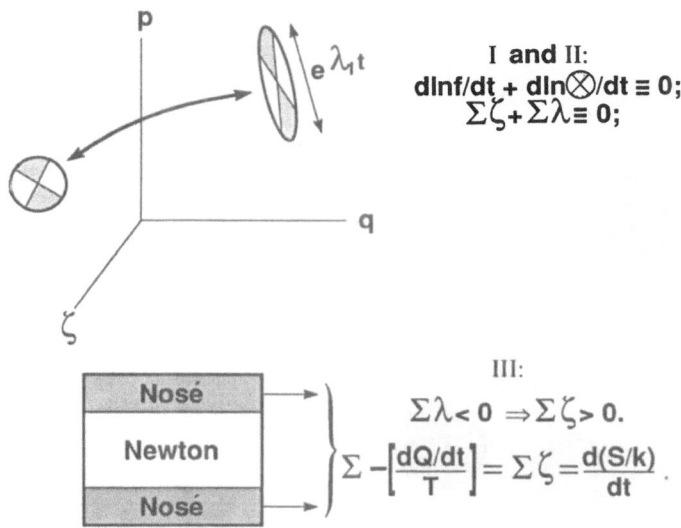

FIGURE 8. Summary of the mechanical basis of the Second Law of Thermodynamics. The upper part of the FIGURE relates probability conservation to the friction coefficients $\{\zeta\}$ and the Lyapunov spectrum $\{\lambda\}$. The lower part indicates the connection of sums of these numbers to the thermodynamic entropy production dS_{TOTAL}/dt.

If we use Nosé mechanics to simulate such a flow we find that the phase-space distribution function $f(q,p,t)$ rapidly collapses, with the ratio $f(t)/f(0)$ diverging as $e^{-\Delta Q/kT}$, with ΔQ negative, onto a zero-volume (decreasing as $e^{+\Delta Q/kT}$) object called a STRANGE ATTRACTOR. Why is the attractor "strange"? First, within the attractor, trajectories separate rapidly from one another, exponentially fast, despite the attractor's having zero volume. Second, there is no probability density on the attractor. The probability per unit volume approaches no limit for smaller and smaller phase-space binnings. The distribution contains irregularities on all spatial scales. This topological behavior is completely unlike the smooth functions to which ordinary calculus applies. Our present understanding of strange-set topology is today still primitive and crude. The existence of these strange phase-space objects depends upon the "Lyapunov instability" of the equations of motion, that is, the tendency for nearby phase-space trajectories to separate from each other exponentially fast, in time.

The shrinking of the occupied phase space and the corresponding collapse and divergence of $f(q,p,t)$ gives rise to a qualitative change in phase-space topology, a loss of phase-space dimensionality. This loss is macroscopic and depends on the departure from equilibrium. For instance, a cubic centimeter of water, sheared slowly at 1 hertz, has a distribution function f which increases e-fold on a picosecond timescale. The loss of phase-space dimensionality for this case is negligible, of order 1 part in 10^{24}. The much stronger gradients in shockwaves can make the Lyapunov-unstable divergence 24 orders of magnitude faster and the corresponding dimensionality loss 24 orders of magnitude greater. In the next section we discuss the detailed characterization of Lyapunov instability and its consequences.

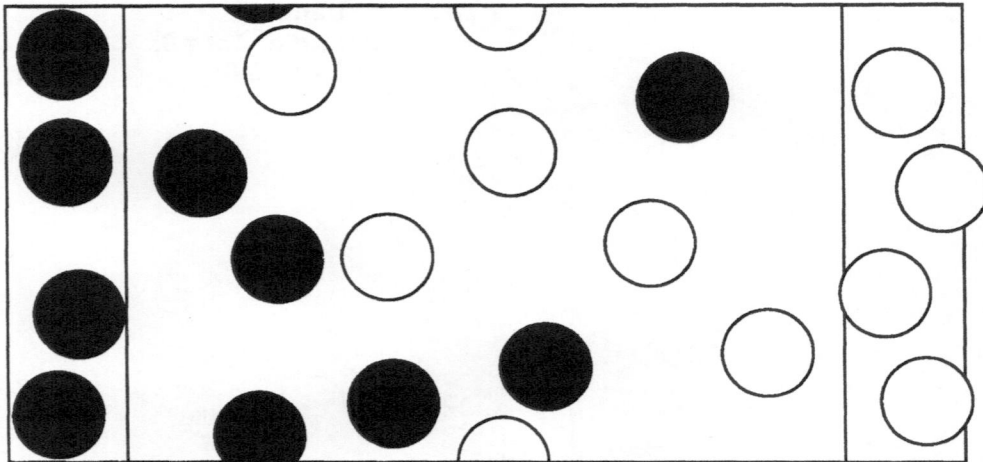

FIGURE 9. A prototypical nonequilibrium problem in which two reservoirs, with characteristic values of the concentration, velocity, or temperature, cause a nonequilibrium mass, momentum, or energy flux through the central Newtonian region.

3. LYAPUNOV INSTABILITY

We have seen that the geometric requirement that stationary states be bounded in phase space breaks the symmetry of the equations of motion, forcing the time development to seek out the direction of phase-space shrinking. In these stationary nonequilibrium continuously-shrinking flows the comoving phase-space hypervolume $\otimes$ exhibits two paradoxical combinations of contraction and expansion properties:

First, the behavior of nearby points in phase space suggests *expansion* rather than contraction, by separating, exponentially fast in time, as the motion goes on. For a simple bouncing-ball illustration[13] of this exponential separation see Figure 10. With each bounce the horizontal offset from the ball's center increases as a geometric series. The resulting exponential instability is best visulized on the semilogarithmic plot shown in the righthand side of the Figure. But, as we have seen, simultaneously with the separating, the comoving *occupied* volume in phase space, $\otimes$, necessarily shrinks, exponentially fast, to zero.

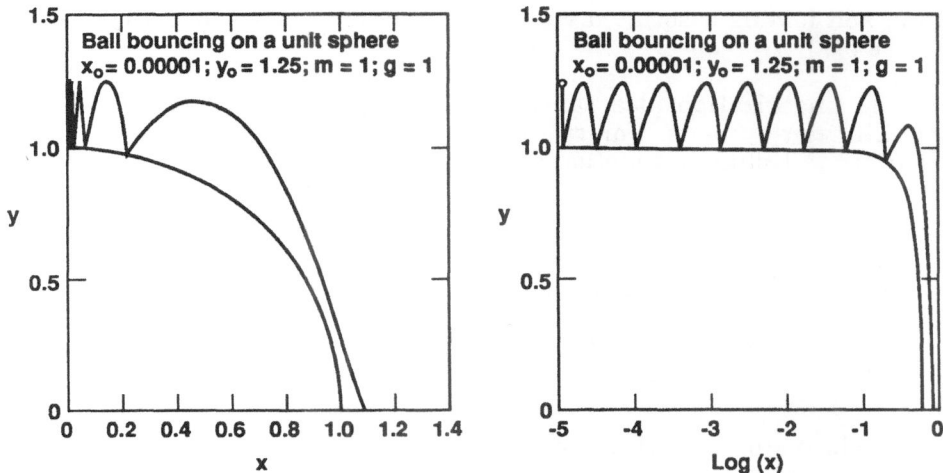

FIGURE 10. A gravity-driven bouncing ball. The Lyapunov-unstable motion causes the horizontal offset from the unstable central fixed point above the ball to increase with each bounce. The exponential character of this instability is illustrated by the logarithmic-scale motion shown on the righthand side of the FIGURE.

Second, at least for the "simple" examples considered here, it appears that the limiting and well-behaved "attractor" set toward which this shrinking motion converges is everywhere arbitrarily close to pathologically unphysical time-reversed "repellor" points which violate the Second Law of Thermodynamics[14].

The first property, expansion and separation, is typical of all strange attractors but the second Second-Law property is not. Most of the maps and flows studied in nonlinear dynamics lack the time reversibility of mechanics. Newtonian and Nosé-Hoover mechanics both exhibit the paradox that to every trajectory obeying the Second Law there is a reversed trajectory which violates it. With this second paradox the zero-volume attractors of nonequilibrium mechanics richly deserve both the appellation "strange" and the study required to make them less so. In this section we first describe the measurement of Lyapunov spectra characterizing attractors. We then review recent results and discuss the link between Lyapunov spectra and static attractor structure. At present the structural analysis of attractors is crude. It is mostly restricted to time-averaged scalar properties of three-dimensional objects and still lacks a classification scheme with which to distinguish attractors from one another.

Among the relatively-crude (because it is time-averaged) ways to analyze the paradoxical strange-attractor motion is the Lyapunov spectrum. To define it we consider the comoving deformation of an infinitesimal phase-space hypersphere, centered on a "reference trajectory" and spanned by the corotating set of infinitesimal orthonormal basis vectors $\{\delta\} = \{(\delta q, \delta p)\}$. In Figure 11 we indicate the rotation of the comoving vectors, shown as arrows, by shading half the hypersphere in which they are embedded. A trajectory segment composed of five equal time steps is shown in Figure 11. Because the comoving and corotating embedding hypersphere is infinitesimal, within it the equations of motion can be linearized relative to those applying at the hypersphere center. This linearization completely avoids the complex bending and folding associated with the Smale horseshoes of the type shown in Figure 5.

The linearization idea can be applied equally well to equilibrium or nonequilibrium systems. For simplicity, consider the linearized equations appropriate to Hamilton's motion equations:

$$d\delta q/dt = \delta p/m; \quad d\delta p/dt = \delta q(dF(q)/dq)$$

or, to simplify and generalize the notation by introducing the <u>dynamical matrix</u> D:

$$d\delta/dt = D \cdot \delta$$

These basis-vector equations of motion, just like those of the reference trajectory, are time-reversible, with the signs of δp and dt changing in a time-reversed trajectory. Such a linearized motion would convert a hypersphere to a rotating hyperellipsoid characterized by the exponential growth and decay rates of its (orthogonal) principal axes. These rates, when time-averaged and ordered from the largest, λ_1, to the smallest, λ_N, in an N-dimensional phase space, constitute the "Lyapunov Spectrum" $\{\lambda_i\}$.

An alternative description of this same growth-rate spectrum can be based on the time-averaged deformation of infinitesimal 1-, 2-, 3-, ... dimensional phase-space objects. To generalize the simple property of exponential divergence of the one-dimensional distance between neighboring trajectories, it is natural next to consider the divergence (or convergence) of infinitesimal two-dimensional phase-space areas, followed by three-dimensional volumes, four-dimensional hypervolumes, and so on. When time-averaged, the successive orthogonal growth rates required are again the Lyapunov exponents. The largest, λ_1, describes the rate of divergence of an infinitesimal one-dimensional *length* δr linking two trajectories:

$$\lambda_1 = d\ln\delta r/dt$$

By adding λ_1 to the next largest exponent, λ_2, we get the sum, $\lambda_1 + \lambda_2$, describing the growth rate for a two-dimensional *area* dA defined by three neighboring trajectories:

$$\lambda_1 + \lambda_2 = d\ln\delta A/dt$$

By adding more dimensions we finally reach the last such relation:

$$\Sigma\lambda_i = d\ln\delta\otimes /dt$$

where the sum includes all N Lyapunov exponents and $\delta\otimes$ is a comoving infinitesimal hypervolume element with the full phase-space dimensionality.

Numerical characterization of the complete spectrum of Lyapunov exponents[12] is computationally intensive. What can be done is severely limited by computational speed. In 1989 it is feasible to compute accurate values of hundreds, but not yet thousands, of Lyapunov exponents, and several such calculations have been carried out. The same basic technique applies for either stationary or time-periodic processes.

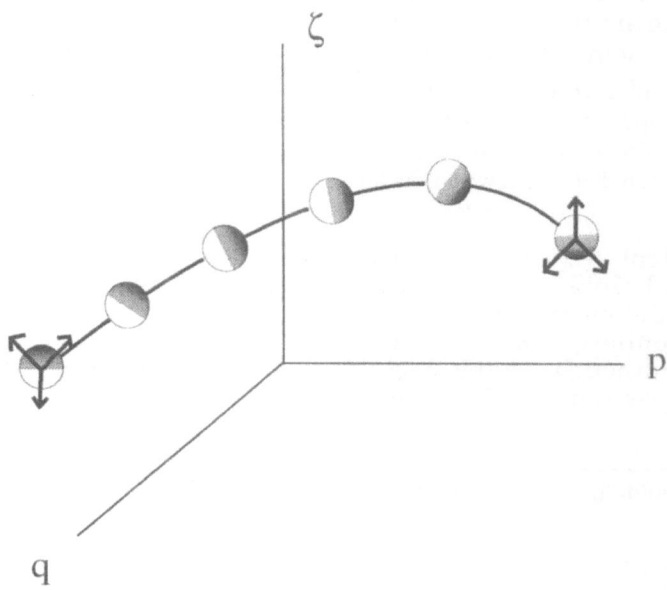

FIGURE 11. The time-development of three comoving and corotating infinitesimal phase-space basis vectors. By measuring the tendency of these vectors to grow, or to shrink, the Lyapunov spectrum can be determined.

Begin by following the construction indicated in Figure 11, erecting a comoving orthonormal set of basis vectors $\{\delta_i\}$, one for each of the N phase-space dimensions, with the origin of the set fixed on an unconstrained moving "reference trajectory". Then the linearized equations of motion of the basis vectors (which can be thought of as describing the motion of N neighboring "satellite trajectories") are solved <u>with the restriction that the basis vectors remain orthonormal</u>. This orthogonality constraint can be repeatedly imposed by Gram-Schmidt orthonormalization. Alternatively, and more elegantly, the constraint can be imposed continuously, by including Lagrange multipliers directly in the equations of motion. In all, N(N+1)/2 multipliers λ_{ij}(or constraints) are required to maintain the orthonormality of N independent basis vectors:

$$d\delta_1/dt = D\cdot\delta_1 - \lambda_{11}\delta_1,$$
$$d\delta_2/dt = D\cdot\delta_2 - \lambda_{22}\delta_2 - \lambda_{21}\delta_1,$$
$$d\delta_3/dt = D\cdot\delta_3 - \lambda_{33}\delta_3 - \lambda_{32}\delta_2 - \lambda_{31}\delta_1,$$
$$\cdots,$$

where the time-averaged values of the diagonal elements $<\{\lambda_{ii}\}>$ of the lower triangular array of Lagrange multipliers,

$$\lambda_{ij} = \delta_i\cdot D\cdot\delta_j + \delta_j\cdot D\cdot\delta_i; \lambda_{ii} = \delta_i\cdot D\cdot\delta_i$$

are the set of Lyapunov exponents $\{\lambda_i\}$:

215

$$\langle\{\lambda_{ii}\}\rangle = \{\lambda_i\}$$

The differential equations determining the Lyapunov exponents through the matrix D turn out to be *odd* in the time so that in the time-reversed motion the most positive Lyapunov exponent becomes the most negative, and *vice vers*a. But, because the rotation of the basis vectors $\{\delta\}$ is unconstrained and depends on the past history of the reference trajectory, there is no simple general relationship between the <u>orientations</u> of the vectors going backward and forward in time.

Typical equilibrium Lyapunov spectra[12] for two- and three-dimensional fluids and solids are shown in Figure 12. Such spectra are much less distinctive and informative than the phonon spectra of solid-state physics. Similarly, the spectra of rotation rates of the vectors show little structure, although, unlike the Lyapunov exponents themselves, the rotation rates increase rapidly with system size.

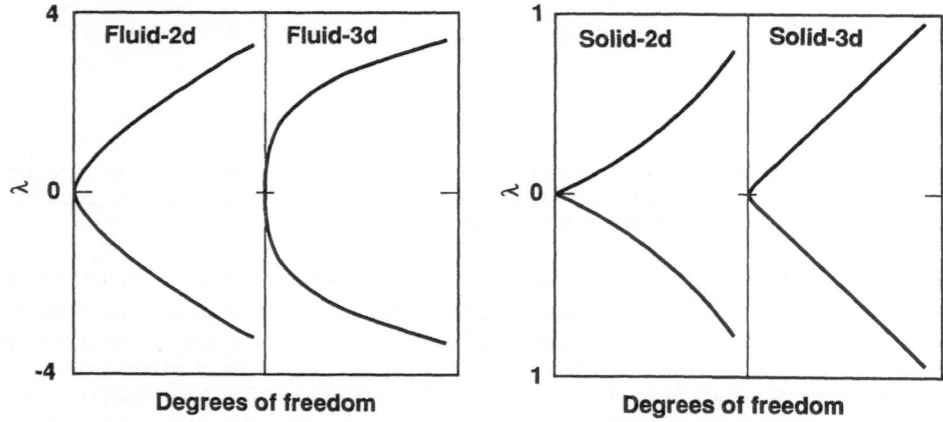

FIGURE 12. Lyapunov spectra for equilibrium two- and three-dimensional fluids and solids. The Lyapunov exponents are arranged in positive-negative pairs, increasing in magnitude from left to right. The number of such pairs is equal to the number of degrees of freedom required to describe the system.

Even *away* from equilibrium the Nosé-Hoover equations for the Lyapunov spectrum are time-reversible. This time symmetry might wrongly suggest that the Lyapunov spectrum remains symmetric about zero. In fact the symmetry is broken because the time-reversed trajectory is even less stable than the original one. Figures 13 and 14 show Lyapunov spectra for two- and three-dimensional boundary-driven steady shear flows. Figure 13 displays spectra for four two-dimensional systems with 4, 9, 16, and 25 Newtonian particles driven by isothermal boundaries. Figure 14 shows two three-dimensional eight-particle systems, one with fixed boundaries, and one with moving boundaries. For all of the five nonequilibrium systems shown in these Figures the Lyapunov exponents are, on the average, negative. Notice that in the equilibrium case shown in Figure 14 the exponents have instead the symmetric distribution suggested by the time symmetry of the equations of motion. This generic nonequilibrium shift away from the symmetric distribution occurs for any nonequilibrium flow involving the dissipation of concentration, velocity, or temperature differences.

Several interesting conclusions can be drawn from such measured spectra and their nonequilibrium shifts. The negative exponents give directly the rates at which <u>past</u> information is <u>destroyed</u> while the positive exponents give the rate at which <u>future</u> information is <u>created</u>, with the sum of the Lyapunov spectrum corresponding directly to the macroscopic thermodynamic dissipation. It is remarkable that the *dynamic* spectrum is also directly linked to a *static* property of nonequilibrium attractors, the "information dimension", through the Kaplan-Yorke conjecture.

The spectrum is also linked to irreversible thermodynamics. To see this connection let us reexamine the result from Section 2 linking the total change in entropy to the heat transfer associated with Nosé-Hoover thermostats:

$$dS_{TOTAL}/dt = \Sigma -(dQ/dt)/T > 0$$

As before, -DQ represents heat extracted from the nonequilibrium system by its surroundings. Because the frictional forces extract heat at a time-averaged rate equal to the product of the frictional force, $-\zeta p$, and the velocity, (p/m):

$$-<dQ/dt> = <\zeta p(p/m)> = <\zeta>kT$$

the time-averaged heat transfer rates, divided by the corresponding temperatures T, are exactly equal to the friction coefficients ζ for the corresponding thermostatted degrees of freedom. Thus

$$dS_{TOTAL}/dt = k\Sigma<\zeta> = k<dlnf/dt> > 0$$

Because probability (conserved by *any* flow) corresponds to the product of the probability density f(q,p,t) and the comoving phase-space hypervolume f, the logarithm of the product, $\ln(f\otimes) = \ln f + \ln\otimes$, is constant, and the comoving phase-space hypervolume $\otimes$ must shrink to zero as given by the summed Lyapunov spectrum:

$$kdln\otimes/dt = k\Sigma\lambda = -kdlnf/dt = -dS_{TOTAL}/dt$$

Thus there is a direct connection between the rate of thermodynamic dissipation, dS_{TOTAL}/dt, and the Lyapunov spectrum $\{\lambda\}$.

We see that comoving low-dimensional phase-space objects *grow* exponentially in time while high-dimensional objects *shrink*. Kaplan and Yorke made the reasonable conjecture[15] that the time-averaged linearly-interpolated dimensionality of a phase-space object which neither shrinks nor grows, so that $\Sigma\lambda$ vanishes, is equal to the "information dimension" of the corresponding strange attractor. The information dimension can be independently defined by determining, for small phase-space bins of size ε, the dependence of the integrated probability density on the bin size.

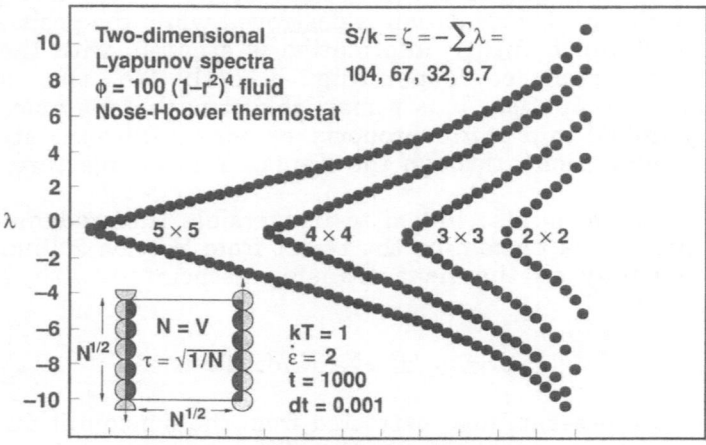

FIGURE 13. Lyapunov spectra for 4-, 9-, 16-, and 25-body two-dimensional boundary-driven steady-state shear flows. Note that the spectra are shifted toward negative values. The sum of the exponents is proportional to the irreversible entropy production.

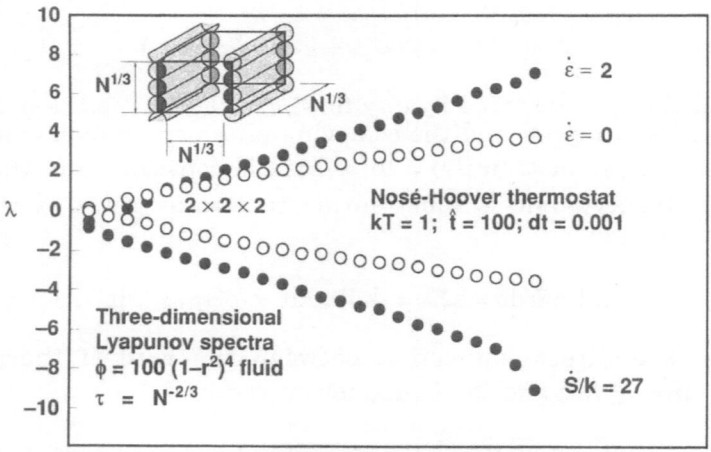

FIGURE 14. Lyapunov spectra for two eight-body three-dimensional systems. The rate-dependence of the spectrum is primarily linear in the strain rate, with an increase in the exponents' magnitudes over those found at zero strain-rate and indicated by open circles (equilibrium). The data also reveal an additional negative shift, approximately quadratic in the strain rate.

In a D-dimensional space an integrated probability density varies as the Dth power of the bin size ε. For singular distributions, in which the probability density approaches zero or infinity as ε vanishes, the variation of bin probabilities with bin size can be determined:

$$P_k(\varepsilon) = N_k(\varepsilon)/N$$

yielding the ratio, $<\ln P(\varepsilon)>/\ln\varepsilon$. For small ε this ratio approaches a measure of dimensionality called the "information dimension". The remarkable connection between the dynamic Lyapunov spectrum and the static information dimension has been verified for many simple models. In the following section we explore the concept of fractal dimension in more detail.

4. FRACTALS

A cornerstone of statistical mechanics is the Liouville Theorem result that f(q,p) is unchanged in equilibrium Hamiltonian flows. Then, because f⊗ corresponds to probability, ⊗ must likewise be unchanged by the motion. Away from equilibrium, with heat transfer included, things are different. In nonequilibrium steady states the density f diverges and the volume ⊗ vanishes.

What do the divergence of the stationary probability density f(q,p) and the vanishing of the phase-space volume ⊗ mean? Both singular behaviors seemed mysterious until the steady-state probability density <f(q,p)> could be measured and displayed graphically, for relatively-simple systems[14,16-19] with only a few degrees of freedom. The divergence of f signals the formation of a new phase-space object with zero volume but with an intricate and singular probability distribution. The singular phase-space object has structure on two scales. On a *small* scale the distribution approaches infinity exponentially fast, as $e^{-\Delta Q/kT} \propto e^{-\Sigma\lambda t}$, while on a fixed *coarse-grained* scale the integrated probability density slowly converges to a singular structure varying locally as a fractional power of the bin size. These coarse-grained distributions have a "fractal" structure. Apparent holes and gaps appear in the phase space and these cease to vanish or simplify no matter how small the scale of observation. At a sufficiently small scale this structure is no longer physics, but mathematics. In practice the uncertainty principle limits observation scales to roughly 17 digits, so that the scaling relation can exist over no more than 17 orders of magnitude.

Chhabra and Jensen[20] showed how to identify the singularities included in the multifractal structure. This can be done by finding the limiting dependence of moments of the {N_k} on the bin size ε. Figure 15 shows the singularity spectrum for a Lorentz gas or "Galton Board" problem detailed in the next section. The spectrum describes the fractal character of two-dimensional "Poincaré-section" cuts through the three-dimensional phase space. The abscissa gives the power-law dependence of the singularity on the bin size and the ordinate the corresponding "bin counting" fractal dimensionality. An ordinary two-dimensional cross section for this problem would correspond to a delta function at the point (2,2). The numerical results show that though most of the distribution corresponds to contraction, with a and f both less than the spatial dimension of two, the nonequilibrium steady-

219

state distribution also contains anomalous expanding regions, with a probability density that vanishes on a small scale (a > 2).

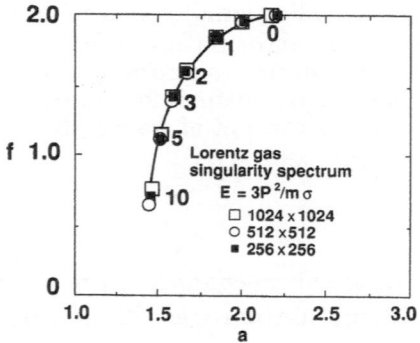

FIGURE 15. The multifractal singularity spectrum for the steady nonequilibrium field-driven Galton Board. The box-counting dimension, f(a), is plotted as a function of the singularity strength a. The singularity strength is the power with which the integrated bin probability depends upon the bin size ε in the small-ε limit.

5. FEW-BODY RESULTS

The study of few-body systems has been useful in understanding the pardoxes associated with irreversibility, thermodynamics, and phase-space structure of nonequilibrium flows. The combination of Runge-Kutta integration with MacIntosh and Stellar Graphics, makes it possible to view results quickly and to preserve them inexpensively on videotape. In this section we describe three such interesting problems: The Galton Board, the Viscous Lorentz Gas, and the Free Expansion in One Dimension.

The Galton Board

For the simplest nonequilibrium steady-state illustration of Nosé-Hoover mechanics consider the Galton Board[14-16] illustrated on the righthand side of Figure 16. The system is a caricature of solid-state diffusion, and involves the field-induced isokinetic (constant-speed) motion of a mass point in a periodic array of scatterers. A parallelogram unit cell is indicated in the Figure. A constant horizontal field tends to drive a moving mass point to the right, parallel to the field direction. The fixed hard-disk scatterers periodically scatter the momentum, with a bias toward head-on collisions. After scattering, the acceleration toward the x direction begins again. A single particle moves through the board, at *fixed* kinetic energy, $p^2/(2m)$, undergoing chaotic mixing collisions with the scatterers. Because the kinetic energy is fixed, the space describing successive collisions can be reduced to a two-dimensional one, spanned by the two angles (a for position and b for momentum) required to describe a collision. See Figure 17 for a definition of the angles a and b and Figure 16 for the corresponding Poincaré section showing, as separate dots, 10,000 successive collisions.

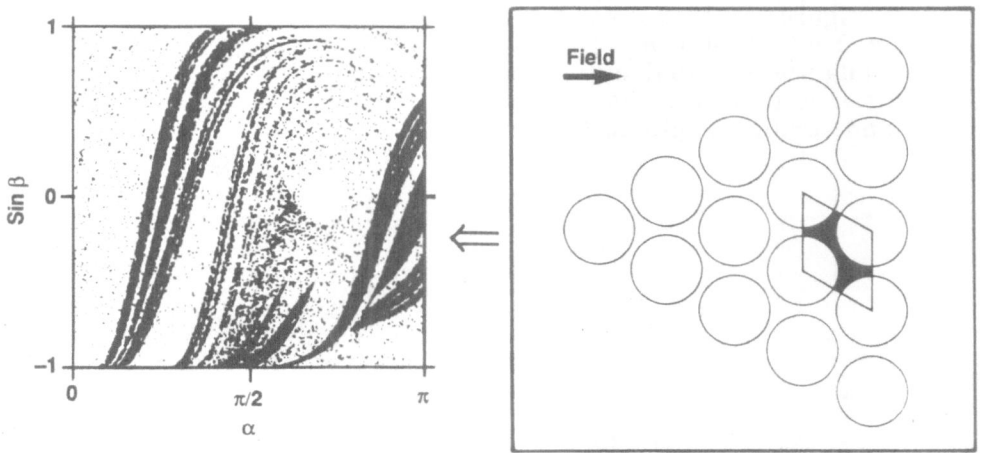

FIGURE 16. The righthand side of the FIGURE indicates the geometry of a finite "Galton Board", in which an external field accelerates a mass point to the right, through the periodic array of scatterers. The values of the two angles, α and β, describing collisions and defined in FIGURE 17, give strange attractors with Poincaré sections of the kind shown on the lefthand side of the FIGURE.

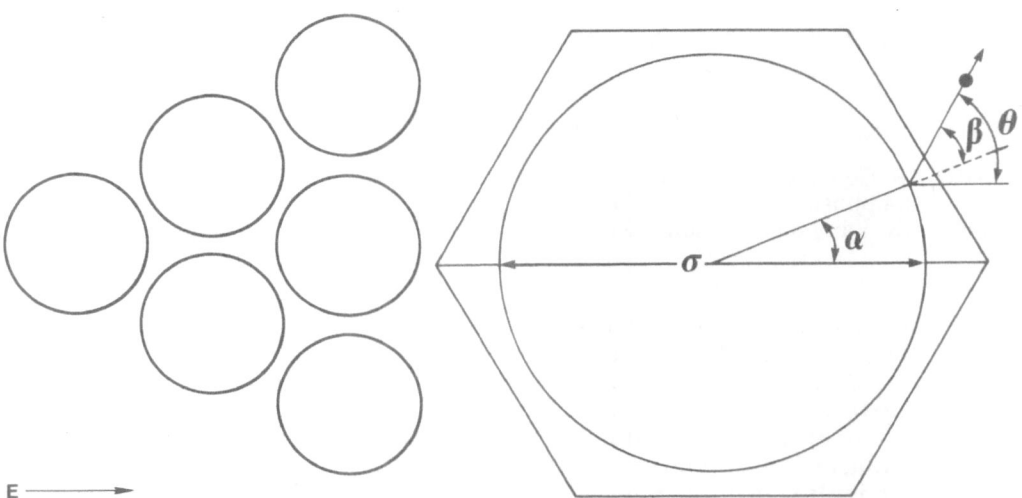

FIGURE 17. Definition of the angles α and β describing Galton-Board collisions.

The equations of motion for the Galton Board problem have simple analytic forms. The basic one, which gives the time variation of the direction of motion between collisions, is

$$d\theta/dt = -pE\sin\theta.$$

A detailed numerical investigation reveals that the resulting flow always behaves irreversibly, as it must, with a positive field-dependent conductivity. The scattering particle moves to lower and lower potential energy at a time-averaged steady rate. The visual appearance of the phase-space distribution converges rapidly to a "strange attractor" with a cross section of the type shown in Figure 16. Visual convergence to a nonequilibrium steady state takes only about 10 collisions. This is shown in Figure 18 by tracing the time history of quadrants of 2500 initial conditions after 1, 2, 3, 5, and 10 collisions each. On a long time scale the intricate multifractal structure penetrates to smaller and smaller length scales.

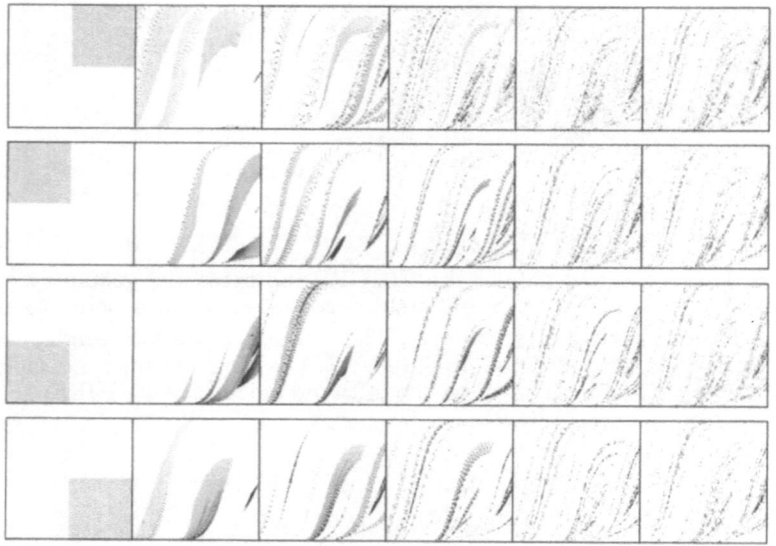

$$E = 3p^2/m\sigma$$

FIGURE 18. Convergence of four sets of 2500 initial conditions, each taking up one-fourth of the phase-space Poincaré section of FIGURE 16, to the steady-state strange attractor. Evolving distributions after 1, 2, 3, 5, and 10 collisions are shown.

The Galton Board fractal objects are fascinating. Successive enlargements show that the structure persists to the smallest feasible scales. See Figure 19 for a series of four enlargements. Such enlargements are used in the logarithmic extrapolation of bin populations to smaller and smaller sizes. The multifractal spectrum of Figure 15 was generated by analyzing a series of enlargements equivalent to ten successive twofold magnifications. At any particular scale numerical work strongly suggests that increasing the number of collisions will eventually fill every bin, no matter how small or improbable. Thus this nonequilibrium steady mixing flow is ergodic, but with the more probable parts of phase-space becoming more and more sparse as the scale of observation is refined. The time-reversibility of the equations of motion implies the existence of a repellor (with the attractor velocities reversed--see Figure 20). It is an amazing consequence of the ergodic nature of the flow that every repellor point lies arbitarily close to attractor points, and *vice versa*. The plausibility of this ergodicity is apparent from the more-detailed Figure 21, which shows 640,000 points distributed over a computer-generated print originally covering an area of one square meter.

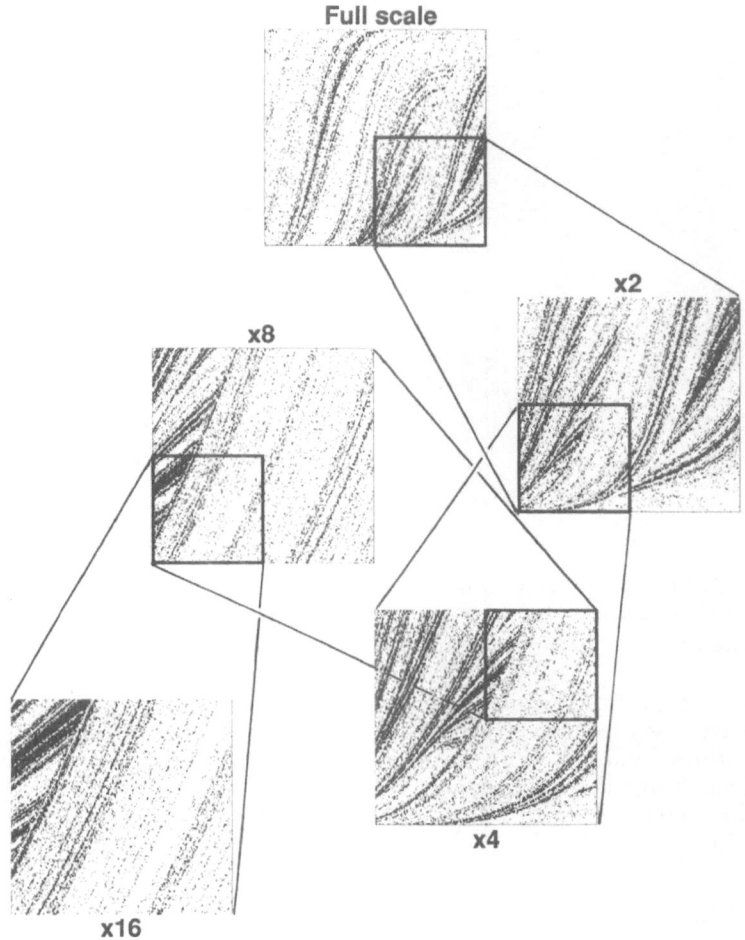

FIGURE 19. Galton Board attractor structure shown at 2-, 4-, 8-, and 16-fold magnification. In each snapshot 10,000 collisions are shown.

The Galton Board motion takes place in a three-dimensional phase space and accordingly has three Lyapunov exponents. One is positive, giving the mean rate of growth in phase-space separation due to scattering off the convex surfaces. One is zero, corresponding to the motion along the trajectory direction. One is negative, sufficiently so that the three-term exponent sum is negative and equal to the rate of total entropy increase $dS_{TOTAL}/dt = -(dQ/dt)/T$. For certain ranges of field strength the positive Lyapunov exponent changes sign and the motion collapses onto a one-dimensional limit cycle with a Poincaré cross section composed of a few discrete dots.

This Galton Board problem provides a nice mechanical illustration of the Second Law of Thermodynamics. The requirement that the phase-space distribution is confined to a finite region of space implies that the time-averaged friction coefficient ζ be positive and that the particle therefore moves from higher to lower potential energy. If the friction were negative the energy would eventually diverge. If the friction were zero there could be no average current.

$$E = 4.00 \ p^2/m\sigma$$

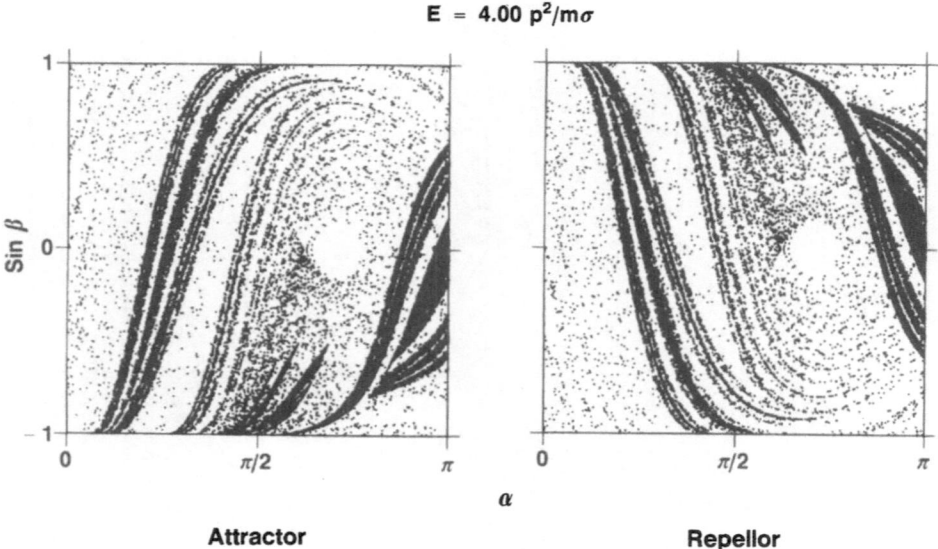

Attractor　　　　　　　　　　**Repellor**

FIGURE 20. Galton Board attractor and repellor sections. Because the equations of motion are time-reversible the reversed (repellor) motion corresponds to a reflection of the attractor about a horizontal line. This changes the sign of the velocity angle β.

Series of related many-body simulations, with half the particles accelerated in one direction by a field and the rest in the opposite direction, gave analogous results. The dimensionality of the many-body strange attractor describing the motion varied quadratically with the field strength (as would be expected from the mirror symmetry linking positive field and negative field results).

The Viscous Lorentz Gas

Numerical analysis of a time-varying problem is complicated by the additional phase-space dimension. Time variation can be introduced through a varying external field or by using time-periodic boundary conditions. A periodic shear (See Figure 22 for a three-dimensional version) is the simplest example, and the two-body "Lorentz Gas" shear flow is the simplest of these. The viscous Lorentz Gas[18] resembles the Galton Board. The viscous flow can be thought of as a symmetric two-body problem, or as an equivalent one-body scattering problem, but now with the boundaries moving at a rate given by the macroscopic constant strain rate. Four unit cells of such a periodically-shearing system are shown, at five equally-spaced times, in Figure 23.

The local motion of a moving particle, relative to the systematic velocity u_x, can be constrained to be isokinetic, by using Gauss' Principle of Least Constraint. If this is done, the resulting equation of motion, *between* collisions, is again simple,

$$d\theta/dt = \gamma\sin^2\theta$$

FIGURE 21. Details of the full attractor Poincaré section of FIGURE 19 representing 640,000 successive collisions. The original computer-generated picture was one square meter in size. Numerical analyses using 1000 times as many points as are shown here are currently feasible.

where γ is the "strain rate",

$$\gamma = du_x/dy$$

The hard-disk scattering analysis can be simplified by considering the large-F limit of a *constant* repulsive central force F. With this assumption the equation of motion *during* collisions,

$$d\theta/dt = (F/p)\sin(\alpha-\theta)$$

can likewise be integrated analytically. By taking the large-F limit and using a geometric condition, that the polar-coordinate integral of dr/dt *through* a collision vanishes, the post-collision direction of motion can be found as the solution of a single transcendental equation. In the large-F limit the angle α, which measures the location of the scattering particle relative to the x axis, is constant during a collision.

This problem is periodic in time, as well as in space, because in a time equal to $1/\gamma$ the system progresses from one checkerboard-like configuration to the next. For an initially-square unit cell, a shear of unity, corresponding to an oblique-cell corner angle of 45°, brings the system back to its original configuration. See again Figure 23 for a four-cell square-corner version of this periodic shear flow. Thus, in addition to the two angles describing a collision, the *phase* of the boundary condition must be specified, making the Poincaré-section description of successive collisions a three-dimensional problem. Nevertheless, a bin structure of $2,097,152 = 2^7 \times 2^7 \times 2^7$ bins is sufficient to determine an accurate multifractal spectrum of singularity strengths. The results are similar to those shown for the Galton Board in Figure 15. We find no trace of the cusps found by Morriss in his analysis of a similar soft-disk problem[19].

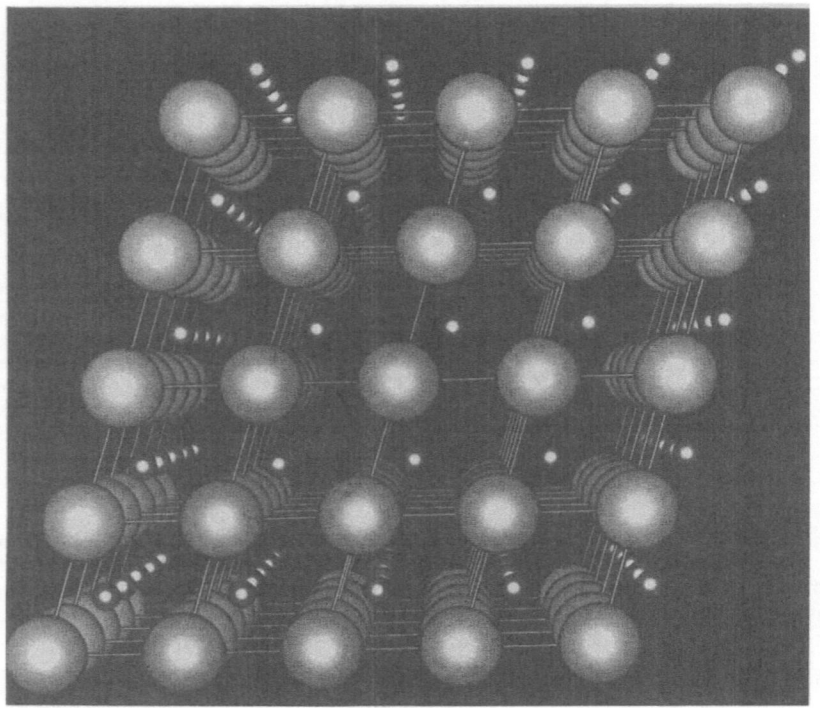

FIGURE 22. A two-body three-dimensional periodic shear flow. For convenience in identifying the unit cell (125 cells are shown in the FIGURE) the two particles are shown as spheres with different diameters.

The distribution of Poincaré-section points in the four-dimensional Lorentz-gas' phase-space, corresponding to the information dimension D_1, is shown in a pair of stereo views in Figure 24. To see this "Poincaré cube" in three dimensions make a copy of this Figure in which the spacing between the centers of the left-eye and right-eye images corresponds to the distance between your own eyes. A viewing distance of about 30 cm is best.

The two angles describing the collisions, α and β, are displayed as a function of the shear (the axis perpendicular to the plane of the paper). This shear-flow problem has a somewhat simpler appearing Poincaré-cube structure than does the diffusive Galton Board problem, despite the extra phase-space dimension. Just as in the Galton Board case the Poincaré cube pictures suggest the filling of all bins with a zero-volume completely-ergodic chaotic attractor.

At sufficiently high strain rates the motion collapses onto a one-dimensional periodic limit cycle. In the resulting Poincaré-cube representation such a cycle appears as a discrete set of dots. In either of these steady nonequilibrium two-body problems, diffusion or shear, the strange-attractor motion develops so as to give positive transport coefficients consistent with the Second Law of Thermodynamics.

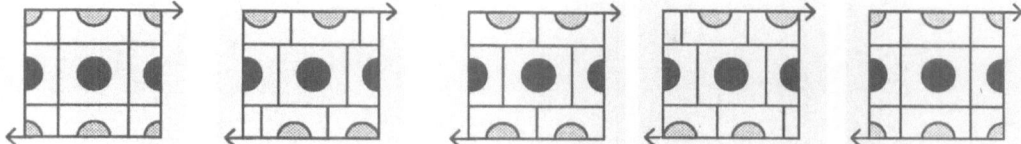

FIGURE 23. Moving boundary particles for the two-dimensional periodic two-body viscous Lorentz gas discussed in Section 5. The square shown represents an area of four unit cells. Only the periodically-repeated Boundary particle is shown. A mass point moves in the field of these particles. Snapshots corresponding to shears of 0, 1/4, 2/4, 3/4, and 1 (equivalent again to 0) are shown.

Free Expansion in One Dimension

There are many paradoxes associated with the thermodynamics of incompressible phase-space flow described by Liouville's Theorem. The best known is an adiabatic free expansion in which the expanding fluid does no external work, accepts no heat, and hence expands at constant energy. The paradox is that the phase-space density, which, according to Gibbs, should vary with entropy, remains unchanged in the process while the thermodynamic entropy increases. This paradox can be resolved, or at least avoided, by using Nosé-Hoover mechanics to embed the paradoxical expansion in a periodic process. This approach emphasizes the importance of interactions with external forces in any process designed to measure entropy. To embed the free expansion in a periodic process consider a three-step periodic cycle:

i)   Expand the volume instantaneously.
ii) Compress the system, at a finite rate, back to the initial volume.
iii) Thermostat the dense hot system to restore the original temperature.

The simplest realization of this cycle is one-dimensional. To represent a system initially confined we consider a thermostatted particle in a simple one-dimensional "box" given by a harmonic Hooke's-Law potential:

$$\phi = x^2/2$$

To start a periodic expansion, compression, and thermalization cycle the potential ϕ changes discontinuously to the new form:

$$\phi = \{(x+1)^2/2 \quad \text{or} \quad 0 \quad \text{or} \quad (x-1)^2/2\}$$
$$\text{for}$$
$$\{x<-1 \quad \text{or} \quad -1<x<1 \quad \text{or} \quad 1<x\}$$

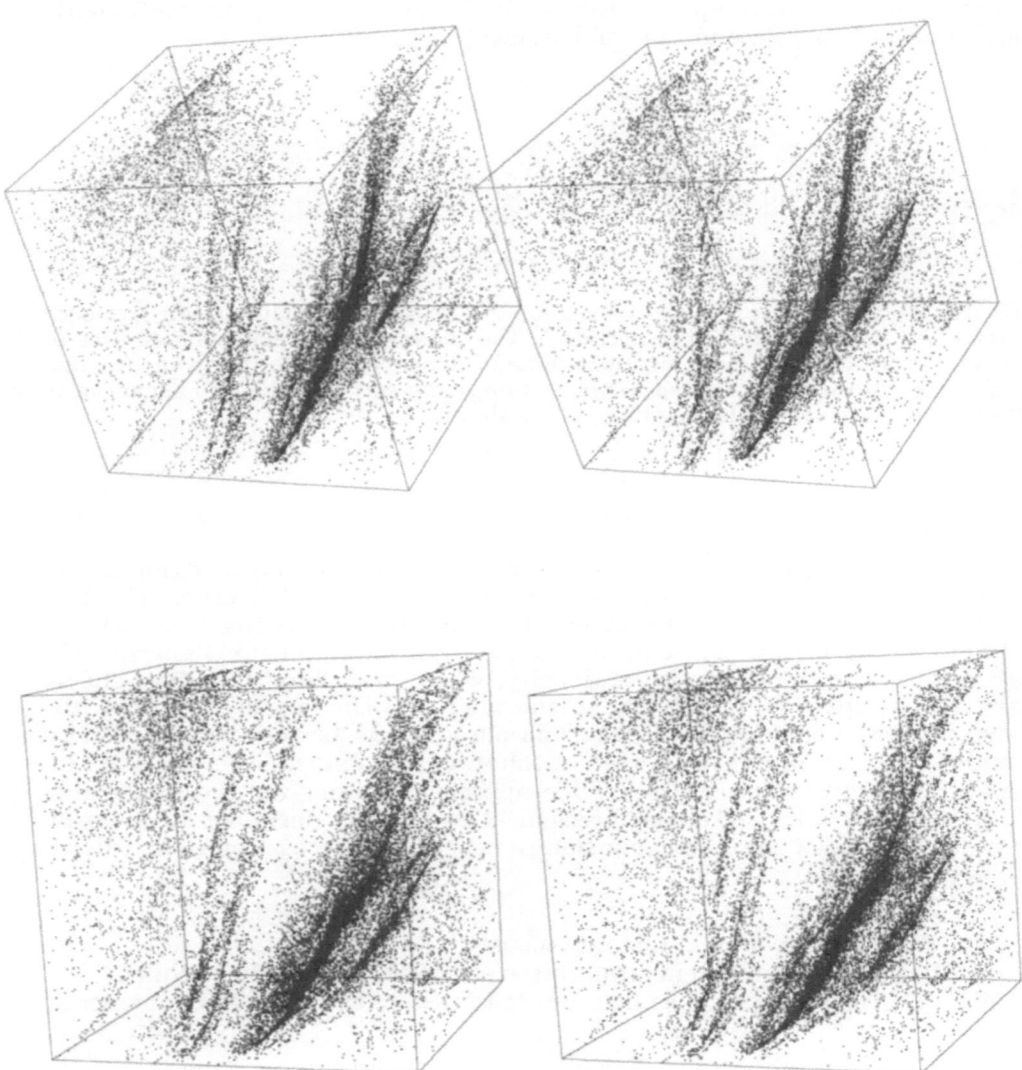

FIGURE 24. Two stereo views of the phase-space Poincaré cube detailing the phase-space distribution of collisions for the two-dimensional shear flow of a Lorentz Gas driven by the periodic boundaries of FIGURE 23. The (mainly) vertical and horizontal axes correspond respectively to $\sin\beta$ and α. The axis (mainly) perpendicular to the plane of the paper represents strain, or equivalently, phase, in the time-periodic boundary motion of FIGURE 23.

Next, over a relatively long period of 1000 fourth-order Runge-Kutta time steps of length dt = 0.01 each, the moving particle is slowly and steadily compressed. This adiabatic compression, at one tenth the thermal velocity, corresponds thermodynamically to a nearly-isentropic process. Finally, the compressed particle is thermostatted, using Nosé-Hoover mechanics, by solving the equations of motion

$$dx/dt = p \qquad dp/dt = -x - \zeta p \qquad d\zeta/dt = (p^2 - 1)/\tau^2$$

where the relaxation time, τ, is 0.1. The thermostatting is applied over the shortest interval such that ζ begins and ends at zero. The particle mass has been chosen equal to one, for convenience.

This three-step irreversible thermodynamic process, incorporating expansion, compression, and thermalization, is perfectly time-reversible, obeying the same equations of motion in either direction of time. Nevertheless the requirement that the occupied phase space not diverge implies that the motion can only converge to a zero-volume strange attractor. By Liouville's Theorem, the phase volume is unchanged during both the expansion and compression steps. Thus the thermalization must, on the average, contract the occupied phase volume.

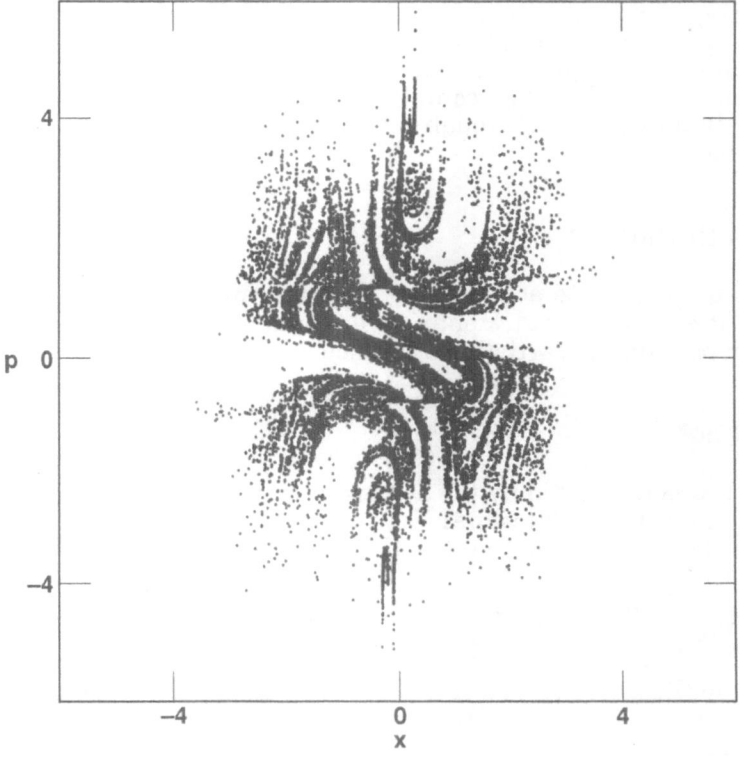

FIGURE 25. Strange attractor Poincaré section corresponding to the one-dimensional free expansion described in the text. After the expansion, compression, and thermalization stages, at the instant just preceding the next expansion, the momentum is plotted as a function of the particle coordinate. The Lyapunov-unstable strange attractor reflects the entropy increase associated with free expansion.

Numerical work confirms this conclusion. Figure 25 shows a Poincaré section of the resulting strange attractor. Just as in the Galton Board and Viscous Lorentz Gas the fractal structure is apparent. It can be verified that the dissipation associated with the thermostatting exactly accounts for the irreversible entropy production associated with the free expansion step. Thus the Nosé-Hoover mechanics makes it possible to analyze transient irreversibility too, by embedding the problem in question in a periodic process.

6. SUMMARY

To describe irreversible processes at an atomistic level it is convenient to apply boundary constraints using Nosé-Hoover mechanics. The inevitable conversion of work to heat, summarized by the Second Law of Thermodynamics, appears in *any* nonequilibrium state, steady, periodic, or transient. With Nosé-Hoover reversible and deterministic mechanics, the direction of the motion is determined by the geometric requirement that the phase-space volume cannot grow with time. By using techniques borrowed from nonlinear dynamics it can be established that the resulting phase-space structures are strange attractors, hosts of chaotic, microscopically reversible but macroscopically irreversible motion. The magnitude of the shrinking within the attractor depends only on the work done by external forces, or, equivalently, the heat extracted by Nosé thermostats. The information dimension of the motion in phase space can be estimated from the Lyapunov spectrum and more highly correlated fractal dimensions can be obtained by bin-counting methods. The complexity of chaotic many-body flows will ultimately be understood through study of few-body caricatures of the type described here.

ACKNOWLEDGMENT

It is a pleasure to acknowledge the continuing counsel and inspiration of Harald Posch and Brad Holian. We thank Kirk T. Hadley and Paul H. Dunlap for painstaking artistic assistance.

REFERENCES

* This work was supported by the Department of Energy and performed at the Lawrence Livermore National Laboratory under the auspices of the University of California pursuant to Contract W-7405-Eng-48.

1. J. Gleick, **Chaos, Making a New Science** (Viking, New York, 1987).
2. J. A. Yorke, "**Chaotic Dynamics**" (NASA Ames, California Lectures of 12 July 1989, unpublished).
3. See the Proceedings of Recent Meetings of the Materials Research Society (November 1988, Boston, for instance) and topical conferences of the American Physical Society such as **Shockwaves in Condensed Matter 1987**, S. C. Schmidt and N. C. Holmes, Eds. (North-Holland, Amsterdam, 1988)
4. E. Meiburg, Phys. Fluids 29, 3107 (1986).
5. M. Mareschal, M. M. Mansour, A. Puhl, and E. Kestemont, Phys. Rev. Letts. **61**, 2550 (1988).
6. E. N. Lorenz, J. Atmos. Sci. **20**, 130 (1963).

7. F. F. Abraham, W. E. Rudge, D. J. Auerbach, and S. W. Koch, Phys. Rev. Letts. **52**, 445 (1984).

8. A. J. DeGroot, S. R. Parker, and E. M. Johansson, **SVD and Signal Processing, Algorithms, Applications, and Architectures** (North-Holland, Amsterdam, 1988)

9. S. Nosé, Mol. Phys. **52**, 255 (1984) and **57**, 187 (1986); and J. Chem. Phys. **81**, 511 (1984). W. G. Hoover, Phys. Rev. **A31**, 1695 (1985).

10. J. Jellinek and R. S. Berry, Phys. Rev. **A38**, 3609 (1988).

11. B. L. Holian, W. G. Hoover, and H. A. Posch, Phys. Rev. Letts. **59**, 10 (1987); W. G. Hoover, Phys. Rev. A37, 252 (1988).

12. H. A. Posch and W. G. Hoover, Phys. Rev. A39, 2175 (1989).

13. W. G. Hoover, **Lecture Notes in Physics #258: Molecular Dynamics** (Springer-Verlag, Heidelberg, 1986).

14. W. G. Hoover and B. Moran, Phys. Rev. **A** (in press, 1989).

15. H. G. Schuster, **Deterministic Chaos** (Physik-Verlag, Weinheim, FRG, 1984).

16. W. G. Hoover, C. G. Tull, and H. A. Posch, Phys. Lett. **A131**, 211 (1988).

17. W. G. Hoover, H. A. Posch, B. L. Holian, M. J. Gillan, M. Mareschal, and C. Massobrio, Mol. Sim. **1**, 79 (1987).

18. B. Moran, W. G. Hoover, and S. Bestiale, J. Stat. Phys. **48**, 709 (1987).

19. G. P. Morriss, Phys. Rev. **A39**, 4811 (1989); G. P. Morriss, Phys. Lett. **122A**, 236(1987); A. J. C. Ladd and W. G. Hoover, J. Stat. Phys. **38**, 973 (1985).

20. A. Chhabra and R. V. Jensen, Phys. Rev. Letts. **62**, 1327 (1989).

THE APPROACH TO EQUILIBRIUM AND MOLECULAR DYNAMICS

Ilya PRIGOGINE[#][§], Eddy KESTEMONT[#], Michel MARESCHAL[#]

[#] Université Libre de Bruxelles, CP231
B1050, Brussels, Belgium

[§] Center of Statistical Mechanics and Thermodynamics
University of Texas at Austin, Texas

1. INTRODUCTION

Some of the first studies in Molecular Dynamics were already concerned with the problem of approach to equilibrium. Alder and Wainwright[1] and later Bellemans and Orban[2] have calculated the time evolution of the Boltzmann $\mathcal{H}$-function for a dilute gas of hard disks and hard spheres put initially in a non-equilibrium state. This was realized by choosing initial velocities of particles having all the same magnitude but pointing in random directions. These simulations have shown that the irreversible behavior predicted by Boltzmann already appears for systems made of a few hundred particles; these experiments have also illustrated the fact that MD trajectories in phase space are accurate over a few collision times only, as can be seen by reversing the particle velocities.

MD simulations which show non-equilibrium behavior have since been used for the study of transport properties and to investigate the connection between the microscopic and macroscopic description of dynamical systems. Some very remarkable results have been obtained in this field, many of them being reported in this volume. The approach to equilibrium ask for precise results and this is quite demanding on cpu time. Besides, some examples have also demonstrated an important N-dependence. Let us only cite the Bellemans and Orban[3] calculations made for a non-equilibrium gas. They computed the contribution of the higher orders terms in the development of the canonical entropy in terms of distribution functions of an increasing number of particles: systems as large as 100 000 particles would be necessary in order to measure the two particles distribution function contribution to the (canonical) entropy.

We come here to some of the basic problems of modern statistical mechanics. How to reconcile time evolution predicted by Boltzmann for his $\mathcal{H}$-function with the time-constancy of the Gibbs $\mathcal{H}$-function, which is a direct consequence of the Liouville equation ? As generally admitted, irreversibility

Microscopic Simulations of Complex Flows
Edited by M. Mareschal, Plenum Press, New York, 1990

is usually associated to an increase of entropy. Every macroscopic system such as dense gas, liquid or solid, presents two aspects, related to order and disorder. In traditional statistical mechanics, we have on one side "disorder" in velocity distribution and on the other side "order" as expressed in pair-correlation functions or in higher-order correlations functions. Our aim here is to show how order is related to irreversible processes. It is quite remarkable that it is possible to introduce initial conditions such that the approach to equilibrium state involves a deviation from Boltzmann $\mathcal{H}$-theorem, and shows the existence of some kind of dialectics between order and disorder. Some years ago, one of us[4] has already introduced the idea of flow of correlations as providing a basic mechanism for the approach to equilibrium, and determining the direction of the arrow of time. We shall not go here into more details[5]. Note however that in this flow of correlations the canonical entropy

$$S_{can} / k_B = - \int f_N \ln f_N \, d \, \mathbf{r}^{(N)} \, d \, \mathbf{p}^{(N)} \qquad (1)$$

remains constant in time essentially because the decay to their equilibrium values of the contributions of the lower order distribution functions, $f_1, f_2, f_3...$ is compensated by the departure from equilibrium of higher order contributions. Hence the name of flow of correlations.

In the paper by Bellemans and Orban, this idea was already investigated. The initial state realized on the computer was controlled at the level of the one particle distribution function, but no correlations existed between the particles. In the course of time, the two particle correlations function are first build and relax to its equilibrium value, then the three particle correlations and so on. After one collision time, the one particle distribution function has already reached its equilibrium form while higher orders correlations have still to be build up. This approach to equilibrium is schematically represented as the sequence

$$f_1 \rightarrow f_2 \rightarrow f_3 \rightarrow .... \rightarrow f_N \qquad (2)$$

In a previous paper, we have already shown how the correlations which are present between colliding particles are destroyed by successive collisions[6]. Here we want to report on simulations made in order to follow the approach to equilibrium of the spatial order between particles, as expressed by 2-body correlations. We shall study the transient behavior of a fluid consisting of an assembly of N particles, put initially out of equilibrium, and sufficiently dense, so that the pair correlation function at equilibrium already shows a few peaks. As initial conditions, we have a Maxwellian velocity distribution, while the particles are put at the nodes of a 2-dimension triangular lattice. So, we start with a system presenting *too much* space correlations. Because of the constancy of the canonical entropy, we expect that while the system evolves to disorder (taking binary correlations into account), it has to go to order (taking velocity distribution into account). This means that velocity distribution would temporarily leave the Maxwellian state. This is a quite remarkable "anti-Boltzmann" behavior. Moreover, this shows a strong coupling between the various degrees of freedom of the system. So, here, at the microscopic level of dynamical description, like in macroscopic physics, irreversibility is both generating order and disorder.

In the following paragraph, we present the system and the simulation procedure. Next, we show the time behavior of the pair correlation function in the simulations done and then the evolution of the moments of the velocity distribution function. We end with a few remarks on the perspectives offered by molecular dynamics in the problem of the approach to equilibrium.

2. THE MODEL

To simplify as much as possible the dynamics of the system, we consider a two-dimensional fluid made of one hundred hard disks. The particles are placed in a rectangular box whose dimensions are chosen to match the desired numerical density. In the units of the molecular diameter, the number density ranges from 0.3 to 0.8 (close-packing corresponds to nearly 1.15 in two dimensions). As usual, periodic boundary conditions are used (a particle leaving the system from one side re-enters on the opposite side with the same velocity).

Particles are followed in time, undergoing binary instantaneous collisions. To obtain a sufficient precision on the calculated observables, one thousand independent (or nearly independent) experiments are performed, all starting from the same spatial configuration but with different velocities for the particles. Quantities of interest are then computed during each run (as a function of time) and averaged over all the experiments.

The initial state is the following: the hard disks are placed at the nodes of a triangular lattice with velocities sampled from a Maxwellian distribution corresponding to a given temperature. In order to make a non-equilibrium ensemble average, many trajectories are run under the same conditions, from one hundred to one thousand. The average over these sets of non-equilibrium states permits to study the time dependence of the pair correlation function, g_2 (r) (hereafter denoted $g(r,t)$), and of the one particle velocity distribution function: in particular, we have measured the moments $<v^2>$ and $<v^4>$ of this distribution function as a function of time, to see the possible disturbance of the initial distribution from equilibrium.

The correlation function $g(r,t)$ is computed as the number of particles at distance r of a given particle per unit volume (in our case unit area). It is to be noted that this computation has a dependence in N^2 hence the choice of a small size for the systems studied.

At the initial time, because the particles are regularly placed, $g(r,0)$ is made of a set of peaks. As time goes by, these peaks smear out to give the usual form of the equilibrium correlation function, with an asymptotic value of 1 for large values of r [7] (in our case, a few molecular diameters).

We also study the behavior of the velocity distribution function when, initially, particles are not randomly located: the chosen density in that case is set equal to 0.6. Here also, the particles are initially placed on a lattice. The initial velocities for each experiment are sampled from a Maxwellian distribution. It is to be noted that this technique implies a fluctuation in the kinetic energy from one experiment to the other, as it should in a canonical distribution; of course, during any given run, the energy and the total momentum are constant. We verified that the expected mean values over all experiments are indeed correct. In order to follow the possible departure from a Maxwellian of the velocity distribution, we compute the time dependent quantity R, defined as

$$R = \frac{<v^2>^2}{<v^4>}$$

(3)

and whose theoretical equilibrium value is 0.5. <...> denotes an ensemble average performed over many different systems. As $<v^2>$ is constant in time, the time dependence of R comes from the $<v^4>$ term. Its departure from 0.5 measures the non gaussian character of the velocity distribution function.

3. SOME RESULTS

Figure 1 shows the correlation functions g(r,t) after 5 mean collision times, for r ranging from 1 to 5 molecular diameters. The three curves correspond to number densities equal respectively to 0.3, 0.6 and 0.8. In each case, the different maxima of the function are, as is well known, reminders of the first, second, ... neighbors of the close-packed structure. For high values of r, the functions tend to the value 1 which corresponds to an homogeneous situation: the probability to find a particle at such distances of a given one becomes constant.

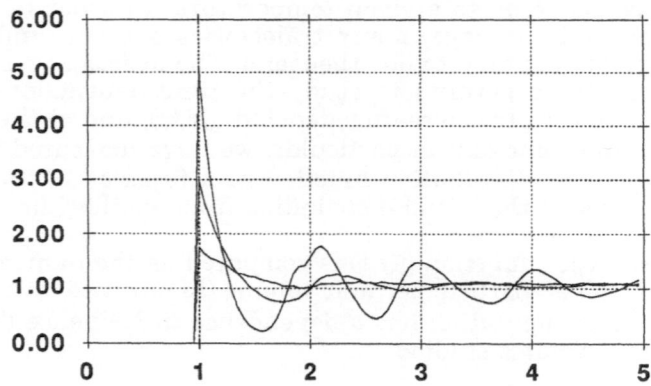

FIGURE 1. g(r) at equilibrium for densities 0.3, 0.6 and 0.8 as functions of r (in molecular diameter units).

Figure 2 gives a three dimensional plot of g as a function of r (in molecular diameter units) and t (in mean collision time units τ computed from an equilibrium molecular dynamics experiment). As can be seen on this graph, even for very short times, characteristic features of g(r,t) do appear, in particular the maximum related to the first neighbors. Of course, the final number of peaks and their intensity are related to the density of the system: for very low values of the density, they nearly vanish while for high densities, many important peaks are present.

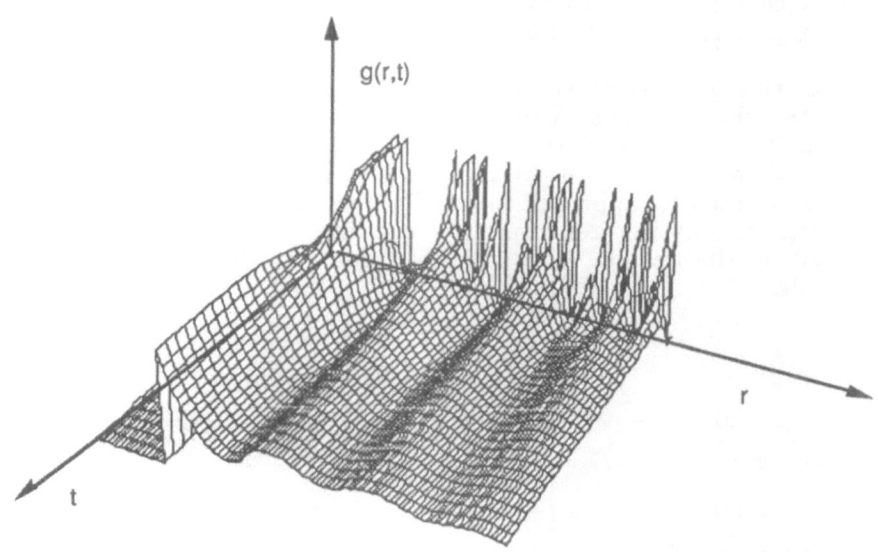

FIGURE 2. pair correlation function, g(r,t), as a function of time. The total time shown on the graph corresponds to three collision times.

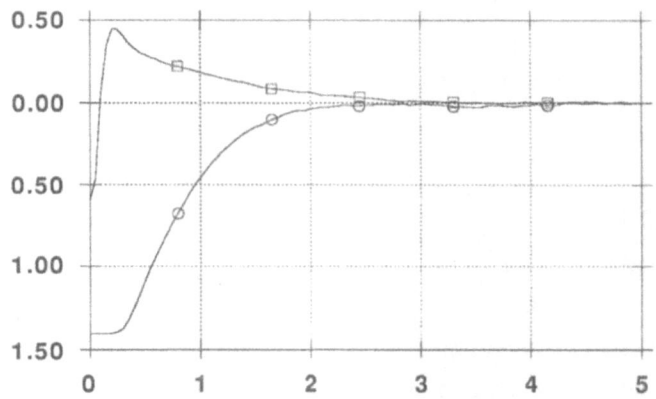

FIGURE 3. $g(r_1)$ (circles) and $g(r_2)$ (squares) as functions of time (density 0.6); time unit: mean collision time τ.

Initially the pair correlation function has high peaks at distances where particles are located and is vanishing in between these points. On a time scale of less than one collision time (i.e. the time between two collisions), one sees that the maxima tend to decrease towards their equilibrium values while the minima increase. The total time shown on the figure corresponds to 3 collision times.

In figure 3, we have plotted the time behavior of g for two particular values of r corresponding to the first (r_1) and second maximum (r_2) of the equilibrium pair correlation function (density = 0.6).

For the very first time steps important transients are present; moreover the choice of the initial condition corresponds to totally arbitrary values of g for these two distances. If one looks at the time behavior of the departures from equilibrium $g(r_1,t) - g(r_1,\infty)$ and $g(r_2,t) - g(r_2,\infty)$ for times such that a systematic approach to the equilibrium values is observed (figure 4), then an interesting feature appears: the first maximum of $g(r)$, is approached by the system more rapidly than the one corresponding to the second shell.

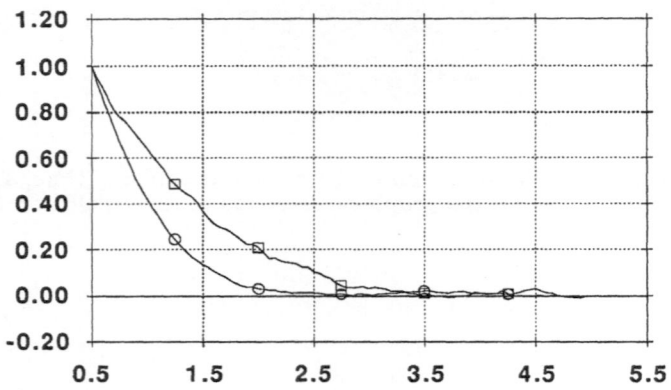

FIGURE 4. $g(r_1,t) - g(r_1,\infty)$ (circles) and $g(r_2,t) - g(r_2,\infty)$ (squares) as functions of time (in units τ); density 0.6

It is as if the process of building the spatial correlations takes more time for larger distances, the equilibrium being reached first at small r values (the initial values of g are arbitrarily normalized to unity to facilitate the comparison of time scales).

To study the behavior of the one particle velocity distribution function and to get some reference values, we followed in time a system in equilibrium to obtain an estimate of the fluctuations of the quantity R. One hundred successive experiments were performed to evaluate this quantity in time. The result is given in figure 5. As expected, the value of R fluctuates near 0.5; however, due to the smallness of the system, standard deviation remains quite important of the order of 0.2 ; the estimated absolute error for these experiments amounts to $\dfrac{0.2}{\sqrt{100}}$ i.e. 0.02.

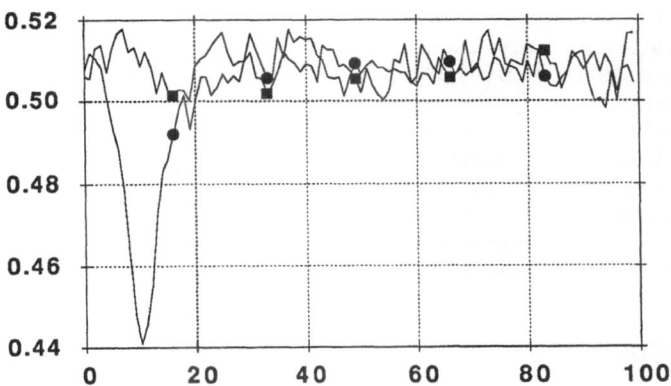

FIGURE 5. $R = <v^2>^2 / <v^4>$ as a function of time (in mean collision times τ)
squares: equilibrium values, circles non equilibrium values

Figure 5 gives also the behavior of R in a non equilibrium situation. As mentioned earlier, the spatial distribution of particles is regular and the initial velocities are picked out of a Maxwellian distribution. The starting value is very close to 0.5 due to the choice of the initial velocities. As time goes by, R departs from its equilibrium value by more than the standard deviation. This is due to the fact that particles are not, as far as space distribution is concerned, in an homogeneous equilibrium situation.

4. CONCLUSIONS

The computations reported here illustrate the mechanism by which a fluid system forced out of equilibrium reaches its final state. The initial situation consists of a state which has an equilibrium velocity distribution but a non-equilibrium spatial distribution: indeed, the particles are put on a lattice, this is an "over-ordered" state for the number density considered. The particles start to build the equilibrium pair correlations, and the measurements made suggest that the time needed for g(r) to reach its equilibrium value increases with r.

During that time, one sees a deviation of the velocity distribution from the initial Maxwellian. This sole observation could make us think that the system is departing from equilibrium: as a matter of fact, the contribution of f_1 to the entropy decreases during that time. Of course the total entropy does not decrease, but Boltzmann's entropy does. Then, once g(r) has reached its final value, the velocity distribution evolves back towards a Maxwellian. The peculiar initial correlations are responsible for this transient evolution.

Not much can be said about higher order correlations as this would require more cpu time and computer memory than is presently available. We hope to come back to these questions in a subsequent paper. The qualitative behavior observed here can be derived from our theoretical approach[8]. A fascinating question is the following: the time necessary to reach equilibrium depends on the distance which we consider. We have noticed that the first maximum of g(r), is approached by the system more rapidly than the one corresponding to the second shell. The time to approach equilibrium for macroscopic distances may become quite long. Moreover, an equilibrium system is characterized by correlations involving arbitrary number of particles. The question may therefore be asked: are there any system at all in nature at equilibrium in the sense of microscopic correlation distributions? It is likely that in the absence of external constraints, all systems evolve rapidly to local equilibrium, but still present evolving processes at the level of higher-order correlation functions, which escape experimental observation. It is amusing that each macroscopic system is perpetuating an arrow of time associated to modern cosmology.

ACKNOWLEDGEMENTS

This research has been supported by the Robert A. Welch Foundation under grant F-365, the US Department of Energy under grant DE-FG05-88ER13897 and the Actions de Recherches Concertées of the Belgian Government.

REFERENCES

1   B. J. Alder, T. Wainwright, in Transport Processes in Statistical Mechanics, I. Prigogine (editor), J. Wiley-Interscience, New York (1958).

2   J. Orban and A. Bellemans, Phys. Lett. **24A**, 620 (1967)

3   J. Orban and A. Bellemans, J. Stat. Phys. **1**, 467 (1969)

4   I. Prigogine, Non-Equilibrium Statistical Mechanics, J. Wiley-Interscience, New York (1960)

5   I. Prigogine and T. Petrosky, Physica **147A**, 461 (1988)
    T. Petrosky and H. Hasegawa, Physica **160 A**, 351 (1989)
    I. Prigogine and T. Petrosky (to appear)

6   I. Prigogine, E. Kestemont and M. Mareschal, in From Chemical to Biological Organization, M. Markus, S. Muller and G. Nicolis editors, Springer-Verlag, Berlin (1989)

7   J. A. Barker, D. E. Henderson, Rev. Mod. Phys. **48**, 587 (1976)

8   I. Prigogine, D. Driebe and T. Petrosky (to appear).

COMPUTER SIMULATION OF COLLAPSING SYSTEMS

H. A. Posch[1], H. Narnhofer[2] and W. Thirring[2]

[1]Institute for Experimental Physics and
[2]Institute for Theoretical Physics
University of Vienna, Boltzmanngasse 5
A-1090 Vienna, Austria

1. INTRODUCTION

In statistical mechanics one is usually concerned with mechanically stable N-particle systems for which the Hamiltonian H_N is assumed to be bounded from below by a constant $\sim N$. Theoretically this property can be proved[1] if (a) the system is treated quantum mechanically, (b) the particles of one sign of charge are all fermions, and (c) gravitation is neglected. Mechanical stability manifests itself in a positive microcanonical specific heat c_V. Systems for which c_V is negative are mechanically unstable, implying that for large N the relativistic Hamiltonian becomes unbounded from below and the system collapses[2]. The gravitational collapse of stars after exhaustion of their nuclear fuel serves as an example which was given considerable attention in the past[3]. In view of the fact that the coupling of such systems to an external thermal bath is weak, the use of a microcanonical ensemble for the description of such phenomena seems to be a natural choice.

Some time ago, a classical cell model was proposed by one of the present authors (W.T.), which is simple enough to be solved exactly but nevertheless retains all the salient features of unstable systems[4,5,1]. In the following section we shall give a brief account of the assumptions and results of this theory which in its original version assumes a small region $V_0 \subset V$ in which the collapse takes place. We shall refer to it as model "A"[5,1]. In an extended and refined theory henceforth referred to as model "B"[4], the system volume V is subdivided into M cells and the most probable distribution of the particles in these cells is obtained by maximizing the microcanonical entropy. For sufficiently low internal energies, it is found that the equilibrium states correspond to a big cluster of particles in one of the cells, with the rest of the particles equally distributed in the remaining $M - 1$ boxes. Thus, the simplifying assumptions of model A are confirmed.

Recently, in a very stimulating article[6], Compagner, Bruin and Roelse made a comparison of the predictions of model B with results obtained from molecular-dynamics simulations of a two-dimensional system of purely attractive particles. They found that the qualitative features of the model are very well reproduced by the simulations. Using the same pair potential as in reference 6 we have performed a series of molecular dynamics simulations of a two-dimensional system of N=400 particles. In addition to the equilbrium properties we are also interested in the transient behaviour of these systems from a highly nonequilibrium initial state to equilibrium. It is the purpose of this paper to report some of our recent results.

Microscopic Simulations of Complex Flows
Edited by M. Mareschal, Plenum Press, New York, 1990

After a short review of the cell model in section 2, the simulation technique is described in some detail in section 3. The results are discussed further in section 4 and a comparison of the equilibrium properties with the cell-model predictions is given.

2. MICROCANONICAL CELL MODEL

The Hamiltonian of our N-particle system is written as

$$H_N(\mathbf{\Gamma}) = \sum_i \mathbf{p}_i^2/2m + \Phi(\mathbf{X}), \tag{1}$$

$$\Phi(\mathbf{X}) = \sum_{i<j} \sum_j \phi(\mathbf{x}_i, \mathbf{x}_j). \tag{2}$$

Here, $\mathbf{\Gamma} = \{\mathbf{X} = (\mathbf{x}_1, \ldots, \mathbf{x}_N); \mathbf{P} = (\mathbf{p}_1, \ldots, \mathbf{p}_N)\}$ denotes a point in $2dN$-dimensional phase space, and d is the dimensionality of the system. In the following we shall consider only $d = 2$. The sum is over all particles, $i = 1, \ldots, N$.

In the original model A[5,1], a small region V_0 is assumed in which the collapse takes place. Two particles i and j have a negative interaction energy $-\epsilon_0$, if both are located in V_0; outside V_0 the particles do not interact and move freely. The pair potential takes the form

$$\phi(\mathbf{x}_i, \mathbf{x}_j) = -\epsilon_0 \chi(\mathbf{x}_i; V_0) \chi(\mathbf{x}_j; V_0), \tag{3}$$

where

$$\chi(\mathbf{x}; V_\alpha) = \begin{cases} 1 & \text{if } \mathbf{x} \in V_\alpha \\ 0 & \text{if } \mathbf{x} \notin V_\alpha, \end{cases} \tag{4}$$

and the energy parameter $\epsilon_0 > 0$ is a constant.

In the extended model B the volume V is subdivided into M cells of volume $V_\alpha, \alpha = 1, \ldots, M$. Only particles within the same cell interact with a constant interaction energy $-\epsilon_0$. Particles belonging to different cells do not interact. The pair potential may be written as

$$\phi(\mathbf{x}_i, \mathbf{x}_j) = -\epsilon_0 \sum_\alpha \chi(\mathbf{x}_i; V_\alpha) \chi(\mathbf{x}_j; V_\alpha) \tag{5}$$

and is totally attractive. To characterize the thermodynamic state we define energy and temperature parameters by

$$\epsilon = 1 + \frac{2U}{N(N-1)\epsilon_0} \tag{6}$$

$$\vartheta = \frac{2k_B T}{N\epsilon_0} = \frac{4E_{kin}}{dN^2\epsilon_0}, \tag{7}$$

where U is the value of H_N, E_{kin} the kinetic energy, T the temperature, and k_B Boltzmann's constant. $\epsilon = 0$ corresponds to the totally collapsed state. In the following one has to distinguish between two limiting cases:

The homogeneous phase: For large ϵ the maximum of the miocrocanonical entropy is obtained for a homogeneous distribution of the particles in all M cells. The temperature is related to the energy by

$$\vartheta = \frac{2}{d}(\epsilon - 1 + \frac{1}{M}), \tag{8}$$

and the heat capacity

$$c_V = 1/(\partial T/\partial U)_V = d(N-1)k_B/2 \tag{9}$$

is that of an ideal gas.

The collapsed phase: At low energies ϵ the most probable states require the formation of a big cluster of N_c particles in a single cell, say $\alpha = C$, with $N - N_c$ particles equally distributed in the remaining $M - 1$ cells forming a gas. If $b = (N - N_c)/(M - 1)N_c$ denotes the ratio of the densities in the gas and in the cluster, the relation between energy and temperature can be written in parametric form:

$$\vartheta(b) = \frac{2(1-b)}{[1 + b(M-1)]\ln(1/b)},$$ (10)

$$\epsilon(b) = 1 + \frac{d}{2}\vartheta(b) - \frac{1 + (M-1)b^2}{[1 + (M-1)b]^2}.$$ (11)

In Figure 1 $\vartheta(\epsilon)$ is depicted for two different numbers M of square cells covering the system volume V: $M = 400$ (dashed curve) and $M = 1600$ (smooth curve). As may be verified from these curves there is a certain range of energies, for which $\vartheta(\epsilon)$ has a negative slope indicating a negative microcanonical specific heat c_V.

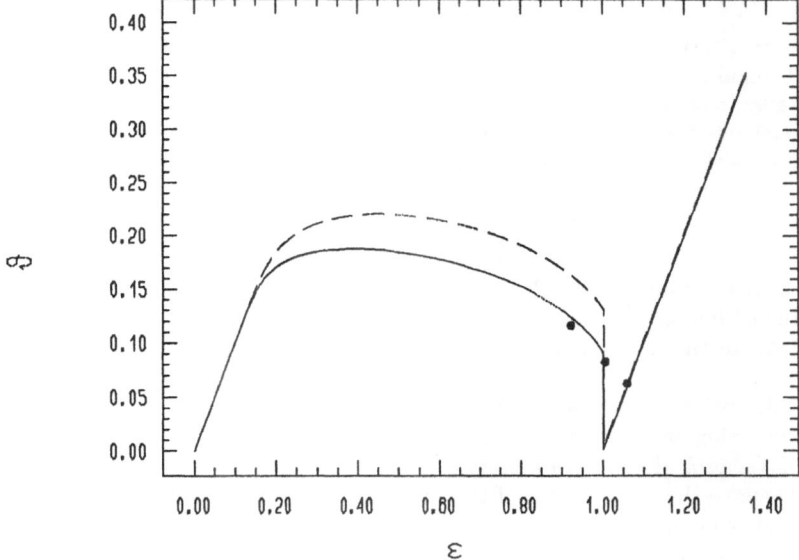

Figure 1. Temperature ϑ versus energy ϵ for a 400-particle system. The curves are for the cell-model: M=1600 (full line) and M=400 (dashed line). The simulation results are indicated by dots.

The transition from the homogeneous phase to the clustered phase occurs near $\epsilon = 1$. However, there is a whole range of energies, in which there are two local maxima of the entropy and in which the two phases coexist. For a certain value of b the two maxima are of equal height. This is taken as the condition for the transition[4,6] and is indicated in Figure 1 by the vertical line connecting the clustered and homogeneous branches of the $\vartheta(\epsilon)$-curve. No latent heat is connected with this transition[4].

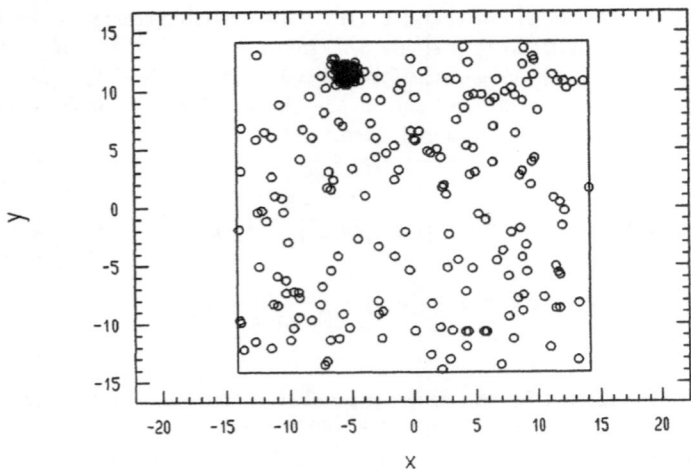

Figure 2. Typical snapshot of an equilibrated configuration for $\epsilon = 0.923$. The large cluster contains $N_c = 211$ particles.

3. MODEL FOR MOLECULAR DYNAMICS SIMULATION

The gravitational potential is not well suited for numerical simulations: it is of long range and singular at the origin. Though we have gravity in mind, we have chosen a potential of short range and regular at the origin. Neither the singularity nor the long range of the gravitational potential is necessary to reproduce the unusual clustering behaviour we are interested in. Following Compagner et. al.[6] we have taken a potential

$$\phi(\mathbf{x}_i, \mathbf{x}_j) = -\epsilon_0 \exp\{-\left(\frac{\mathbf{x}_i - \mathbf{x}_j}{\sigma}\right)^\delta\}, \tag{12}$$

with $\delta = 2$ for our simulation. A larger value for δ would make the resemblance between (12) and the model potentials (3) or (5) more pronounced but would possibly require a smaller time step Δt for the simulation.

The definition of a cluster in such a system is somewhat arbitrary. It turns out, however, that the clustering properties are insensitive to the particular definition. We take as a typical interaction range D the inflection point of the potential (12), $D = \sigma/\sqrt{2}$. Two particles are said to belong to the same cluster, if their separation is less than D. During the simulation run every quarter of a unit time a complete cluster analysis is performed and the cluster properties are averaged over many such configurations. As an example we show in Figure 2 a snapshot of a typical equilibrium configuration for a system of 400 particles in a box of volume $V/\sigma^2 = 800$. The main cluster contains 211 particles, whereas the remaining 181 particles are uniformly distributed. Since the diameter of the disks in Figure 2 is D, two overlapping disks belong to the same cluster. In agreement with the theoretical prediction of the cell model only one big cluster is observed.

A Gear fourth-order predictor-corrector algorithm in the N-representation is used for the integration of the equations of motion[7]. Periodic boundary conditions are employed. In the following, including the figures, we use reduced units, where the particle mass m as well as the potential parameters ϵ_0 and σ are unity.

The system chosen consists of $N = 400$ particles in two dimensions, $d = 2$. Two different types of initial conditions are used:

- an equal distribution of all N particles in the square box with volume V, and

- a totally condensed system with all N particles on top of each other.

In both cases the velocity components v_α are taken as equally distributed in $-v_0 < v_\alpha < v_0$, $v_0 = \sqrt{3 k_B T / m}$. For low enough total energies ϵ, the large cluster starts growing in the first case, whereas in the second case the particle number of the main cluster is rapidly decreasing by evaporating off single particles or small droplets of a few particles. After equilibration the average number of particles $\langle N_c \rangle_0$ of the big cluster is independent of the initial condition, if the simulations are performed with the same total energy ϵ. To achieve equilibrium very long runs of at least two million timesteps of length $\Delta t = 0.0025$ are required. During this time the total energy is constant to within 0.1%.

4. EQUILIBRIUM AND TRANSIENT PROPERTIES OF CLUSTERING SYSTEMS

Simulations were carried out for two different system volumes V corresponding to the densities $\rho = N \sigma^3 / V = 0.5$ and 0.125, respectively, and for various internal energies. The results for the equilibrium temperature parameters ϑ are indicated by the dots in Figure 1. The points agree rather well with the cell-model predictions for $M = 1600$. A similar qualitative agreement between computer simulation and theory has been found previously by Compagner et al.[6].

For energies $\epsilon \sim 1$ close to the phase transition energy the average number $\langle N_c \rangle_0$ of particles in the main cluster under equilibrium conditions may be estimated from the cell model. In lowest order the model predicts[6,8]

$$\langle N_c \rangle_0 = dN / \ln M. \tag{13}$$

With $d = 2$ and $N = 400$ one finds $\langle N_c \rangle_0 = 134$ and 108 for $M = 400$ and 1600, respectively. The simulation result for $\rho = 0.5$ is $\langle N_c \rangle_0 = 140$. The disagreement is due to the differences in the pair potentials used for the simulation and the cell model.

In the following we shall discuss some properties of the cluster both in its nonequilibrium transient stage and after reaching equilibrium. To obtain appropriate averages for the dynamical variables of interest for a particular number N_c of cluster particles, the system state was analyzed after every 100 time steps and particles belonging to the main cluster were determined. If the instantaneous cluster-particle number happened to be equal to N_c, the particles belonging to this cluster were renumbered, $i = 1, \ldots, N_c$, and this state was taken to contribute to the evaluation of time averages for this particular N_c. Nonequilibrium averages obtained in the transient stage are denoted by $\langle \ldots; N_c \rangle$, equilibrium averages by $\langle \ldots \rangle_0$. During one run a whole range of N_c-values was analyzed simultaneously.

The instantaneous cluster-particle number fluctuates around its average value $\langle N_c(t) \rangle$, which increases with time in the nonequilibrium transient regime, if the initial condition is taken to be a random distribution, until it finally reaches the equilibrium value $\langle N_c \rangle_0$. Similarly, $\langle N_c(t) \rangle$ decreases with time from N to $\langle N_c \rangle_0$, if a totally condensed state is taken for initial condition.

As an example in Figures 3 to 6 we consider a system with density $\rho = 0.5$, with an internal energy $\epsilon = 1.0059$ close to the phase transition but still in the clustering phase, and with a homogeneous initial particle distribution.

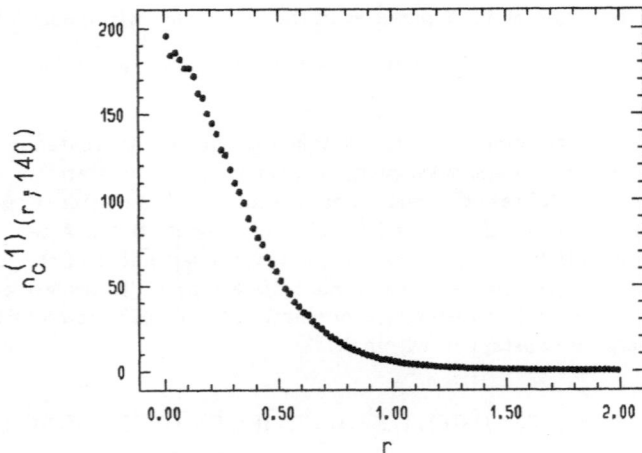

Figure 3. Local particle density around the center of mass of an equilibrium cluster containing 140 particles.

The local particle density within the cluster is defined by

$$n_c^{(1)}(r; N_c) = \langle \sum_{i=1}^{N_c} \delta(\mathbf{r} - \mathbf{x}_i - \mathbf{R}_c); N_c \rangle, \tag{14}$$

where $\mathbf{R}_c$ is its center of mass. This function is normalized such that

$$\int_V n_c^{(1)}(r; N_c) d\mathbf{r} = N_c. \tag{15}$$

In Figure 3 the local particle density is shown for a cluster with $N_c = 140$ equal to the equilibrium value $\langle N_c \rangle_0$. Its shape is almost a Gaussian and should be obtainable from a Vlasov equation for this problem[8]. The local density of the surrounding "gas" is uniformly distributed over the remainder of the system volume.

To study the time evolution of the cluster size during its nonequilibrium transient growth, we define a size parameter

$$s(N_c) = \frac{2}{N_c(N_c - 1)} \langle \sum_{i=1}^{N_c-1} \sum_{j=i+1}^{N_c} (\mathbf{x}_i - \mathbf{x}_j)^2; N_c \rangle, \tag{16}$$

where the sums are over all particles belonging to the cluster with a particle number N_c. As inferred from Figure 4, the surprising result is that the cluster size and consequently also its diameter is hardly affected by the transient condensation process.

Another important property is the kinetic energy of the cluster particles. In Figure 5 we show the kinetic energy per cluster particle defined by

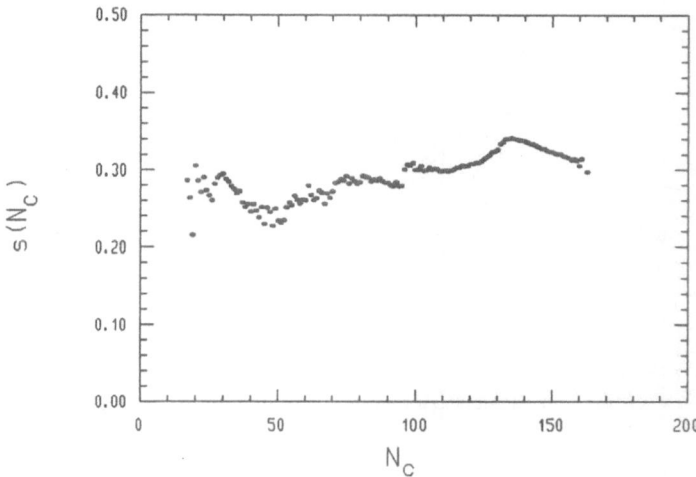

Figure 4. Size parameter of the condensing cluster as a function of its particle number.

$$\left\langle \frac{E_{kin}}{N_c} \right\rangle = \frac{1}{N_c} \left\langle \sum_{i=1}^{N_c} \frac{m}{2} (\mathbf{v}_i - \mathbf{V}_c)^2 ; N_c \right\rangle, \tag{17}$$

where $\mathbf{v}_i$ is the particle velocity and $\mathbf{V}_c$ the center-of-mass velocity of the cluster. All transient states of the growing cluster are shown. The error bars are given by the standard deviation and in essence indicate the number of configurations encountered with a particular N_c during the simulation run. In the transient regime it is a measure of the speed with which the mass of the cluster grows. The equilibrium particle number is 140, the largest cluster found during

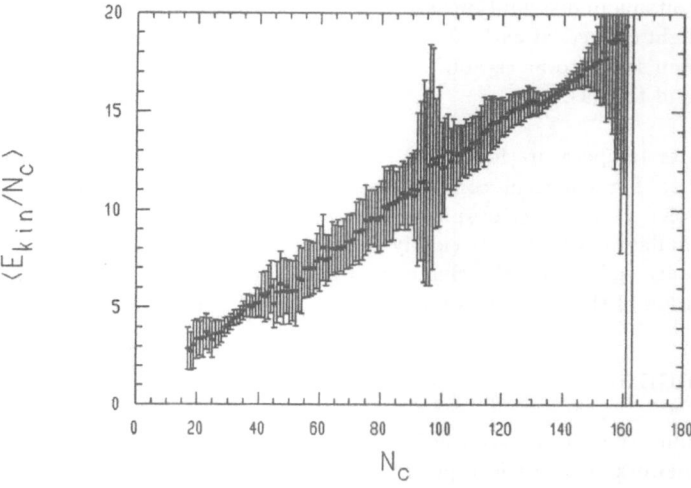

Figure 5. Kinetic energy per particle for the condensing cluster as a function of the cluster-particle number.

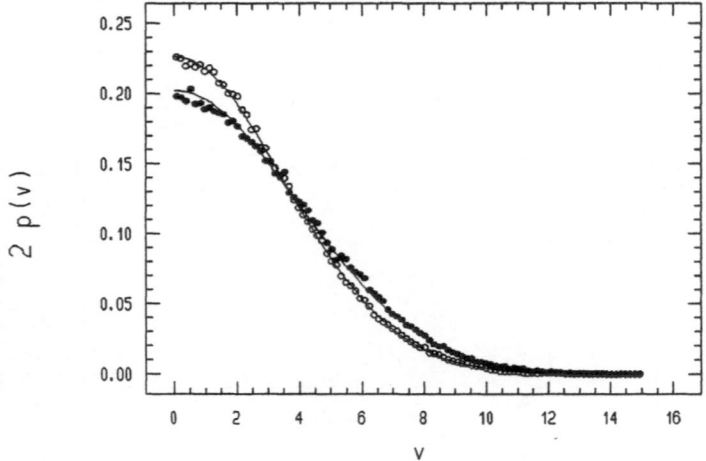

Figure 6. Normalized velocity distribution of the cluster particles (full circles) and of the gas particles (open circles) for a transient nonequilibrium system with a cluster-particle number 130. The smooth curves are calculated from a Maxwell-Boltzmann distribution.

this run over 3 million time steps is 162. As indicated by Figure 5, $\langle E_{kin}/N_c \rangle$, and therefore the local cluster temperature, is directly proportional to N_c. The more particles the cluster contains the hotter it is.

The notion of a "cluster temperature" is meaningful only if the cluster particles are Maxwell - Boltzmann distributed even in nonequilibrium transient states. This condition is examined in Figure 6 for a transient 130-particle cluster, where the probability distributions $p(v)$ for the cluster particles and for the remaining gas particles are shown. In both cases the Maxwell-Boltzmann distributions are well obeyed. This is true also for even smaller N_c. Only for very light clusters at early stages of the run significant deviations may be observed[8]. Local equilibrium seems to be established very quickly in spite of the different temperatures of the cluster and the gas.

The cluster temperature for the state in Figure 6 is 15.6, the temperature of the gas particles is 12.4. For transient condensing systems one always finds a hot cluster slowly floating in a cooler atmosphere. In spite of the somewhat artificial nature of the pair potential used for the simulations the results closely resemble the situation for a collapsing star. Only after approaching equilibrium the cluster and gas temperatures become identical. A more detailed discussion of this point and further results will be published separately[8].

ACKNOWLEDGEMENTS

The computations were performed at the computing center of the University of Vienna within the framework of IBM's European Supercomputing Initiative. We gratefully acknowledge the generous allocation of computer time. We also thank Dr. John Hague, IBM-UK, for his assistance with the vectorization of some of the programs. We are grateful to the authors of reference 6 for informing us of their results prior to publication.

REFERENCES

1. W. Thirring, *A Course in Mathematical Physics*, Vol.4: *Quantum mechanics of large systems*, Springer Verlag, Wien, New York (1983).

2. W. Thirring, Naturwissenschaften, **73**, 605 (1986).

3. D. Lynden-Bell, in *Gravitation in Astrophysics*, edited by B. Carter and J.B. Hartle, NATO ASI Series B: Physics Vol. 156, Plenum Press, New York (1987).

4. P. Hertel and W. Thirring, Ann. of Physics, **63**, 520 (1971).

5. W. Thirring, Z. Physik, **235**, 339 (1970).

6. A. Compagner, C. Bruin and A. Roelse, preprint.

7. W.F. van Gunsteren and H.J.C. Berendsen, Molec.Phys., **34**, 1311 (1977).

8. H.A. Posch, H. Narnhofer and W. Thirring, in preparation.

LYAPUNOV SPECTRA IN HAMILTONIAN DYNAMICS

Stefano Ruffo

Istituto Nazionale di Fisica Nucleare, Sezione di Firenze
L.go E.Fermi, 2 50125 FIRENZE, *ITALY*
and Facoltá di Scienze MFN, Universitá della Basilicata
Potenza, *ITALY*

Introduction

Let us consider a system of N classical interacting particles. The corresponding Hamilton equations of motion are:

$$\dot{q}_i = p_i,$$
$$\dot{p}_i = -\frac{\partial H}{\partial q_i} \tag{1}$$

with:

$$H = \sum_{i=1}^{N} \frac{p_i^2}{2} + V(\{q_i\})$$

In order to integrate this set of ordinary differential equations on a computer we need a time discretization algorithm. Let us choose the Verlet algorithm [1] which, written in Hamiltonian form, reads as follows:

$$q_i(t + \Delta t) = q_i(t) + p_i(t)\Delta t$$
$$p_i(t + \Delta t) = p_i(t) - \frac{\partial V}{\partial q_i}(\{q_i(t + \Delta t)\})\Delta t \tag{2}$$

The nice property of this algorithm is that it preserves the symplectic structure of the Hamiltonian flow.Let us explain this fact in a way which will result useful afterwards. According to algorithm (2) the variations of coordinates and momenta $(\delta q_i, \delta p_i)$ evolve in time:

$$\delta q_i(t + \Delta t) = \delta q_i(t) + \Delta t \delta p_i(t)$$
$$\delta p_i(t + \Delta t) = -\sum_j \delta_{ij}\delta p_j - \frac{\partial V}{\partial q_i \partial q_j}\Delta t\left(\delta q_j + \delta p_j \Delta t\right)$$

In matrix form the $2N$ vector $\vec{\psi} = (\delta\vec{q}, \delta\vec{p})$ evolves in time according to the following linear transformation:

$$\vec{\psi}(t + \Delta t) = \mathbf{S}_t \vec{\psi}(t) \qquad (3)$$

where

$$\mathbf{S}_t = \begin{pmatrix} \mathbf{I} & \mathbf{I}\Delta t \\ \mathbf{A}_t & \mathbf{I} + \mathbf{A}_t \Delta t \end{pmatrix} \qquad (4)$$

$\mathbf{I}$ being the $N \times N$ identity matrix and:

$$(\mathbf{A}_t)_{ij} = -\frac{\partial V}{\partial q_i \partial q_j}\big|_t \Delta t$$

The matrix $\mathbf{S}_t$ is a symplectic matrix, i.e. $\mathbf{S}_t^T \mathbf{J} \mathbf{S}_t = \mathbf{J}$ where:

$$\mathbf{J} = \begin{pmatrix} \mathbf{O} & \mathbf{I} \\ -\mathbf{I} & \mathbf{O} \end{pmatrix}$$

A consequence of being symplectic is that $det\mathbf{S}_t = 1$. This property imply that algorithm(2) is in fact an area preserving map, but it has other important consequences on the spectrum of Lyapunov exponents, as we will comment in a while. This linear transformation (3) can be readily solved at the n-th time step

$$\vec{\psi}(t + n\Delta t) = \prod_{i=1}^{n} \mathbf{S}_{t+(i-1)\Delta t} \vec{\psi}(t) \equiv \mathbf{M}_n \vec{\psi}(t) \qquad (5)$$

Matrix $\mathbf{M}_n$ is the fundamental ingredient in the computation of Lyapunov spectra.

Liapunov spectra

In a famous theorem Oseledec [2] proved the existence of the following limit:

$$\Lambda = \lim_{n \to \infty} \left(M_n^T M_n \right)^{1/2n}$$

The logarithms of the 2N eigenvalues λ_i of Λ are the Lyapunov exponents:

$$\mu_i = log(\lambda_i) \quad i = 1, \cdots 2N$$

In practice one computes the Lyapunov exponents using a well known algorithm [3] which is based on the evaluation of the rate of growth of hypervolumes spanned by independent initial vectors $\vec{\psi}_i(0)$. An important dependence of the $\mu_i's$ is on the initial phase-space point. We will consider in the following only examples where this dependence is not present, or it is weak; this is the case for instance of highly chaotic dynamics with only one ergodic component. The Liapunov exponents μ_i can be ordered from large to small

$$\mu_1 \geq \mu_2 \cdots \geq \mu_{2N}$$

252

admitting some possible degeneracy. For symplectic (Hamiltonian) dynamics they obey a symmetry property:

$$\mu_i = -\mu_{2N+1-i} \quad i = 1, \cdots, 2N \tag{6}$$

i.e. Lyapunov exponents are equal in modulus and opposite in sign in pairs. We sketch in a few lines the proof of this property. Let Sp be the ensemble of Lyapunov exponents, that we call the spectrum:

$$Sp(\mathbf{M}) \equiv (\mu_1, \cdots, \mu_{2N}) \tag{7}$$

Property (6) amounts to say that changing sign to all the exponents leave the ensemble (7) unchanged:

$$Sp(\mathbf{M}) = -Sp(\mathbf{M}) \tag{8}$$

Let us prove this property in two steps. First of all we observe that contracting (expanding) directions become expanding (contracting) under time inversion:

$$Sp((\mathbf{M}^T)^{-1}) = -Sp(\mathbf{M}) \tag{9}$$

But $\mathbf{M}$ is symplectic, then $-\mathbf{JMJ} = (\mathbf{M}^T)^{-1}$ and since $\mathbf{J}^2 = -\mathbf{I}$ we obtain:

$$Sp((\mathbf{M}^T)^{-1}) = Sp(\mathbf{M}) \tag{10}$$

Equations (9) and (10) prove property (8). Due to this symmetry property in Hamiltonian dynamics one needs to compute only the positive Lyapunov spectrum.

Following a suggestion of Ruelle [4] we have proposed in [5] the existence of the following thermodynamic limit for highly chaotic Hamiltonian dynamics:

$$\phi(x) = \lim_{N \to \infty} \mu_{xN}(N) \tag{11}$$

Of course we are assuming that the dependence on the initial condition can be neglected. The relevance of the thermodynamic limit (11) can be exhibited using many arguments, here we mention two of them. The area below $\phi(x)$ $[0 \leq x \leq 1]$ is the Kolmogorov-Pesin- Sinai entropy density, which characterizes the rate of information production of the chaotic dynamics.

If $\phi(x)$ is invertible, then:

$$\eta(\phi) = 1 - x(\phi) \tag{12}$$

is the density of exponents such that $\bar{\phi} \leq \phi$, i.e. the integrated density of the spectrum, a quantity recalling an analogy with Schrödinger operator theory, which we will make more explicit in the following. A list of papers which have dealt with the existence of the limit (11), using both analytic and (mainly) numerical techniques is given in Ref. [6]. All the properties discussed in this section remain valid for symplectic matrices of different form than matrix (4). For instance Eckmann and Wayne [7] have recently proposed to approximate the dynamics of systems of coupled oscillators at large energy per oscillator with the following matrix:

$$\hat{S}_t = exp\left(\tau \begin{pmatrix} O & -\Omega_t \\ I & O \end{pmatrix}\right) \tag{13}$$

with Ω_t tridiagonal and τ a "time" mesh parameter. An interesting and fruitful idea, which allow the construction of sistematic approximations of the true dynamics (5), consists in replacing the matrix product in (5), which is ordered along the orbit, with a product of random matrices (some degree of correlation may also be permitted). This class of approximations establish a link with analogous results obtained on Lyapunov spectra of products of random matrices [8], but also lead to analytic results both for the Lyapunov spectra themselves [7] and for the scaling properties deriving from small perturbations of integrable cases [9].

Liapunov vectors

It is interesting to look at the dynamics of the vectors $\vec{\psi}$ in cases where the dynamics is chaotic or random. Initially one might expect that no new insight is achieved by looking at them, since applying random matrices to a random vector can produce only a random vector, perhaps with some interesting probability distribution. On the contrary one observes numerically that Lyapunov vectors show a very interesting localization phenomenon. For instance, after a few time steps, the vector corresponding to the maximal Lyapunov exponent have only a few components which are significantly different from zero (and a similar effect is present for vectors corresponding to smaller exponents). A justification of this behavior is possible for matrices of the form (4) where A_t is a tridiagonal symmetric matrix with random elements. From now on we put, for simplicity $\Delta t = 1$. The application of matrix S_t defines a "random" map:

$$\begin{cases} \vec{x}(t+1) = \vec{x}(t) + \vec{y}(t) \\ \vec{y}(t+1) = \vec{y}(t) + A_t\vec{x}(t+1) \end{cases} \tag{14}$$

where $\vec{x}, \vec{y}$ are N dimensional components of $\vec{\psi} : \vec{\psi} = (\vec{x}, \vec{y})$.

Let us now perform the following linear transformation:

$$\begin{cases} \vec{p} = \vec{x} \\ \vec{q} = \vec{x} + \vec{y} \end{cases}$$

then the map (14) can be rewritten as a second order difference equation:

$$(\triangle - \mathbf{A}_t)\vec{p}(t) = 0 \qquad (15)$$

where $\triangle$ is the discrete operator:

$$\triangle \vec{p}(t) = \vec{p}(t+1) - 2\vec{p}(t) + \vec{p}(t-1)$$

Eq. (15) is some sort of Schrödinger equation in a random potential and therefore one expects localization (in some cases) of the N-component "wave- function" $\vec{p}$ on the basis of Anderson theory [10]. A better understanding of this localization phenomenon is obtained introducing the following ratios:

$$R_t^k = \frac{p^k(t)}{p^k(t-1)} \qquad T_t^k = \frac{p^k(t)}{p^{k-1}(t)} \qquad (16)$$

where the index k denotes the k-th component of the vector $\vec{p}$. For the sake of simplicity we will further specialize the model to the case where the sum of the elements in each row of $\mathbf{A}_t$ is zero, i.e. $\sum_j (\mathbf{A}_t)_{ij} = 0 \quad i = 1, \cdots, N$, so that the dynamics is fully determined by extracting a random N component vector $\vec{\epsilon}_t$ at each time step. This corresponds to imposing the conservation of momentum in the case of coupled oscillator models, acquiring some physical justification. Given these restrictions Eq. (15) rewrite in terms of the ratios in (16) as:

$$R_{t+1}^k = 2 - \gamma_t^k - 1/R_t^k$$
$$T_{t+1}^k = \frac{T_t^k R_t^k}{2 - \gamma_t^{k-1} - 1/R_t^k} \qquad (17)$$

with $\gamma_t^k = \epsilon_t^k(1 - 1/T_t^k) + \epsilon_t^{k+1}(1 - T_t^{k+1})$ where again the index k denotes the k-th component of the vectors involved. This formulas generalize the mapping studied in Ref [11], corresponding to a one dimensional discrete Schrödinger equation in a random potential. They describe an "intermittency" process where the components of $\vec{p}$ stay near a constant value for a randomly varying latency period, which is then followed by a "burst" resulting in a sudden variation of their values. It should be noticed that this intermittency is present both in space and in time. A detailed study of this intermittency process would allow better estimates of the Lyapunov spectrum than the ones contained in Ref [7], which are based on the isotropy of the motion of the Lyapunov vector on the N-dimensional unit radius hypersphere.

Acknowledgements

I thank Ricardo Lima, Roberto Livi, Antonio Politi and Angelo Vulpiani for the innumerable discussions on the subject of this paper and S. Isola for a careful reading of this manuscript.

References

[1] L.Verlet: Phys. Rev.,**159**,89 (1967);

[2] V. I. Oseledec: Trans. Moscow Math. Soc., **19**, 197 (1968);

[3] G. Benettin, L. Galgani,A. Giorgilli and J. M. Strelcyn: Meccanica ,March,**9**, 21 (1980);

[4] D. Ruelle: Comm. Math. Phys., **87**,287 (1982);

[5] R. Livi, A. Politi, S. Ruffo: J. Phys., **A 19**, 2033 (1986);

[6] P. Grassberger, H. Kantz: Phys. Lett.,**A123**,437 (1987); J. L. Pichard, C. André: Europhys. Lett.,**2**,477 (1986); B. Derrida,J.L. Pichard, K. Mecheri: Journ. de Physique (Paris),**48**,733 (1987);
B. Derrida,N. Zanon: J. Stat. Phys.,**50**,509 (1988);

[7] J.P. Eckmann,E.C. Wayne: J. Stat. Phys.,**50**,853 (1988);

[8] C. M. Newman: Comm. Math. Phys.,**103**, 121 (1986);

[9] G. Parisi and A. Vulpiani: J. Phys.,**A19**, L425 (1986); R. Lima and S. Ruffo: J. Stat. Phys.,**52**, 259 (1988);

[10] P.W. Anderson: Phys. Rev.,**109**,1492 (1958);

[11] B. Derrida, E. Gardner: J. de Physique (Paris),**45**,1283 (1984);

PART IV

APPLICATIONS (II): TRANSPORT AND MODE COUPLING THEORIES

TIME CORRELATION FUNCTIONS AND HYDRODYNAMIC MODES

FOR LATTICE GAS CELLULAR AUTOMATA

James W. Dufty

Department of Physics
University of Florida
Gainesville, FL 32611

INTRODUCTION

The explicit construction of lattice gas cellular automata with fluid symmetry at the Navier-Stokes level has stimulated a great deal of research on this and related discrete dynamical systems.[1] The primary objective has been the description of complex hydrodynamic flows by simulation as a potentially more efficient method than numerical solution to the continuum mechanics equations.   In this context the automaton is merely a tool for constructing the macroscopic dynamics of interest (Navier-Stokes equations), and attention to its microscopic properties is motivated only by the need to determine the parameters of the macroscopic equations (e.g., transport coefficients).   The successful simulation of complex flows by this method and current limitations are reviewed by Boon in this volume.[2]

More recently, a quite different interest in lattice gas cellular automata has developed whereby they are viewed as models for the statistical mechanics of many-body fluids.[3]   Since the equilibrium statistical mechanics is rather trivial for these models, the primary objective is to understand how the macroscopic dynamics results from its underlying microdynamics.   Although the collision rules are only very crude representations of scattering processes in real fluids, nevertheless it is expected that many important qualitative features of many-body dynamics are shared with real fluids.   For example, a mean field Boltzmann approximation appears to be quite accurate for some automata at low densities.[4]   An analysis of conditions for the validity of the Boltzmann equation and the corrections required for higher densities is expected to be more tractable for cellular automata than for real fluids.   Similarly, long-standing questions about asymptotic slow decay of time correlation functions ("long time tails")[5] and the related problem of non-standard hydrodynamics below critical dimensionality, d=2, perhaps can be studied in more detail for these discrete dynamical systems.

The discussion here emphasizes this second role of cellular automata, and reviews briefly some results obtained from the generalization of statistical mechanical methods developed for continuous fluid systems to a broad class of discrete models.[6-8]   Attention is limited to a description of dynamical excitations near equilibrium, and consequently the interesting nonlinear flows described elsewhere in this volume are

Microscopic Simulations of Complex Flows
Edited by M. Mareschal, Plenum Press, New York, 1990

excluded. However, a quite complete description of the linear dynamics is possible in terms of equilibrium time correlation functions of the local conserved densities (e.g., mass, momentum). Experience with continuous fluid systems has shown that such time correlation functions are particularly useful for analysis by both theory and simulation.

In the next section the application of linear response methods[9] is given to relate nonequilibrium dynamics __near__ equilibrium to the dynamics of fluctuations __at__ equilibrium. For long wavelength excitations at long times, this relationship allows determination of the hydrodynamic modes from equilibrium time correlation functions. The explicit form of the hydrodynamic modes at the Navier-Stokes level is given in section 3. The transport coefficients are expressed as Green-Kubo relations similar to those for continuous fluids[10] with some important differences. In section 4 it is shown how these Green-Kubo expressions can be evaluated approximately from a Boltzmann-Enskog kinetic equation. The calculation of correlation functions in the hydrodynamic limit is discussed in section 5, with emphasis on sub-critical dimensionality. Finally, some concluding remarks are offered.

LINEAR RESPONSE

Macroscopic hydrodynamics is associated with the nonequilibrium average of the local conserved densities,

$$<A_\alpha(\vec{r},t)> = \sum_{\{n\}} P(\{n\})A_\alpha(\vec{r},t) \quad , \quad A_\alpha(\vec{r},t) = \sum_c a_\alpha(c)n(\vec{r},\vec{c},t) \tag{1}$$

The state of the system at some initial time is specified by a distribution function (ensemble) defined over the occupation numbers, $n(\vec{r},\vec{c},t)$, for each lattice site $\vec{r}$ and velocity state $\vec{c}$. The time dependence is generated from local collision rules of the form,

$$n(\vec{r}+\vec{c},\vec{c},t+1) = n(\vec{r},\vec{c},t) + \Delta(\vec{c},\vec{r},t) \tag{2}$$

where Δ is a function of the occupation numbers for all states, $\vec{c}$, at site $\vec{r}$ and time t. The functions $a_\alpha(c)$ in (2) characterize the local conserved densities by the property,

$$\sum_c a_\alpha(c)\Delta(\vec{c},\vec{r},t) = 0 \tag{3}$$

A key problem for simple discrete dynamical systems is the identification of __all__ local conserved densities. In general there are spurious conservation laws in addition to those expected by analogy to continuous fluids.[11] These must be included in the analysis since they will contribute also on the same long time, large wavelength scales.

Consider a special initial state corresponding to a specified value for the average conserved density near its equilibrium value,

$$P(\{n\}) = P_o(\{n\})\left[1 + \sum_{\vec{r}} y_\alpha(\vec{r})\left(A_\alpha(\vec{r}) - <A_\alpha(\vec{r})>_o\right)\right] \tag{4}$$

where $P_o(\{n\})$ denotes the stationary equilibrium state. Here and below a summation over repeated Greek labels is implied. The function $y_\alpha(\vec{r})$ represents the initial deviation of $<A_\alpha(\vec{r})>$ from its initial equilibrium value $<A_\alpha(\vec{r})>_o$. More generally, substitution of (4) into (1) gives the

258

linear response to such deviations for $t \geq 0$,

$$\delta<A_\alpha(\vec{r}, t)> = \sum_{\vec{r}'} G_{\alpha\beta}(\vec{r}-\vec{r}', t) \, \chi_{\beta\gamma}^{-1} \, \delta<A_\alpha(\vec{r}')> \tag{5}$$

where the time correlation matrix G and the susceptibility matrix χ are defined by,

$$G_{\alpha\beta}(\vec{r}-\vec{r}', t) = <A_\alpha(\vec{r}, t)[A_\beta(\vec{r}') - <A_\beta(\vec{r}')>_0]>_0 \tag{6}$$

$$\chi_{\alpha\beta} \, \delta(\vec{r}-\vec{r}') = G_{\alpha\beta}(\vec{r}-\vec{r}', t=0) \tag{7}$$

Translational invariance among lattice sites has been used. The local form of (7) follows from the fact that the equilibrium distribution factors into a product over all sites. The result (5) links the nonequilibrium dynamics of $\delta<A_\alpha(\vec{r}, t)>$ to the equilibrium time dependent fluctuations of $\{A_\alpha(\vec{r}, t)\}$. In particular, the hydrodynamic equations for $\delta<A_\alpha(\vec{r}, t)>$ are reflected in the spectrum of equilibrium fluctuations.

HYDRODYNAMIC MODES AND GREEN-KUBO EXPRESSIONS

The hydrodynamic modes are most conveniently identified from a Fourier representation of $\delta<A_\alpha(\vec{r}, t)>$,

$$\delta<\hat{A}(\vec{k}, t)> \equiv V^{-1} \sum_{\vec{r}} e^{-i\vec{k}\cdot\vec{r}} \, \delta<A(\vec{r}, t)> \tag{8}$$

Here and in the following an obvious matrix notation is used (e.g., $A \leftrightarrow \{A_\alpha\}$). The hydrodynamic equations are identified from the long time, long wavelength (small k) limit of (8) in the form,

$$\left(\frac{\partial}{\partial t} + L(\vec{k})\right)\delta<\hat{A}(\vec{k}, t)> = 0 \tag{9}$$

The time derivative now appears because (9) is valid only for times large compared to the discrete time step, so that t is effectively a continuous variable in this limit. The matrix L is the hydrodynamic matrix whose eigenvalues and eigenvectors give the hydrodynamic modes, or elementary excitations about equilibrium.

To determine L we use the fact that the Fourier transformed equilibrium time correlation matrix also obeys the hydrodynamic equations for long times and long wavelengths. The Fourier representation of $G(\vec{r}-\vec{r}', t)$ is defined by,

$$\hat{G}(\vec{k}, t) \equiv <\hat{A}(\vec{k}, t)|\hat{A}(\vec{k})> \equiv V^{-1}<\hat{A}(\vec{k}, t)[\hat{A}^*(\vec{k}) - <\hat{A}^*(\vec{k})>_0]>_0 \tag{10}$$

The second equality defines the scalar product notation. To simplify the notation, the caret (^) over Fourier-transformed variables will be suppressed. The hydrodynamic limit of $G(\vec{k}, t)$ and the form of the matrix L can be identified in a convenient way using the conservation laws,

$$\Delta_t A(\vec{k}, t) + i\vec{k}\cdot\vec{B}(\vec{k}, t+1) = 0 \tag{11}$$

where $\Delta_t A(t) = A(t+1) - A(t)$ and the fluxes, $\vec{B}_\alpha$, are defined by,

$$\vec{B}_\alpha(\vec{k},t) = \sum_c \vec{b}_\alpha(\vec{c},\vec{k})\hat{n}(\vec{c},\vec{k};t) \; , \quad \vec{b}_\alpha(\vec{c},\vec{k}) = \vec{c}a_\alpha(\vec{c})[e^{i\vec{k}\cdot\vec{c}}-1]/1\vec{k}\cdot\vec{c} \quad (12)$$

An exact equation for the correlation matrix of conserved densities can now be written in the form,

$$[\Delta_t + \mathcal{L}(\vec{k},t)]\, G(\vec{k},t) = 0 \tag{13}$$

$$\mathcal{L}(\vec{k},t) \equiv i\vec{k}\cdot\langle\vec{B}(\vec{k},t+1)|A(\vec{k})\rangle\, G^{-1}(\vec{k},t)$$

$$= i\vec{k}\cdot\langle A(\vec{k},t)|\vec{B}(\vec{k})\rangle\, G^{-1}(\vec{k},t) \tag{14}$$

The second equality follows from stationarity of $G(\vec{k},t)$ and the conservation laws.[6] To further simplify this result, apply the identity,

$$\mathcal{L}(\vec{k},t) = \mathcal{L}(\vec{k},0) + \sum_{\tau=0}^{t-1} \Delta_\tau \mathcal{L}(\vec{k},\tau) \; , \quad t \geq 1 \tag{15}$$

Then, using stationarity of the equilibrium correlation matrices and the conservation law (11) one more time, $\mathcal{L}$ is expressed in terms of the autocorrelation function for the fluxes. This is the form suitable for identifying Green-Kubo relations.

The long wavelength limit is obtained by expanding in powers of k. Similarly, for long times the difference operator has a differential representation,

$$\Delta_t G(k,t) \to [\frac{\partial}{\partial t} + \frac{1}{2}\frac{\partial^2}{\partial t^2} + ..]\, G(k,t) \; , \quad t \gg 1$$

$$\to [\frac{\partial}{\partial t} + \frac{1}{2}\mathcal{L}^2 + ..]\, G(k,t) \; , \quad t \gg 1 \text{ and } k \ll 1 \tag{16}$$

The elimination of the second time derivative in terms of $\mathcal{L}$ is similar to that in the Chapman-Enskog expansion, and is justified for small k by the fact that $\mathcal{L}$ is of first order in k.[4] Use of (16) in (13) allows determination of the hydrodynamic matrix in (3.2) to any desired order in k. The calculation to Navier-Stokes order (k^2) is straightforward and only the results are given,

$$[\frac{\partial}{\partial t} + L(k)]\, G(\vec{k},t) = 0 \; , \quad L(k) = ik\Omega + k^2\Lambda \tag{17}$$

The matrix Ω characterizes the Euler order linear hydrodynamics and is defined by,

$$\Omega = \hat{k}\cdot\langle\vec{B}|A\rangle\chi^{-1} \; , \quad \chi \equiv \langle A|A\rangle \tag{18}$$

The transport or Navier-Stokes matrix, Λ, is given by the long time limit of the Green-Kubo relation,

$$\Lambda(t) = \sum_{\tau=0}^{t} \phi(\tau) - \frac{1}{2}\phi(0) \tag{19}$$

where $\phi(t)$ is the autocorrelation matrix for the "subtracted" fluxes,

$$\phi(t) = \langle J(t)|J\rangle\chi^{-1} \; , \quad J = \hat{k}\cdot\vec{B} - A\Omega^+ \tag{20}$$

Closer inspection shows that J is that part of $\hat{k}\cdot\vec{B}$ which is orthogonal to the conserved densities. The matrix elements of Λ determine the transport coefficients for the Navier-Stokes hydrodynamic equations. Equations (17) – (20) are the primary results of the linear response analysis. They provide an exact description of the hydrodynamic excitations near equilibrium with explicit expressions for the transport coefficients in terms of equilibrium correlation functions.

BOLTZMANN-ENSKOG APPROXIMATION

The evaluation of the susceptibility and Euler matrices, χ and Ω, is easily performed for a given CA since the occupation numbers for different states $\{\vec{c},\vec{r}\}$ at equal times are uncorrelated in the equilibrium state.[12]

$$\langle \delta n(\vec{c},\vec{r})\delta n(\vec{c}',\vec{0})\rangle = K\ \delta_{\vec{r},\vec{0}}\ \delta_{\vec{c},\vec{c}'}\ ,\qquad K = f(1-f) = \rho(b-\rho)/b^2 \qquad (21)$$

Here f and ρ = bf are the average occupation per state $\{\vec{c},\vec{r}\}$ and per site $\vec{r}$ respectively, where b is the number of velocity states per site. The relation (21) guarantees that the susceptibility matrix, χ, and the Euler matrix, Ω, are k-independent,

$$\chi_{\alpha\beta} \equiv \langle A_\alpha(\vec{k})|A_\beta(\vec{k})\rangle = K\ \sum_c\ a_\alpha(c)a_\beta(c) \qquad (22)$$

$$(\Omega\chi)_{\alpha\beta} = \hat{k}\cdot\langle\vec{B}_\alpha(\vec{k})|A_\beta(\vec{k})\rangle = K\ \sum_c\ \hat{k}\cdot\vec{c}a_\alpha(c)a_\beta(c) \qquad (23)$$

Further reduction of χ and Ω requires specification of the CA.

Evaluation of the matrix Λ for the transport coefficients is more difficult and involves a detailed analysis of the many-body dynamics. The theoretical analysis of time correlation functions is not well-developed at this point, although the simplest approximation (Boltzmann-Enskog) provides a surprisingly good first estimate for the transport coefficients. In this section the Boltzmann-Enskog approximation is introduced as the exact short time limit of the linear kinetic theory for time correlation functions.

The fundamental phase space density (Fourier representation) is introduced by,

$$G(c,c',\vec{k};t) \equiv \langle n(\vec{c},\vec{k},t)|n(\vec{c}',\vec{k},0)\rangle \qquad (24)$$

Then the correlation functions of conserved densities and their fluxes can be calculated directly from $G(c,c',k;t)$. For example,

$$\langle A_\alpha(\vec{k},t)|A_\beta(\vec{k})\rangle = \sum_{c,c'}\ a_\alpha(c)a_\beta(c')G(c,c',\vec{k};t) \qquad (25)$$

A formal kinetic equation for $G(c,c',\vec{k};t)$ can be identified in a manner analogous to that used in the last section to obtain the hydrodynamic equations. First, the conservation law (3) is applied to obtain,

$$\left\{\Delta_t + (1-e^{-i\vec{k}\cdot\vec{c}})\right\}G(c,c',\vec{k};t) = \langle\Delta(c,\vec{k},t)|n(c',\vec{k},0)\rangle \qquad (26)$$

The second term in the curly brackets on the left side reduces to $i\vec{k}\cdot\vec{c}$ for small k and describes the free streaming of particles. The term on the right side describes collisions, where $\Delta(c,\vec{k},t)$ is defined by,

$$\Delta(c,\vec{k},t) \equiv \sum_r e^{i\vec{k}\cdot(\vec{r}-\vec{c})} \Delta(\vec{c},\vec{r},t) \qquad (27)$$

A formal kinetic theory is now introduced by defining a collision "operator" (matrix in velocity space), $I(c,c',\vec{k};t)$, according to,

$$\left\{\Delta_t + (1-e^{-i\vec{k}\cdot\vec{c}})\right\}G(c,c',\vec{k};t) = \sum_{c''} I(c,c'',\vec{k};t)G(c'',c',\vec{k};t) \qquad (28)$$

Comparison with (26) gives this formally exact collision operator as,

$$\sum_{c''} I(c,c'',\vec{k};t)G(c'',c',\vec{k};t) \equiv \langle\Delta(c,\vec{k},t)|n(c',\vec{k},0)\rangle \qquad (29)$$

The Boltzmann-Enskog approximation results from a Markov approximation: successive collisions are uncorrelated so that the dynamics at each time step is completely determined by the state at the previous time step. In this case, the collision operator in (29) is approximated for all times by its value at t=0 after one collision,

$$I_E(c,c',k) = K^{-1}\langle\Delta(c,\vec{k},0)|n(c',\vec{k},0)\rangle \qquad (30)$$

Since the average on the right side of (30) involves only products of occupation numbers at the same time, it is easily evaluated once the collision rules are specified. Then $G(c,c'\vec{k},t)$, and consequently the correlation functions, can be calculated directly from the solution to (28). For example, the current autocorrelation functions, (20), can be written,

$$\phi_{\alpha\beta}(t) = \sum_{c,c'} j_\alpha(c)j_\beta(c')G(c,c',\vec{k};t)$$

$$= K\sum_c j_\alpha(c)\not{j}_\beta(c,t) \quad , \quad j_\alpha(c) \equiv \hat{k}\cdot\vec{b}_\beta(c,\vec{0}) - a_\sigma(c)\Omega^+_{\sigma\beta} \qquad (31)$$

where $\not{j}_\beta(c,t)$ is the solution to the homogeneous (k=0) Boltzmann-Enskog equation,

$$\Delta_t \not{j}_\beta(c,t) = \sum_{c'} I_E(c,c',\vec{0})\not{j}_\beta(c',t) \quad , \quad \not{j}_\beta(c,0) = j_\beta(c) \qquad (32)$$

Here $a_\alpha(c)$ and $b_\alpha(c)$ are the phase functions characterizing the conserved densities and their fluxes, defined in Eqs.(1) and (12) respectively. Similarly, the Green-Kubo expression for the transport matrix in the Boltzmann-Enskog approximation is found to be,

$$(\Lambda_E)_{\alpha\beta} = K\sum_c j_\alpha(c)[\not{J}_\sigma(c) - \frac{1}{2}j_\sigma(c)]\chi^{-1}_{\sigma\beta} \qquad (33)$$

where $\not{J}_\sigma(c)$ is the solution to,

$$\sum_{c'} I_E(c,c',\vec{0})\not{J}_\sigma(c') = j_\sigma(c) \qquad (34)$$

The conservation laws and the definition of I_E imply,

262

$$\sum_c a_\alpha(c) I_E(c, c', \vec{0}) = 0 \tag{35}$$

Therefore, solutions to (34) exist only if $j_\sigma(c)$ is consistent with (35), i.e.,

$$K \sum_c a_\alpha(c) j_\sigma(c) = 0 = \langle J_\alpha(\vec{0}) | A_\beta(\vec{0}) \rangle \tag{36}$$

This "Fredholm alternative" is satisfied as a consequence of the definition of the subtracted flux, J, in Eq.(20).

HYDRODYNAMIC LIMIT AND MODE COUPLING

The correlation matrix for the conserved densities, $G(\vec{k}, t)$, can be calculated for large time and small k directly from the hydrodynamic equations (17). The result can be expressed in terms of the hydrodynamic modes, defined by the eigenvalues and eigenvectors of the matrix, L(k),

$$L(\vec{k}) \psi^\mu(\vec{k}) = \omega^\mu(k) \psi^\mu(\vec{k}) \quad , \quad \omega^\mu(k) = ic_\mu k + \lambda_\mu k^2 \tag{37}$$

The eigenvalues have been expressed (to Navier-Stokes order) in terms of the sound speeds, c_μ, and the transport coefficients, λ_μ. The eigenvectors are linear combinations of the conserved densities, chosen such that $\{\psi^\mu\}$ form a bi-orthogonal set in the sense,

$$\langle \psi^\mu_\alpha | \psi^\nu_\alpha \rangle = \delta_{\mu\nu} \tag{38}$$

(See reference 8 for further details of this construction). The correlation matrix, $G(\vec{k}, t)$, in the hydrodynamic limit then becomes,

$$G(\vec{k}, t) \rightarrow \sum_\mu M_\mu e^{-\lambda_\mu k^2 t} \quad , \quad M_\mu = \langle A(\vec{k}) | \psi^\mu(\vec{k}) \rangle \langle \psi^\mu(\vec{k}) | A(\vec{k}) \rangle \tag{39}$$

This is an important result because it shows that the hydrodynamic modes can be measured directly by simulation of the equilibrium correlation functions. Most attention has been given to simulation of the Green-Kubo (k=0) correlation functions, but there is clearly more potential information contained in $G(\vec{k}, t)$. The hydrodynamic form (40) requires small k (large system size) and large t (long runs).

There is an additional, stronger, motivation for simulating the hydrodynamic limit of $G(\vec{k}, t)$ for CA in dimensions less than or equal to 2. In this case it is expected (by analogy with continuous fluids) that the long time limit of $\Lambda(t)$ in (19) does not exist. This means that the small k expansion of $L(\vec{k})$, leading to Navier-Stokes order hydrodynamics, does not exist. Instead, either $L(\vec{k})$ must have some non-analytic dependence on k, or the time dependence of the coefficients in the k-expansion must be retained. The CA promise to provide a rich testing ground for existing phenomenological theoretical models to describe this process.

The basis for problems with transport in low dimensional systems is the slow decay of coupled hydrodynamic modes over a broad band of wavevectors. A phenomenological description of these processes is provided by mode coupling theory.[6,13] In the present context it is sufficient to restrict attention to the flux correlation functions, $\phi(t)$, occurring in the Green-Kubo expression. The slowest components of the microscopic fluxes J(t) are expected to arise from projections along the

263

conserved densities, $A(\vec{k})$, since fluctuations in the latter decay slowly for small k. However, since the fluxes are orthogonal to A by (20), the leading contribution comes from bilinear products of the A's. Let $B(\vec{k}, \vec{k}')$ denote the part of the direct product of two A's that is orthogonal to linear combinations of A,

$$B_{\alpha\beta}(\vec{k}, \vec{k}') = A_{\alpha}(\vec{k})A_{\beta}^{*}(\vec{k}') - \langle A_{\alpha}(\vec{k})A_{\beta}^{*}(\vec{k}')\rangle$$

$$-V^{-1} \sum_{k'', \gamma, \nu} \langle A_{\alpha}(\vec{k})A_{\beta}(\vec{k}')[A_{\gamma}(\vec{k}'') - \langle A_{\gamma}(\vec{k}'')\rangle]^{*}\rangle \chi_{\gamma\nu}^{-1}[A_{\nu}(\vec{k}'') - \langle A_{\nu}(\vec{k}'')\rangle] \quad (40)$$

The slow component of J is now identified as its projection along B,

$$J(\hat{k}) = \sum_{\vec{k}_1, \ldots, \vec{k}_4} \langle J(\hat{k})|B(\vec{k}_1, \vec{k}_2)\rangle g^{-1}(\vec{k}_1, \vec{k}_2; \vec{k}_3, \vec{k}_4)B(\vec{k}_3, \vec{k}_4) + \ldots \quad (41)$$

$$g(\vec{k}_1, \vec{k}_2; \vec{k}_3, \vec{k}_4) = \langle B(\vec{k}_1, \vec{k}_2)|B(\vec{k}_3, \vec{k}_4)\rangle \quad (42)$$

The dots at the end on the right side denote contributions that decay more rapidly at long times. The flux autocorrelation function, $\phi(t)$, now becomes,

$$\phi(t) = \phi_o(t) + \phi_{mc}(t) \quad (43)$$

Here, $\phi_{mc}(t)$ denotes the "mode coupling" part of $\phi(t)$ arising from the contributions to J from the first term on the right side of (42). The remainder, $\phi_o(t)$, is assumed to decay on a shorter time scale. The mode coupling contribution is given by,

$$\phi_{mc}(t) = \sum_{\vec{k}_1, \ldots, \vec{k}_4} V(\vec{k}_1, \vec{k}_2)G(\vec{k}_1, \vec{k}_2, t; \vec{k}_3, \vec{k}_4)V^{*}(\vec{k}_3, \vec{k}_4) \quad (44)$$

with

$$V(\vec{k}_1, \vec{k}_2) \equiv \sum_{k_3, k_4} \langle J(\hat{k})|B(\vec{k}_3, \vec{k}_4)\rangle g^{-1}(\vec{k}_3, \vec{k}_4; \vec{k}_1, \vec{k}_2) \quad (45)$$

$$G(\vec{k}_1, \vec{k}_2, t; \vec{k}_3, \vec{k}_4) \equiv \langle B(\vec{k}_1, \vec{k}_2, t)|B(\vec{k}_3, \vec{k}_4)\rangle \quad (46)$$

The mode coupling theory rests on two approximations. the first is that the bilinear contributions to J shown in (42) dominate for long times. The second approximation refers to the corresponding dominant part of $G(\vec{k}_1, \vec{k}_2, t; \vec{k}_3, \vec{k}_4)$. A cumulant representation in terms of the variables A shows that there is a "disconnected" part depending on products of $G(\vec{k}, t)$:

$$G(\vec{k}_1, \vec{k}_2, t; \vec{k}_3, \vec{k}_4) = \frac{1}{2} G(\vec{k}_1, t)G(\vec{k}_2, t)\left[\delta_{\vec{k}_1, \vec{k}_3}\delta_{\vec{k}_2, \vec{k}_4}\right.$$

$$\left. + \delta_{\vec{k}_1, \vec{k}_4}\delta_{\vec{k}_2, \vec{k}_3}\right] + \ldots \quad (47)$$

Retaining only this disconnected contribution, the self-consistent mode coupling approximation is finally given by,

$$\phi_{mc}(t) = (2V)^{-1} \sum_{\vec{k}} \mathcal{V}(\vec{k}) G(\vec{k}, t) G(-\vec{k}, t) \mathcal{V}^*(\vec{k}) \qquad (48)$$

$$\mathcal{V}(\vec{k})\delta_{k,k'} \equiv 2V^{1/2} \mathcal{V}(k, k') \qquad (49)$$

The calculation of (48) for dimensions greater than 2 is straightforward. First $G(\vec{k}, t)$ is calculated for long times from the hydrodynamic form (40), using the Boltzmann-Enskog values for the transport coefficients. Next, the mode coupling expression (48) is used to calculate $\phi(t)$, leading to a slow algebraic decay $\sim t^{-d/2}$. However, $\phi(t)$ is still summable in the Green-Kubo formula for $t \to \infty$ so the primary effect of mode coupling is simply to renormalize the Boltzmann-Enskog transport coefficients. The form of the hydrodynamic modes is unchanged. The procedure can be iterated to obtain self-consistent transport coefficients.

Now consider the case of $d \leq 2$. The above method of calculation fails since it leads to a mode coupling contribution to the Green-Kubo expression that diverges for $t \to \infty$, which is inconsistent with the assumed form of the hydrodynamic equations (39). Consequently, for low dimensional fluids $G(\vec{k}, t)$ and $\phi(t)$ must be determined self-consistently, not simply the transport coefficients. Such a calculation could be initiated as above, but on second iteration $G(\vec{k}, t)$ must be calculated from (17) using $\Lambda \to \Lambda(t)$. The resulting hydrodynamic limit for $G(\vec{k}, t)$ clearly will not be of the exponential form shown in (39). This is the main point of the discussion in this section: at or below critical dimension equal to two hydrodynamics is anomalous and its proper form is unknown. Simulation of $G(\vec{k}, t)$ in one and two dimensions should shed some light on this question and the validity of the mode coupling theory.

DISCUSSION

Two aspects of the linear response analysis can be high-lighted. First, it represents a straightforward extension of statistical mechanical methods developed for continuous fluids to the case of discrete dynamical systems. It is expected that further application of many-body techniques to these simple systems may indicate more precisely both their limitations and conditions for validity. A second aspect of this analysis is the demonstration that a complete description of hydrodynamic excitations can be obtained from simulation of appropriate equilibrium time correlation functions. This eliminates complexities associated with the introduction of special boundary conditions to generate specific flows. However, it is not clear which approach to simulations is most efficient for a particular problem, such as measurement of the viscosity.

A quantitative estimate of the transport coefficients and hydrodynamic modes can be calculated easily from the Boltzmann-Enskog approximation. The Enskog aspect of this approximation is due to the Fermi exclusion rule for CA, leading to a nonlinear dependence of the collision operator on the density. Otherwise, the linear Boltzmann-Enskog operator is simpler than for continuous fluids since it is simply a finite dimensional matrix in the space of discrete velocities. For self-dual collision rules[12] the Boltzmann-Enskog approximation can be a good approximation at both low and high densities. However, it has been demonstrated recently that CA with substantial "backscattering" can invalidate the Boltzmann-Enskog approximation.[14]

The self-consistent mode coupling analysis given here to describe low dimensional fluids leads to a hydrodynamic representation of the form,

$$G(\vec{k}, t) \rightarrow \sum_\mu M_\mu \exp[ic_\mu kt - \int_o^t d\tau\lambda_\mu(\tau)k^2] \qquad (50)$$

Simulation of a particular hydrodynamic mode for long times at fixed k will determine the deviation from exponential decay. Clearly, the time scale for this simulation is of the same order as that needed in the simulation of $\phi(t)$ to see the algebraic "long time tail". Although this seems out of reach for current simulations, Frenkel has recently developed an algorithm for the related tagged particle transport problem that can access this long time domain.[15] The tails are quite clearly seen.[16] An alternative characterization of hydrodynamics for low dimensional fluids is based on an analysis in terms of the Laplace transformed correlation functions. For fluids above critical dimension equivalent results are obtained. Otherwise, this alternative description replaces the hydrodynamic differential equations by appropriate dispersion relations.[17] An eigenvalue problem like (37) is obtained, although the eigenvalues are no longer analytic in k. The details of this approach are given in reference 8. It is of particular use in the analysis of mode coupling effects for steady state simulations; some theoretical support for the theory in this context has been described recently.[18,19]

REFERENCES

1. For an overview of recent developments see Complex Systems 1, no. 4 (1987); "Discrete Kinetic Theory, Lattice Gas Dynamics, and Foundations of Hydrodynamics", Torino, Italy (World Scientific, Singapore, 1989).
2. J. P. Boon, this volume.
3. For a recent review, see M. Ernst in "Cellular Automata and Modelling of Complex Physical Systems", P. Manneville, editor (Springer, Berlin, 1989).
4. S. Wolfram, J. Stat. Phys. 45, 471 (1986).
5. Y. Pomeau and P.Resibois, Phys. Reports 19C, 69 (1975).
6. J. W. Dufty and M. H. Ernst, J. Phys. Chem. (1989).
7. M. H. Ernst and J. W. Dufty, Phys. Letts. 138, 391 (1989).
8. M. H. Ernst and J. W. Dufty, J. Stat. Phys. (January, 1990).
9. J.A. McLennan, "Introduction to Non-Equilibrium Statistical Mechanics", chapter 9 (Prentice Hall, NJ, 1988).
10. M.S. Green, J. Chem. Phys. 20, 1281 (1952); 22, 398 (1954); R. Kubo, J. Phys. Soc. Japan 12, 570 (1957).
11. G. Zanetti, Phys. Rev. A40, 1539 (1989).
12. U.Frisch, D. d'Humière, B. Hasslacher, P. Lallemand, Y. Pomeau and J. P. Rivet, Complex Systems 1, 649 (1987).
13. L. P. Kadanoff and J. Swift, Phys. Rev. 166, 89 (1968).
14. M.H. Ernst, G.A. van Velzen and P.M. Binder, Phys. Rev. Lett., (to be published).
15. D. Frenkel, this volume.
16. D. Frenkel and M. H. Ernst, Phys. Rev. Lett., submitted.
17. M.H. Ernst and J.R. Dorfman, J. Stat. Phys. 12, 311 (1975).
18. G. Zanetti, this volume.
19. L.P. Kadanoff, G.R. McNamara and G. Zanetti, Complex Systems 1, 797 (1987); and Phys. Rev. A (to be published).

Shear–induced ordering revisited

Werner Loose and Siegfried Hess

Institut für Theoretische Physik, Technische Universität Berlin

PN 7–1, Hardenbergstr. 36, D–1000 Berlin 12, West Germany

I. Introduction

The simulation of a stationary homogeneous shear flow is probably the most thoroughly investigated application of nonequilibrium molecular dynamics (NEMD). Well–probed algorithms are at hand which are discussed in several excellent reviews [1] and conference proceedings [2]. Even for the nonlinear regime their validity was established by means of nonlinear response theory for transient phenomena [3] and by comparing the steady–state response of a sheared gas with predictions of the Boltzmann equation [4,5]. Conversely, for the homogeneous heat flow a kinetic theory analysis was recently used to prove that no homogeneous driving force exists which generates the correct nonlinear heat conductivity [6].

Scepticism about the method was usually raised in view of the constant–temperature constraint of the simulation; e.g., the influence of the thermostatting forces on the transport properties is, in general, difficult to predict *a priori*. Shear viscosities for homogeneous systems with and without viscous heating, however, were found to exhibit the same qualitative behavior [7]. Problems with the very definition of the (kinetic) temperature arise for simulations at high shear rates and will be discussed in this article. These ambiguities are intimately related to the phenomenon of shear–induced ordering as observed in high shear rate simulations of dense simple fluids [8–11]. In the next section we will briefly recall its 'history' and the debate induced by the early work.

As already indicated, the objections focussed on the temperature control and led to the concept of a *profile–unbiased thermostat* (PUT) introduced by Evans and Morriss [12]. They conclude that the ordering transition is an artefact of the conventional *biased* thermostatting. In section III the PUT is introduced and results obtained with our version of the PUT are presented in section IV for 2 and 3 dimensional systems. Again, shear–induced ordering is observed. We interpret the ordering transition as a consequence of a hydrodynamic instability and, in section V, sketch out a simple linear stability analysis which seems to support our findings. Open questions will be addressed in the concluding section as well as implications for the experimental detection of these phenomena.

Microscopic Simulations of Complex Flows
Edited by M. Mareschal, Plenum Press, New York, 1990

II. A brief history of the ordering transition

The formation of strings of particles was first observed for a dense hard–sphere fluid by Erpenbeck [8]. These strings parallel to the stream lines form a perfect hexagonal pattern when projected onto a plane orthogonal to the streaming direction. Similar observations for soft–sphere systems were reported by Hess [9] and Woodcock [11]. The tendency of layering in gradient direction, a precursor of the 'string phase', was observed by Heyes *et al.* [13] long before one dared to study the high shear rate regime.

Why should one ever want to further increase the shear rate if even the lowest typical NEMD shear rates, if scaled with the parameters of an atomic fluid, e.g. argon, are already orders of magnitude beyond the capabilities of current laboratory facilities? Of course, the transport properties of simple fluids under extreme conditions are of interest in their own right. The connection to the 'real world', however, can be made by noting that if the interaction of 'macro fluids' like colloidal suspensions is modelled by the same two–particle interaction potential, the corresponding shear rates are readily accessible. This is due to the fact that the nonlinear flow behavior is determined by the dimensionless product of the shear rate and a characteristic relaxation time whose order of magnitude is typically given in picoseconds for atomic liquids and milliseconds for colloidal suspensions. This analogy is supported by comparing the distorted structure of a sheared colloidal suspension with simulation results of a soft–sphere fluid [14]. In addition, agreement with predictions of a kinetic theory equation for the pair–distribution function of a simple fluid [15] was found. Interestingly enough, observations of shear–induced ordering in a concentrated suspension have already been reported as early as 1974 by Hoffman [16]. However, until recently further experimental evidence was lacking and we will comment on this in the concluding section. The quest for a better understanding of the complex flow behavior of dense colloidal suspensions (of spherical particles) is certainly one *raison d'être* for the high shear rate simulations of simple fluids[1].

So far, we were refering to 3–d systems. In 2–d a similar ordering was observed [10], viz. the formation of strings along the streaming direction. In addition, highly compressed strings orthogonal to the former appeared in snapshots of the particle positions. These orthogonal strings are periodically build and destroyed by particles being temporarely trapped between members of adjacent layers. A typical snapshot is shown in Fig. 1: X is the streaming direction and the shear rate is denoted by $\gamma = \partial v_x / \partial y$. The 2–d soft–disk system is essentially identical to the choice of Heyes, Morriss and Evans [10], i.e. density $n = 0.9238$ and temperature $T = 1.0$. All quantities are given in standard reduced units. Note, the velocity profile is perfectly linear.

Despite of the homogeneous velocity gradient, ordered and amorphous phases were found to coexist at lower shear rates. Evans and Morriss [12] argued that the coexistence of two phases with different shear viscosities in a region with constant shear rate is not in accord with the condition for stability which requires the shear stress to be homogeneous. They conclude that stabilizing forces are exerted by the thermostat which is *biased* by assuming a linear velocity profile. They introduced a *profile–unbiased* thermostat which makes no assumptions about the profile and does not even assume the existence of a stationary flow profile. With their PUT the string phase was found to vanish and its earlier observations attributed to the use of an inproper thermostat.

---

[1]The 'challenge' was formulated by Ackerson and Clarke in 1982 in the final sentences of their article in Ref.[2b]

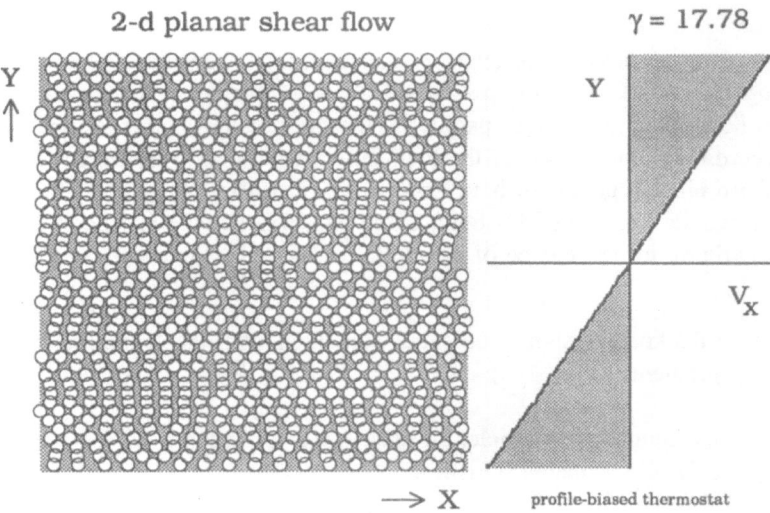

Fig. 1. The 'string phase' for a 968 particles soft–disk fluid (see text). With the use of the conventional (profile–biased) thermostat a perfectly linear velocity profile is observed.

III. Profile–unbiased thermostatting

Any NEMD thermostat is based on the (instantaneous) kinetic temperature calculated as an average of the peculiar velocities squared,

$$T \sim \frac{1}{N} \sum_{i=1}^{N} \left[\dot{r}^i - v(r^i) \right]^2 \quad .$$

Usually, the average streaming velocity at the position of particle i, $v(r^i)$, is *expected* to obey the linear relation: $v(r^i) = \gamma \, y^i \, e^x$, where e^x is the unit vector in streaming direction. Possible deviations from this linear profile are interpreted to be thermal and intuitively one expects that noticable deviations would be suppressed by the temperature control which, in the most simple way, can be accomplished by rescaling the peculiar velocities.

The idea of an *unbiased* thermostat is to actually calculate the streaming velocity $v(r)$ at each time step. To do that one has to divide the basic cell into bins. The local streaming velocity is then obtained as the average velocity of the particles in each bin. Originally, the PUT was introduced for the 2–d soft–disk system mentioned before and the 896 particles were sorted into square area elements with an area such that on average they contain 2 (!) particles [12]. We suspect that this is too fine a separation and that the resulting disappearance of any order might be attributable to the fluctuations induced by particles leaving or entering these tiny bins. And how, if at all, do those particles contribute to the temperature which find themselves alone in their bin? In later implementations [17] the bins contributed according to a weight determined by the respective number of particles.

Previously, we had realized that the breakdown of the (biased) thermostat observed at very high shear rates is connected with the tendency of the fluid particles to move in blocks (plug–like flow). For this reason we already considered it necessary to evaluate the average flow velocity in layers parallel to the flow rather than to assume a linear flow profile. Accordingly, our version of the PUT differs from that of Evans and Morriss in the choice of these bins. It is 'somewhat biased' insofar as we employ a local description for the gradient direction only; i.e., for both, 2 and 3–d systems, the basic cell is divided into layers. Two criteria for the choice of the thickness of the layers are reasonable:

- the layer thickness b should be small compared with the lengthscale given by the velocity gradient: $\gamma b \ll 1$;

- the average number of particles in each layer (of volume $b L^{d-1}$) should be large enough to have reasonable statistics: $n b L^{d-1} \gg 1$.

For all simulations the relation of the particle diameter and the layer thickness was roughly as sketched in Fig. 2. To obtain smoothed velocity and density profiles, the actual size of a particle was taken into account, cf. Fig. 2. The resulting mass density and momentum profiles are only used for the graphical display, not for the actual dynamics. The average velocity of the typically 10 to 20 particles in each of the layers is calculated and used to evaluate the local contribution to the kinetic temperature. *Ad hoc* rescaling of the peculiar velocities is used to control the temperature although a 'Gaussian method' [1] can also be applied in conjunction with a PUT [17]. Evans' least–squares adjustment of the velocity field [18] was found to be convenient to guarantee for a well–defined prescribed (average!) shear gradient. The Lees–Edwards periodic boundary conditions [1] alone, however, would be sufficient.

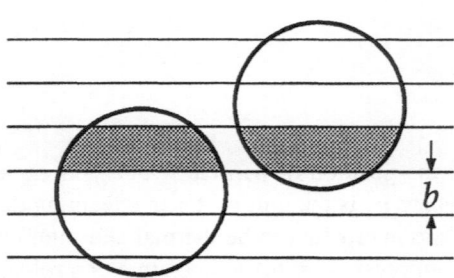

Fig. 2. Typical relation between layer thickness b and particle diameter. For the graphical display the density and velocity profiles are smoothed by taking into account the 'size' of the particles. The weight of the particles of neighbouring layers is determined by their (average) contribution to the mass density (gray shaded).

IV. Results

The 3–d results are for 512 particles interacting via a purely repulsive potential: a soft–sphere core is supplemented by a term which approximates a screened Coulomb potential. Density and temperature are $n = 0.84$ and $T = 0.25$, respectively. A more detailed account of these results can be found elsewhere [19]. A PUT with 32 layers was implemented as described in the preceding section.

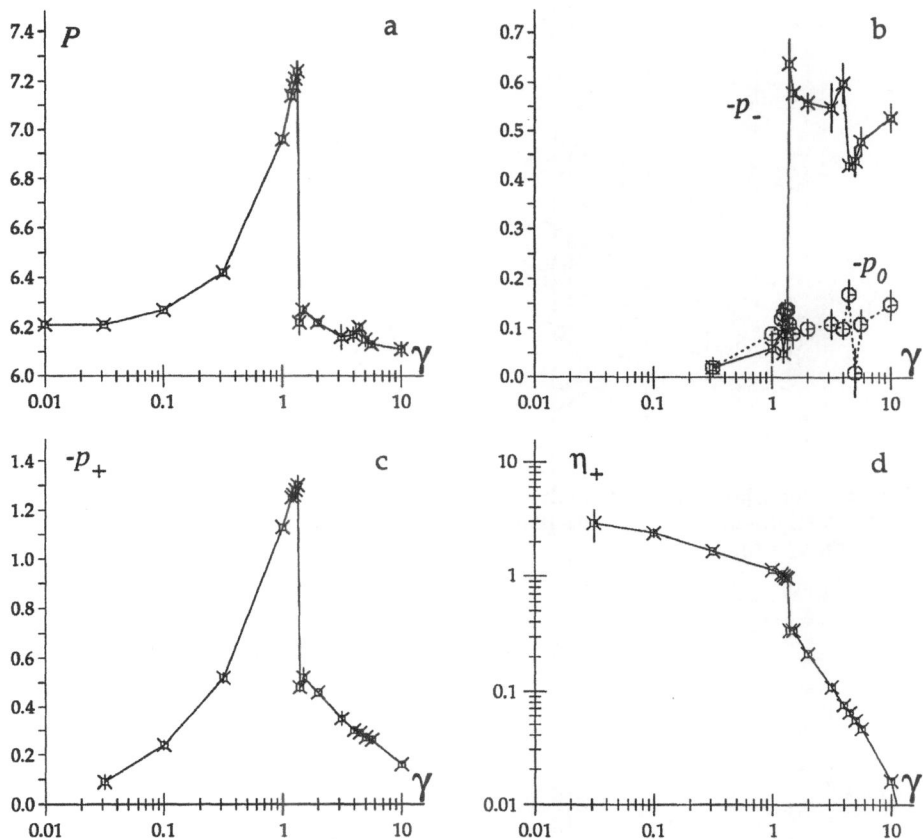

Fig. 3. The shear rate dependence of the components of the pressure tensor (a–c), as defined in the text, and of the shear viscosity η_+ (d).

Fig. 3 displays the shear rate dependence of the pressure tensor: P is its trace, $p_+ = \frac{1}{2}(P_{yx} + P_{xy})$ the friction pressure, and $\eta_+ = -p_+/\gamma$ the shear viscosity; p_- and p_0 are normal pressure differences defined by:

$$p_- \equiv \frac{1}{2}\left(P_{xx} - P_{yy}\right) \qquad ; \qquad p_0 \equiv \frac{1}{2}\left[P_{zz} - \frac{1}{2}\left(P_{xx} + P_{yy}\right)\right] \quad .$$

One observes a 'discontinuous' decrease of P and the stress $(-p_+)$ at a shear rate between $\gamma = 1$ and $\gamma = 2$ which goes along with a steep increase of the normal pressure difference $(-p_-)$. Note, shear thinning (i.e. decreasing viscosity) goes along with an increase of the hydrostatic pressure, often called dilatency. This is an important difference of NEMD simulations performed at constant volume and experiments where usually the pressure is constant.

The snapshots shown in Fig. 4 are for the shear rates $\gamma = 1.35$ and $\gamma = 1.4$, respectively, i.e. close to the pressure jump. The characteristic hexagonal pattern is observed for $\gamma = 1.4$. Fig. 5 displays the velocity and density profiles for four different shear rates. For $\gamma \leq 1.35$ the velocity field is perfectly linear while at $\gamma = 1.4$ a kink–like profile is observed which is even more pronounced at higher shear rates. Note, the layering transition is preceded by density fluctuations on the lengthscale of a particle diameter.

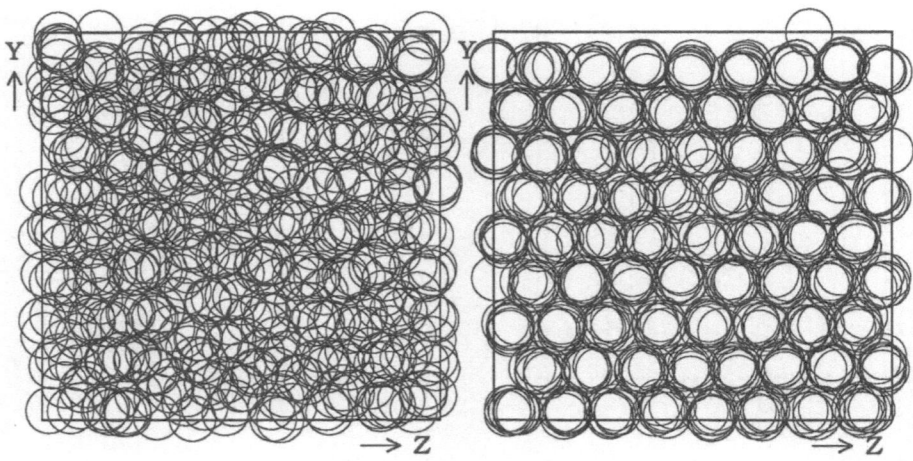

Fig. 4. 'Snapshots' of the particle configuration at $\gamma = 1.35$ (left–hand side) and $\gamma = 1.4$ (right–hand side). The particles are projected onto the YZ–plane (flow in X–direction).

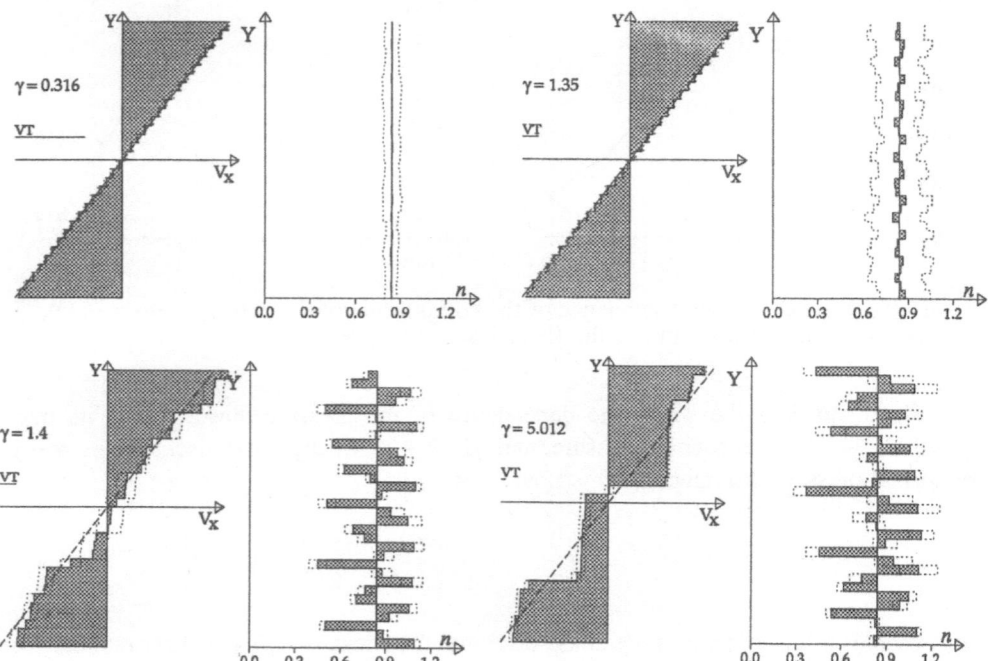

Fig. 5. Velocity and density profiles for 4 characteristic shear rates (3–d). They were obtained as averages over several runs of 2000 time–steps each and the dotted lines are a measure for the standard deviations. For the density profiles, deviations from the prescribed average density $n = 0.84$ are gray–shaded. The line denoted by 'VT' indicates the mean thermal speed.

In 2–d the ordering transition is not as sharply localized. The results are for essentially the same system as used in Refs.[10,12], however, a modified PUT with 128 layers for the 968 particles was used. Fig. 6 displays the friction pressure as function of the shear rate. A broad transition regime is marked by the two vertical lines. The structural changes, viz. the layering can be read off a plot of the maximum of the static structure factor

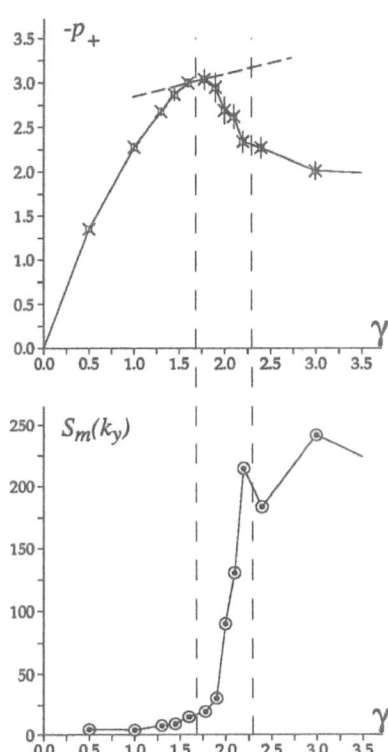

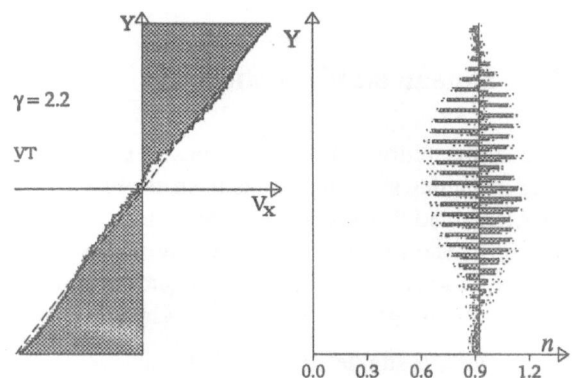

← **Fig. 6.** Shear rate dependence of the friction pressure of a 2–d soft–disk fluid (upper part). A broad transition region is identified by a plot of the maximum of the static structure factor $S(k_y)$. The dashed line stems from predictions of a stability analysis (cf. section V).

$\gamma = 2.2$

↑ **Fig. 7.** Velocity and density profiles for $\gamma = 2.2$ (2–d). They result from averaging over several runs and the dotted lines represent the standard deviations. The layering is superposed by a disturbance of a wavelength corresponding to the box size.

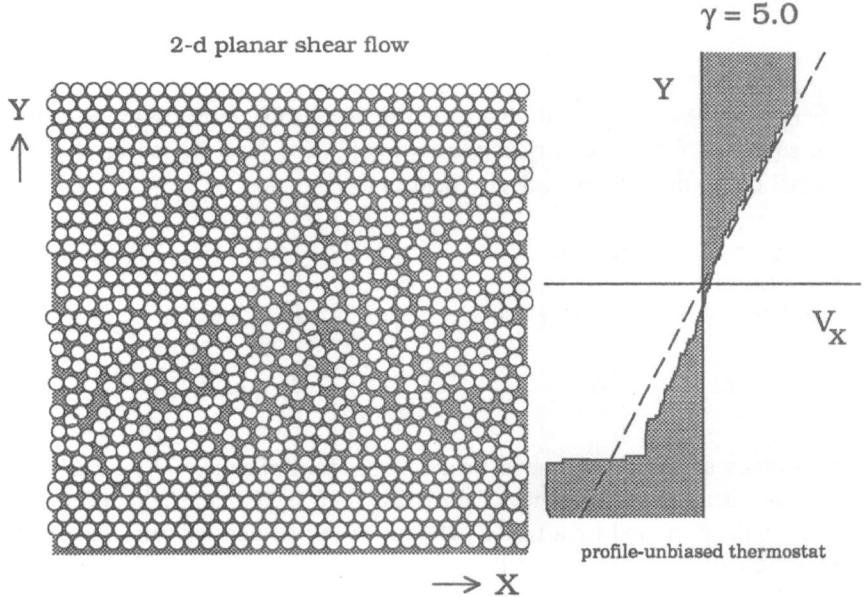

2-d planar shear flow

$\gamma = 5.0$

profile-unbiased thermostat

Fig. 8. Coexistence of ordered and amorphous regions for the 2–d soft–disk fluid. Notice the different local velocity gradients in contrast to Fig. 1.

evaluated along the gradient direction. In view of the almost discontinuous transition in 3–d, additional mechanisms seem to damp the transition in 2–d. Fig. 7 hints at possible system–size effects. For the shear rate $\gamma = 2.2$, i.e. in the transition regime, the velocity and density profiles are modulated by a wavelength of the order of the box size. At higher shear rates the scenario is again similar to the 3–d case, cf. Fig. 8. Again, the velocity profile is kink–like and a perfectly ordered crystalline phase moves as one block. This coexistence seems to contrast the almost perfect order found in the 3–d simulations, cf. Fig. 4. However, coexistent 'solid and fluid–like' regions were also observed in 3–d when the basic box was stretched in gradient direction [20].

V. Linear stability analysis

The applicability of hydrodynamics to phenomena on the length and time scales of molecular dynamics simulations has been established by impressive simulations of classical flow problems and this volume will serve as a perfect reference for this subject. Even hydrodynamic instabilities can be studied by microscopic simulations. This was demonstrated for the two–dimensional Rayleigh–Bénard instability [21] where the results were found to be in very good agreement with the corresponding macroscopic hydrodynamics [22].

Here, a simple linear stability analysis is sketched which suggests that the ordering transition is a consequence of a hydrodynamic instability. The method is standard with the exception that no constitutive laws are employed for the pressure tensor. On the contrary, a criterion is derived which allows one to decide whether a certain flow law will lead to this kind of instability.

The small deviations from the stationary solution are assumed to depend on the y–coordinate (gradient direction) only,

$$n = n_0 + \tilde{n}(y) \quad ; \quad v_x = \gamma_0 y + u(y) \quad ; \quad v_y = v(y) \quad .$$

The stationary solutions are denoted by a subscript zero. This ansatz is to be inserted into the balance equations for the density and momentum. The pressure tensor is assumed to be a functional of the density and velocity fields,

$$P_{\mu\nu} = P_{\mu\nu} \left[n(y), \boldsymbol{\nabla} \boldsymbol{v}(y) \right] \quad .$$

It is then straightforward to derive a criterion for an instability [19],

$$\frac{\partial(-p_+)}{\partial \gamma} \leq K \frac{\partial P_{yy}}{\partial \gamma} \quad ; \quad K \equiv \frac{\partial(-p_+)}{\partial n} \bigg/ \frac{\partial P_{yy}}{\partial n} > 0 \quad .$$

All the derivatives are to be evaluated in the stable flow regime. The criterion gives a threshold for the shear rate dependence of the stress $(-p_+)$. The threshold is determined by the dilatancy of the fluid and by a factor determined by the density dependences of these pressure tensor elements. Note, $\partial P_{yy}/\partial \gamma \approx \partial P/\partial \gamma$ is large and positive in the simulations, cf. Fig. 3a, which is not the case in experiments.

Consistency between NEMD results and the instability criterion would support the relation between the ordering transition and a flow instability. In Fig. 9 the stress versus shear rate curve is plotted on a linear scale. The threshold slope obtained from NEMD data is also shown and reasonable agreement is found. For the 2–d case the corresponding comparison is made in the upper part of Fig. 6. Definite conclusions, however, are obstructed by the broadness of the transition regime.

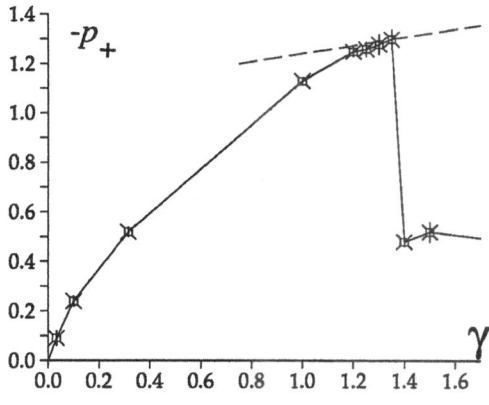

Fig. 9. The shear rate dependence of the stress. A flow instability is predicted if the slope of the tangent is less than 1/6 (dashed line). Consistency with the actual results of the simulation is observed.

The stability analysis gives some interesting phase relations between the different disturbances which are plotted in Fig. 10. If a cosine–like density perturbation (with wave vector k) is assumed the velocity fields are,

$$n \sim \cos(ky) \quad ; \quad v \sim -\frac{\lambda_k}{k} \sin(ky) \quad ; \quad u \sim -\frac{1}{k} \sin(ky) \quad .$$

Note, λ_k changes sign at the instability and becomes positive. Fig. 10 is for positive λ_k, i.e. *after* the instability. It is obvious how the velocity variation in gradient direction amplifies

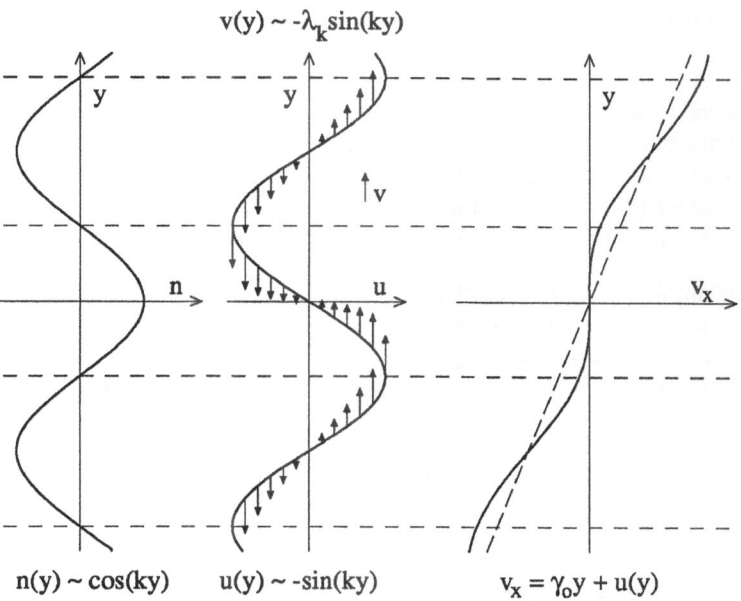

Fig. 10. The relative phases of the density and velocity perturbations in the *unstable* flow regime. An initial density perturbation with a wave vector k is assumed. The arrows denote y–component of the velocity which amplifies the initial disturbance. The resulting velocity profile $v_x(y)$ is also shown. A minimal gradient is found to coincide with maximal density.

the density perturbation; in the stable flow regime the arrows would have the opposite direction. The velocity profile is distorted in a way that a shear minimum goes along with maximal density which corresponds qualitatively to observations made in the NEMD simulations.

The idea to relate the ordering transition to a hydrodynamic instability is not new and, in fact, Kirkpatrick and Nieuwoudt [23] as well as Lutsko, Dufty and Das [24] have substantiated this by more elaborate microscopic theories. It is interesting that density fluctuations of the wavelength of the particle size play an essential role in their instability scenario.

VI. Concluding Remarks

The shear–induced ordering transition of dense simple fluids has been addressed and new support for its occurence has been given. It appears to be induced by a hydrodynamic instability, although additional work is required to make the conclusions more definite. Larger systems should be studied in order to identify possible system–size effects.

The simple hydrodynamic stability analysis leads to an intuitive instability scenario and gives some insight into the necessary requirements to be fulfilled in order to observe shear–induced ordering. The relevance of the constant–volume constraint of the simulation has already been pointed out by Woodcock [25] and is reemphasized by our analysis. If the pressure is constant, as in the usual experimental setup, a steeper decrease of the viscosity curve is required.

The recent light scattering experiments (on dense suspensions) by Ackerson and Pusey [26] veryfied the correlation of the ordering with shear–thinning behavior. The structures they observed are similar but less complete than the NEMD snapshots would suggest. This is in accord with the small angle neutron scattering data of Ottewill and Lindner [27]. The coexistence of ordered and amorphous phases, as seen in some of our simulations, could be an explanation for the relatively weak intensity maxima observed in the scattering experiments. The direct evaluation of $S(k)$ in the NEMD simulation analogous to a small angle multidetector [28] facilitates further comparisons with experiments where, however, also the formfactor of the particles has to be taken into consideration.

Concerning the NEMD temperature control it was shown that the conventional (profile–biased) thermostat may indeed lead to spurious results, viz. a long–range order which is stabilized by the thermostat, cf. Fig. 1. Hence, the concept of unbiased thermostatting as introduced by Evans and Morriss [12] is indispensible for the avoidance of these artefacts. Obviously, there is some freedom of choice for the PUT which may affect the results even qualitatively and, hence, requires more attention in the future[2].

---

[2]Giovanni Ciccotti suggested to evaluate the velocity distribution in each bin of the PUT. It should be Maxwellian if the concept of a temperature, based on the local equilibrium assumption, is meaningful at all.

Acknowledgements

We thank B. J. Ackerson and P. Lindner for a fruitful exchange of ideas and a steady stream of information on scattering experiments performed with sheared systems. Partial financial support by the Deutsche Forschungsgemeinschaft via the SFB "Anisotropic Fluids" is gratefully acknowledged as well as a grant by the "Studienstiftung des deutschen Volkes" (W. L.).

References

[1] W. G. Hoover, Annu. Rev. Phys. Chem. **34**, 103 (1983);
    D. J. Evans and G. P. Morriss, Comp. Phys. Rep. **1**, 297 (1984);
    D. J. Evans and W. G. Hoover, Ann. Rev. Fluid Mech. **18**, 243 (1986);
    M. P. Allen and D. J. Tildesley, *Computer Simulation of Liquids* (Clarendon, Oxford, 1987).

[2] G. Ciccotti and W. G. Hoover (eds.), *Molecular Dynamics Simulation of Statistical Mechanical Systems* (North–Holland, Amsterdam, 1986);
    H. J. M. Hanley (ed.), *Nonlinear Fluid Behavior* (North–Holland, Amsterdam, 1983).

[3] D. J. Evans and G. P. Morriss, Phys. Rev. A **30**, 1528 (1984).

[4] S. Hess, H. J. M. Hanley, and N. Herdegen, Phys. Lett. **105A**, 238 (1984);
    A. J. C. Ladd and W. G. Hoover, J. Stat. Phys. **38**, 973 (1985).

[5] W. Loose, Phys. Lett. **128A**, 39 (1988); W. Loose and S. Hess, Phys. Rev. Lett. **58**, 2443 (1987); Phys. Rev. A **37**, 2099 (1988).

[6] W. Loose, Phys. Rev. A **40**, 2625 (1989).

[7] J. W. Dufty, A. Santos, J. J. Brey and R. F. Rodriguez, Phys. Rev. A **33**, 459 (1986); see also the article of Dufty, Brey and Santos in [2a].

[8] J. J. Erpenbeck, Phys. Rev. Lett. **52**, 1333 (1984).

[9] S. Hess, Int. J. Thermophys. **6**, 657 (1985); J. Méc. Théor. Appl. (special issue) 1 (1985).

[10] D. M. Heyes, G. P. Morriss, D. J. Evans, J. Chem. Phys. **83**, 4760 (1985).

[11] L. V. Woodcock, Phys. Rev. Lett. **54**, 1513 (1985); Chem. Phys. Lett. **111**, 455 (1984).

[12] D. J. Evans and G. P. Morriss, Phys. Rev. Lett. **56**, 2172 (1986).

[13] D. M. Heyes, J. J. Kim, C. J. Montrose, T. A. Litovitz, J. Chem. Phys. **73**, 3987 (1980).

[14] H. J. M. Hanley, J. C. Rainwater, N. A. Clark, B. J. Ackerson, J. Chem. Phys. **79**, 4448 (1983).

[15] S. Hess, Phys. Rev. A **22**, 2844 (1980);
     S. Hess and H. J. M. Hanley, Phys. Rev. A **25**, 1801 (1982);
     H. J. M. Hanley, J. C. Rainwater, and S. Hess, Phys. Rev. A **36**, 1795 (1987).

[16] R. L. Hoffman, J. Coll. Int. Sci. **46**, 491 (1974).

[17] H. J. M. Hanley, G. P. Morriss, T. R. Welberry, D. J. Evans, Physica **149A**, 406 (1988).

[18] D. J. Evans, Mol. Phys. **37**, 1745 (1979).

[19] W. Loose and S. Hess, Rheol. Acta **28**, 91 (1989).

[20] S. Hess and W. Loose, **in:** D. Axelrad and W. Muschik (eds.), *Constitutive Laws and Microstructure* (Springer, Berlin, 1988).

[21] M. Mareschal and E. Kestemont, Nature (London) **329**, 427 (1987); J. Stat. Phys. **48**, 1187 (1987); D. C. Rapaport, Phys. Rev. Lett. **60**, 2480 (1988).

[22] M. Mareschal, M. Malek Mansour, A. Puhl, E. Kestemont, Phys. Rev. Lett. **61**, 2550 (1988).

[23] T. R. Kirkpatrick and J. C. Nieuwoudt, Phys. Rev. Lett. **56**, 885 (1986).

[24] J. W. Lutsko and J. W. Dufty, Phys. Rev. Lett. **57**, 2775 (1986); J. W. Lutsko, J. W. Dufty, S. P. Das, Phys. Rev. A **39**, 1311 (1989).

[25] L. V. Woodcock, J. Non–Newtonian Fluid Mech. **19**, 349 (1986).

[26] B. J. Ackerson and P. N. Pusey, Phys. Rev. Lett. **61**, 1033 (1988).

[27] R. H. Ottewill, P. Lindner, private communication, see also: Annual Report of the Institute Laue–Langevin, Grenoble (1987) p. 76.

[28] O. Hess, W. Loose, T. Weider, and S. Hess, Physica B **156/157**, 505 (1989).

A TEST OF MODE-COUPLING THEORY[1]

D. Frenkel and M. van der Hoef

FOM Institute for Atomic and Molecular Physics
PO Box 41883, 1009 DB Amsterdam, The Netherlands

1   LONG-TIME TAILS

1.1   Introduction

In the history of the kinetic theory of fluids, 1969-1970 was a crucial year. In that year Alder and Wainwright [2] published a paper in which they demonstrated the breakdown of the 'Molecular Chaos' assumption. The Molecular Chaos assumption, originally introduced by Boltzmann as the 'Stoßzahlansatz', states that the collisions experienced by a molecule in a fluid are uncorrelated. One consequence of this assumption is that the velocity auto-correlation function (VACF) of a tagged particle in fluid should decay exponentially. What Alder and Wainwright found is that the VACF of a particle in a moderately dense fluid of hard spheres or hard disks does not decay exponentially but algebraically. These algebraic long-time tails are the consequence of coupling between particle diffusion and shear modes in the fluid.

The Alder-Wainwright simulations caused a complete overhaul of the kinetic theory of dense fluids. The subsequent theoretical analyses of algebraic long-time tails were either based on an extension of kinetic theory [3] or on mode-coupling theory [4]. For a review, see ref. [5] . In the mode-coupling theory by Ernst, Hauge and van Leeuwen [4], it is assumed that the long-time tail is the consequence of coupling between particle diffusion and shear modes in the fluid. To a first approximation the leading term in the long-time tail of the velocity ACF is given by:

$$< v_x(0)v_x(t) > \approx \frac{D-1}{D} \frac{1}{\rho(4\pi(D_0 + \nu_0)t)^{D/2}} \equiv \frac{d_0}{t^{D/2}} \, , \qquad (1)$$

where ρ is the number density, D_0 the 'bare' self-diffusion constant, ν_0 the kinematic viscosity and D the dimensionality.

Following this theoretical work, simulations were performed by Levesque and Ashurst [6] and, most extensively, by Erpenbeck and Wood [7,8] with the aim to verify the validity

---

[1]This paper is based on material that has either been published elsewhere or has been submitted for publication.

of eqn. 1. For three-dimensional fluids these simulations are extremely expensive because very long simulations on very large systems must be performed. Even so, the system sizes studied in the simulations of Erpenbeck and Wood were such that it was essential to apply finite-size corrections to the corresponding mode-coupling theory before a meaningful comparison with the simulations could be made. Following such an approach, Erpenbeck and Wood found agreement between their simulation results for the VACF and a finite-size mode-coupling theory for a number of different densities. Nevertheless the statistical accuracy of their data was such that it was not meaningful to verify either the value of the exponent of the algebraic tail or the functional form of the density-dependent tail coefficient independently.

In the case of two-dimensional fluids there is another problem. Ever since the discovery of hydrodynamic tails, it has been realized that a consistent description of mode-coupling effects in a two-dimensional fluid would result in a long-time tail that decays faster than t^{-1}, because in $2D$ the self-diffusion constant itself diverges. In fact, de Schepper and Ernst [9] computed the coefficient (d_1 in eqn. 2) of the first correction to the t^{-1} tail for a system of hard disks. They predicted that this correction should be negative and proportional to $\log(t/t_0)/t$ (where t_0^{-1} is the initial decay rate of the velocity ACF):

$$< v_x(0)v_x(t) > \approx d_0/t + d_1 \log(t/t_0)/t + ...\text{for } t/t_0 \gg 1. \tag{2}$$

However, this prediction is only expected to hold for times that are not too long. Forster et al. [10] argued that as $t \to \infty$, the tail should be renormalized to $1/(t\sqrt{\log t})$.

Thus far it has not been possible to compare these predictions directly with computer simulation data. The reason is that such a comparison requires accurate knowledge of the velocity ACF for very long times (many tens to hundreds of collision times). This requires very long simulations on very large systems (to avoid problems due to spurious correlations caused carried by sound waves in periodic systems). Thus far such calculations have remained beyond the power of presentday computers.

1.2 Lattice gases

In the remainder of this paper we shall indicate how computer simulations on lattice gases can be used to test theoretical predictions about 'long-time tails' in the velocity autocorrelation function of tagged particle in a fluid.

One of the attractive features of lattice gas models is that they are ideally suited to serve as a testing ground for concepts in kinetic theory. The reason is twofold: on the one hand the very simple structure of most lattice-gas models often makes it possible to work out in closed form the consequences of a particular approximation scheme in kinetic theory. On the other hand, lattice-gas models are ideally suited for computer simulation. Thus approximate schemes in kinetic theory can be tried out on lattice gas models before applying them to more realistic models of fluids or solids. Below we show how Lattice Gas Cellular Automata (LGCA's) can be used to verify in great detail the theoretical predictions about long-time tails in the the velocity autocorrelation function of tagged particles.

Lattice gas cellular automata have recently received a lot of attention because these model systems may provide a 'cheap' alternative to simulate the hydrodynamic behaviour of simple fluids [11]. Because of their simple structure which makes it comparatively easy to work out the consequences of a particular approximation scheme, LGCA's are, in principle, ideally suited to serve as a testing ground for concepts in kinetic theory. It is therefore only natural that lattice gas models were considered as promising systems to study the long-time behaviour of the tagged-particle VACF. This approach has been tried by Boon and Noullez [12] and Binder and d'Humières [13] for 2D systems. However, somewhat disappointingly, the statistical accuracy of these (long) numerical simulations is poor. Hence the presence of a long-time tail was hidden in the statistical noise and a special analysis was required to demonstrate that the simulation data are, in fact, compatible with the presence of an algebraic tail of approximately the expected amplitude [14].

However, using a technique that was recently developed by Frenkel [15,16] it is now possible to compute tagged particle VACF's in lattice gases with an efficiency that is a factor 10^6 to 10^{10} higher than the earlier 'brute force' schemes. In the present paper we shall discuss the application of this new technique to the study of long-time tails.

1.3 Lattice Gas Cellular Automata

In the present paper we shall focus on the properties of LGCA's on a $2D$ triangular lattice (FHP-model [11,20] and on a $3D$ face-centered hyper-cubic (FCHC) lattice.

In LGCA's, the particles are constrained to move along the bonds joining the lattice sites. No two particles can move along the same bond in the same direction. The state of the lattice is completely specified by indicating which links are occupied and which are empty. This implies that lattice-gas particles are indistinguishable.

The time evolution of the system is governed by the following rules:

1. Propagation: all particles move in one time step (for convenience we choose $\Delta t = 1$) from their initial lattice position (say $\mathbf{X}$) to a new position ($\mathbf{X}' = \mathbf{X} + \mathbf{c}_\alpha$; where $\mathbf{c}_\alpha$ is the velocity of species α). The velocities $\mathbf{c}_\alpha$ are such that at the end of the propagation steps all particles are once more positioned at lattice sites.

2. Collision: the particles at all sites on the lattice undergo a collision that converves the total number of particles and the total momentum at each site. Usually, these local collision rules are deterministic.

Provided that the lattice has a sufficiently high symmetry (e.g. triangular in 2 dimensions) and the collision rules are sufficiently isotropic (for a discussion, see [17]), it can be shown that the equation that governs the time evolution of the distribution function of such a lattice gas becomes equivalent to the Navier-Stokes equation for an incompressible fluid in the limit that the flow velocity is much less than the particle velocity, and all spatial variations in the system occur on a scale that is large compared to the mean free path of the lattice gas particles. In this respect LGCA's model atomic fluids.

When attempting to compute the velocity correlation function of a particle in a lattice gas CA, one is immediately confronted with a conceptual problem. As all lattice gas particles are indistinguishable, the velocity correlation function of 'a particle' is ill defined. As soon as a particle has collided it is no longer possible to identify any of the outgoing particles as the original particles whose velocity ACF we are attempting to compute. To avoid this problem, the particle under consideration must be labeled differently from the rest (say, a 'blue' particle in a sea of 'red' particles). Once the collision rules for all particles have been specified we can then compute the velocity ACF of a single tagged particle. This is the approach that has been pursued by Boon and Noullez for the FHP model [12] and by Binder and d'Humières for the HPP model [13]. As mentioned above, this approach yields poor statistics because we must solve the dynamics of all N lattice-gas particles in order to follow the time-evolution of 1 tagged particle.

An alternative approach that effectively side-steps the problem referred to above has been followed by Colvin, Ladd and Alder [24]. These authors compute not the velocity ACF of a tagged particle but the autocorrelation function of the fluid velocity at a lattice site. This method yields somewhat better statistics and made it possible to observe a long-time tail in the site-velocity ACF. However, the site-velocity ACF does not contain the same mode-coupling contributions as the tagged-particle velocity ACF and is therefore, from a theoretical point of view, of less interest.

Fortunately, the matrix method described in the previous section can be extended in such a way that it becomes possible the compute the velocity ACF of a tagged particle in a lattice-gas cellular automaton with high accuracy. In order to do so we have to impose one restriction on the rules of the lattice gas automaton, namely that the collision rules for a tagged particle with untagged particles result in the occupation of the same output states as in the case of collisions between untagged particles. And, most importantly, the tagged particle has equal probability to be in any of the occupied output states. Hence, for the tagged particle the collision rules are stochastic, although for a 'colour-blind' observer, the rules are deterministic. Note also that even in 'collisions' that have the same input and output states the tagged particle may still 'collide', i.e. it may change its velocity state.

With these rules, it is obvious that the average velocity of a tagged particle after a collision at site $\mathbf{X}$, depends only on the (colorless) state at that site. In particular, it depends in no way on where the tagged particle was coming from. We can thus define for every (non-empty) site of the lattice the average post collisional velocity that a tagged particle at that site would have at that time:

$$\mathbf{v_X}(t) = \frac{1}{N_{occ}(\mathbf{X})} \sum_{\alpha=1}^{N_{occ}(\mathbf{X})} \mathbf{c}_\alpha \tag{3}$$

where $N_{occ}(\mathbf{X})$ is the total number of particles at site $\mathbf{X}$ and $\mathbf{c}_\alpha$ are the velocities corresponding to the occupied links. At first sight, eqn. 3 may look similar to the expression for the site velocity in a Lorentz gas. The important difference is that for a Lorentz gas the site velocity is fixed by the (time-independent) distribution of random bonds. In contrast, for the LGCA model the site-velocity for a tagged particle changes with every time step. However, apart from this modification, we can use basically the same techniques that worked for the Lorentz gases, to compute the velocity ACF of a tagged particle in a lattice gas cellular automaton.

Consider a tagged particle that is moving at site $\mathbf{X}$ at time $t = 0$, moving with a velocity $\mathbf{c}_\alpha$. At time $t = 1$, the tagged particle will have collided at site $\mathbf{X} + \mathbf{c}_\alpha$ and its average post collisional velocity will be $\mathbf{v_{X+c_\alpha}}(t = 1)$. Clearly,

$$< \mathbf{v}(0).\mathbf{v}(1) > = \frac{1}{N_{tot}} \sum_{\alpha, \mathbf{X}} s_\alpha(\mathbf{X}, t = 0) \mathbf{c}_\alpha . \mathbf{v_{X+c_\alpha}}(t = 1) \tag{4}$$

where $s_\alpha(\mathbf{X}, t = 0) = 1(0)$ if link α at site $\mathbf{X}$ is occupied (empty) at $t = 0$.

To compute $< \mathbf{v}(0).\mathbf{v}(2) >$, we simply propagate the tagged particle that is at site $\mathbf{X} + \mathbf{c}_\alpha$ to all sites that will be reached from this site in the next time step. For all these sites (denoted by $\{\mathbf{X}(t = 2)\}$ we compute the average post collisional velocity, average it over the set $\{\mathbf{X}(t = 2)\}$ and this yields the average velocity of the tagged particle after two time-steps.

For the actual implementation of this technique the recipe is then the following:

1. compute the average post collisional velocity of the tagged particle at site $\mathbf{X}$, at time $-t$, for all possible $\mathbf{X}$ on the lattice;

2. propagate this average to all sites that will be reached in one cycle from site $\mathbf{X}$;

3. for every site thus reached, compute the average of all the averages that have been propagated to it.

4. iterate steps 2 and 3 until t iterations have been made in total.

Note that in order to compute the velocity correlation function of a tagged particle, use was made of all possible starting positions and trajectories that such a particle could have, compatible with the (deterministic) dynamics of the underlying 'uncolored' lattice gas. The only additional averaging is over all possible time origins and over independent initial conditions. The latter averaging would not be necessary if the lattice gas cellular automaton were strictly ergodic. But as the models studied in this paper are known to have spurious invariants [25], averaging over a number of independent initial conditions is advisable.

1.4 Models

Simulations of tagged particle diffusion in two-dimensional fluids were carried out for a lattice gas model ('FHP-III', 6 speed-1 particles, one rest-particle, defined in ref. [20]). System sizes of up to 500×500 lattice points were studied at densities varying from 5% to 75% occupancy. The simulations were either performed on a CYBER 205 vector-computer or, for the larger systems, on a NEC-SX2 super-computer. As a model for $3D$ fluids we used, the face-centered hyper-cubic(FCHC) lattice gas model [17,21]. Whereas FHP-III is discussed extensively elsewhere [17,20] we should explain the exact version of the 3D FCHC-model that we used in our simulations. In this lattice there are 24 possible velocities, so a collision would require a 2^{24}-word lookup table, which requires a very large shared memory [18]. In the algorithm used in the present paper the 24-bit state is split into two 12-bit sub-states [19], which requires only a small 12-bit lookup table. This splitting can be done in 6 different ways, one of which is choosen randomly at every collision.

The $3D$ simulations were carried out on systems of up to $60 \times 60 \times 60$ lattice points. In all cases correlations were only computed for time intervals less then the shortest time in which any particle could cross the periodic box. This is in contrast to corresponding simulations of long-time tails in atomic fluids [8] where time intervals up to 5 times the acoustic wave traversal time had to be used. In the present simulation the VACF is calculated for different densities varying from $d = 0.05$ to $d = 0.90$, where d is defined as the average number of particles per link per node ($d = \rho/b$). In order to estimate the statistical error of the VACF 5 to 10 independent simulations per density were performed.

1.5 Results

1.5.1 Three dimensional model

Let us first look at the results for one density $d = 0.1$. In figure 1 we show the VACF for this density. For shorter times we see that the decay is approximately exponential, and for longer times we clearly observe the algebraic decay, which appears in the $\log - \log$ plot as a straight line. A convenient way to present the results for a range of densities is to multiply the VACF with $t^{-3/2}$, see figure 2. If the decay is algebraic with the predicted exponent $-3/2$ these functions should reach a constant value in the limit $t \gg t_0$. This behavior is indeed observed. These plateau values should equal the amplitude of the algebraic tail and can be compared with the prediction of mode-coupling theory, see figure 3. This figure shows that there is essentially quantitative agreement between the simulated and the predicted amplitudes for all densities. We wish to stress that there are *no* adjustable parameters in the comparison of theory and simulation.

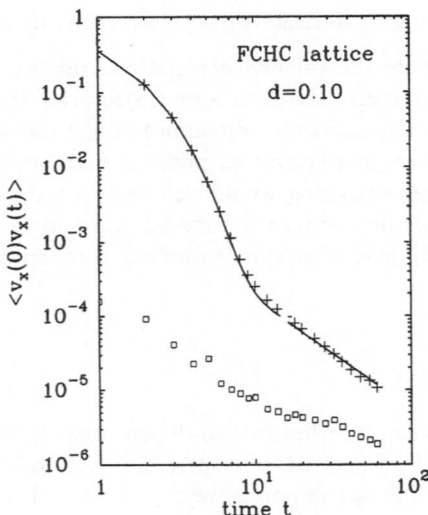

Figure 1. Log-log plot of the normalized velocity autocorrelation function of a 3-dimensional lattice gas cellular automaton on a FCHC lattice for density $d = 0.10$. The density d is in units of average number of particles per link. The time is in units of the time between successive updates of the lattice gas cellular automaton $\Delta t(=1)$. The solid line is the prediction of mode-coupling theory, the crosses are the simulation data. Note that the estimated error (open squares) decreases with increasing t to a value of order 10^{-6}.

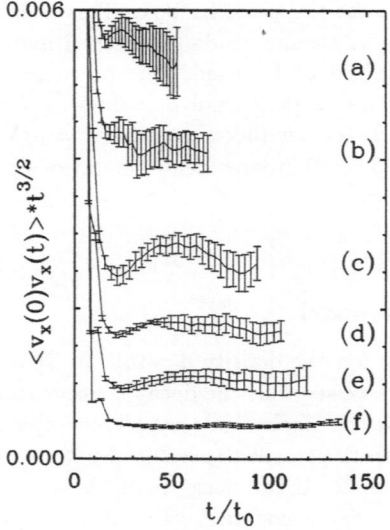

Figure 2. Velocity autocorrelation function multiplied by $t^{3/2}$ for the following densities: (a) $d = 0.2$, (b) $d = 0.3$, (c) $d = 0.5$, (d) $d = 0.6$, (e) $d = 0.7$, (f) $d = 0.8$. The time is in units of mean free time t_0.

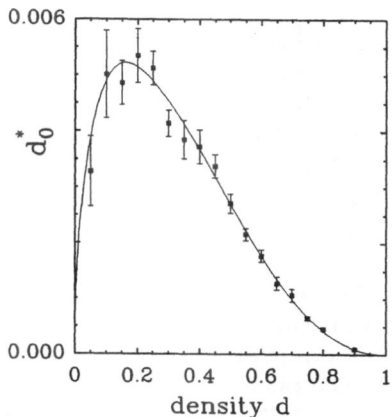

Figure 3. The normalized tail coefficient d_0^* of the tagged-particle VACF for the 3D FCHC system The solid curve is the mode-coupling prediction, the dashed curve is a weighted fit to the simulation data.

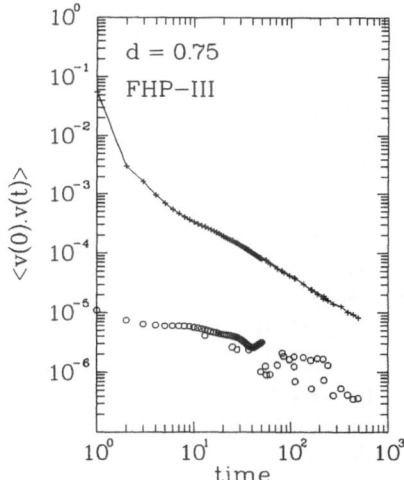

Figure 4. Example of the normalized velocity autocorrelation function of a tagged particle in the in a 2D lattice-gas cellular automaton (FHP-III) at a density $d=0.75$. Note that after an initial rapid decay (and overshoot), the velocity ACF approaches a power-law decay with an exponent -1. The estimated error (open circles) decreases with increasing t to a value of order 10^{-6}. In ref. [12] the statistical error in the long-time tail was of order 0.310^{-2}.

1.5.2 Two dimensional model

Simulations of tagged particle diffusion were carried out for a $2D$ lattice gas model ('FHP-III', 6 speed-1 particles, one rest-particle, with alternating collision rules for odd and even timesteps, see ref. [20]). System sizes of up to 500×500 lattice points were studied at densities d ranging from 0.05 to 0.75 (i.e. 5% to 75% occupancy per velocity state).

Figure 4 shows the velocity autocorrelation function of a tagged particle in the $2D$ lattice gas at a density $d = 0.75$. The velocity ACF has been normalized to 1 at $t = 0$. Here, and in all other cases shown, correlations were only computed for time intervals less than the shortest time in which any particle could cross the periodic box. In many of our calculations the statistical error is of order 10^{-6}, which is about a factor 10^3 lower than has been achieved with conventional techniques. Note that such an error reduction corresponds to a gain of 10^6 in computer time. Initially the decay of the velocity ACF is approximately exponential. The characteristic decay time t_0 of this exponential ranges from 5.6 at $d = 0.05$ to 0.33 at $d = 0.75$. There is a surprisingly large time-interval where the decay is no longer exponential but not yet algebraic. However, after some 30 collision times the decay appears to become algebraic. In order to see this latter effect more clearly, fig. 5 shows the function $t < \mathbf{v}(0).\mathbf{v}(t) >$ for densities $0.1 \leq d \leq 0.75$. If the velocity ACF decays as t^{-1} then the curves in fig. 5 should approach a constant value as $t \to \infty$. Such behavior is indeed observed. This in itself is maybe not surprising, but it is reassuring as it has been argued that the hydrodynamic long-time tails observed in computer simulation on continuous systems may be due to a propagation of numerical errors [26]. In the present simulation the discrete dynamics of the lattice gas is solved exactly, hence propagation of numerical errors is ruled out as a factor affecting either the power-law tails or, for that matter, any corrections to the latter.

We have compared the measured amplitude of the t^{-1} tail with the predictions of mode-coupling theory [22] adapted to the LGCA. As it turns out, the expression for the amplitude of the t^{-1}-tail in a $2D$ LGCA is equal to eqn. 1 multiplied by a factor $(1 - d)$, where the density ρ for continuous fluids must be interpreted as $7d/v_0$, the number density per unit area for the FHP-III model (7 velocity states per site, volume of the unit cell of the triangular lattice $v_0 = \sqrt{3}/2$). The factor $(1 - d)$ is a consequence of the Fermi statistics and guarantees that the state occupied by the tagged particle contains no fluid particle. It should be stressed that the applicability of eqn. 1 to the FHP model is not self-evident because in the FHP-model there exist unphysical hydrodynamic modes (associated with the staggered momentum density, see ref. [25]). These modes can couple to the microscopic stress tensor, thereby affecting the amplitude of the long-time tail of the stress-stress ACF. However, to leading order in $1/t$ the same staggered modes do *not* couple to the tagged particle current.

In figure 6 we compare the simulation results for the amplitude of the d_0 with the predictions of mode coupling theory. As can be seen from the figure, the mode coupling predictions are very close to the simulation results. The remaining discrepancy of a few percent is comparable to that found by Kadanoff et al. [25].

Next, consider the behavior of $< \mathbf{v}(0).\mathbf{v}(t) >$ for $t/t_0 \gg 1$. Velocity correlations were studied for times up to $t = 500$, which corresponds to values of t/t_0 ranging from 10^2 to over 10^3. In this time regime, which has never before been studied numerically, we would expect to observe a $\log(t)/t$ correction to the $1/t$ tail similar to the one predicted by de Schepper and Ernst for the hard-disk model [9]. The relative importance of these logarithmic corrections is grows linearly with the gas density in continuous fluids. If we

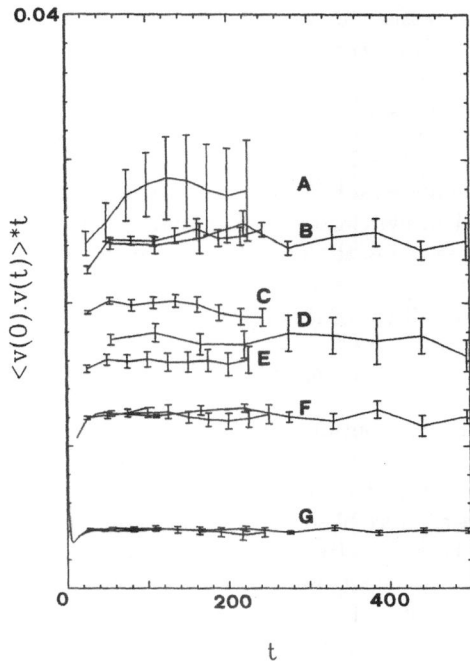

Figure 5. In order to detect possible deviations from the t^{-1} decay in the velocity ACF of a tagged particle in a $2D$ LGCA at long times, this figure shows $t < \mathbf{v}(0).\mathbf{v}(t) > / < v^2 >$ as a function of time t. The letters refer to the density: $A - G$ correspond to $d = 0.1, 0.2, 0.3, 0.35, 0.4, 0.5$ and 0.75 For some densities more than one simulation result is plotted. Within the statistical accuracy of the present calculations (the error bars have a length of 2 standard deviations) no systematic deviations from the t^{-1} decay can be detected. However, our error estimate is probably too conservative because a fit to the simulation data shows that for all 7 points with $d \geq 0.3$ the VACF decays with an effective exponent $\beta > 1$ (up to 3%).

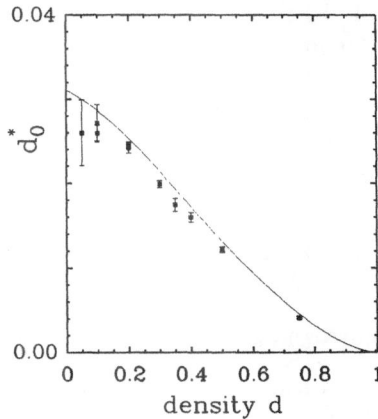

Figure 6. Density dependence of the amplitude of the t^{-1} algebraic tail of the tagged-particle VACF for the two-dimensional FHP-III model. Points: computer simulation results. Drawn curve: mode-coupling theory.

would assume that the expression for d_1 (see eqn. 2) given in ref. [9] also applies in the case of a lattice gas, then we find the following density dependence of the ratio d_1/d_0:

$$d_1/d_0 = -d_0\{(\nu_0 + D_0)^{-1} + (4\nu_0)^{-1} + (8(\nu_0 + \zeta_0))^{-1}\} = -0.247d , \qquad (5)$$

where ζ_0 is the bulk viscosity and $\rho\zeta_0 = 1/14$ for $\rho = 0$ [20]. A ratio d_1/d_0 of this magnitude should be easily observable in the present simulations. However, as can be seen from figure 5 there is no clear evidence for faster than t^{-1} decay at any density. In fact, a $\log - \log$ fit to the long-time tail of the velocity ACF allows us to estimate the ratio d_1/d_0 as a function of density. If we assume a linear density dependence of d_1/d_0 we can put the following bounds on the amplitude of this term:

$$d_1/d_0 = -0.02(3)d , \qquad (6)$$

The kinetic theory estimate for continuous fluids exceeds this value by 7 standard deviations. This suggests that the expression for d_1/d_0 given in ref. [9] does not apply to lattice gases. It seems unlikely that this discrepancy is due to the fact that the simulations do not extend to long enough times. More likely, eqn. 5 is not correct for lattice gases, because the the density effects of the Fermi statistics have not been accounted for and the staggered momentum modes that do not contribute to d_0 are expected to affect d_1. It should be added that even though we compute $< \mathbf{v}(0).\mathbf{v}(t) >$ for $10^2 - 10^3$ mean-free times, we are still well removed from the asymptotic regime where the correlation functions is expected to decay as $1/(t\sqrt{\log t})$.

At present, the lattice-gas equivalent of the prediction for d_1/d_0 given in ref. [9] is still lacking. Clearly, such a theoretical result is highly desirable, as it would allow us to decide whether the suppression of significant corrections to the t^{-1} tail is peculiar to lattice gases or if it is indicative of the behavior of $2D$ fluids in general.

Acknowledgments

We would like to thank Matthieu Ernst and Tony Ladd for stimulating discussions and the steady exchange of information. This work is part of the research program of the Stichting voor Fundamenteel Onderzoek der Materie (Foundation for Fundamental Research on Matter) and was made possible by financial support from the Nederlandse Organisatie voor Zuiver Wetenschappelijk Onderzoek (Netherlands Organization for the Advancement of Research). Computer time on the NEC-SX2 at NLR was made available through a grant by the NFS(Nationaal Fonds Supercomputers).

References

[1] D.Frenkel in: **Liquids, Freezing and the Glass Transition**, J.P.Hansen and D.Levesque (editors), North-Holland, Amsterdam, 1990.

[2] B. J. Alder and T. E. Wainwright, Phys. Rev. **A1**:18(1970).

[3] J.R. Dorfman and E.G.D. Cohen, Phys. Rev. Lett. **25**:1257(1970), Phys. Rev. **A6**:776(1972),**A12**:292(1975).

[4] M.H. Ernst, E.H. Hauge and J.M.J. van Leeuwen, Phys. Rev. **A4**:2055(1971).

[5] Y. Pomeau and P. Résibois, Physics Reports **19**:63(1975).

[6] D. Levesque and W.T. Ashurst, Phys. Rev. Lett. **33**:277(1974).

[7] J.J. Erpenbeck and W.W. Wood, Phys. Rev. **A26**:1648(1982).

[8] J.J. Erpenbeck and W.W. Wood, Phys. Rev. **A32**:412(1985).

[9] I. de Schepper and M. H. Ernst, Physica **87A**:35(1977)

[10] D. Forster, D. R. Nelson and M. J. Stephen, Phys. Rev. **A16**:732(1977).

[11] U. Frisch, B. Hasslacher and Y. Pomeau, Phys. Rev. Lett. **56**:1505(1986).

[12] J.-P. Boon and A. Noullez in: Proceedings of workshop on **Discrete kinetic theory, lattice gas dynamics and foundations of hydrodynamics**, World Scientific, Singapore, 1989.

[13] P.M. Binder and D. d'Humières, Los Alamos preprint LA-UR-1341(1988) and in ref. [23].

[14] J. -P. Boon and A .Noullez, to be published.

[15] D.Frenkel in: Proceedings of workshop on **Cellular Automata and Modelling of Complex Physical Systems**, P. Manneville(editor), Springer, Berlin, 1989.

[16] D.Frenkel and M.H.Ernst, Phys. Rev. Lett. **63**:2165(1989).

[17] U. Frisch, D. d'Humières, B. Hasslacher, P. Lallemand, Y. Pomeau and J.-P. Rivet, Complex Systems **1**:649(1987).

[18] J.-P. Rivet, M. Hénon, U. Frisch and D. d'Humières, Europhys. Lett. **7**:231(1988).

[19] A.J.C. Ladd and D. Frenkel in: Proceedings of workshop on **Cellular Automata and Modelling of Complex Physical Systems**, P. Manneville(editor), Springer, Berlin, 1989.

[20] D. d'Humières and P. Lallemand, Complex Systems **1**:599(1987).

[21] D. d'Humières, P. Lallemand and U.Frisch, Europhys. Lett. **2**:291(1986).

[22] M.H. Ernst in: Proceedings of Les Houches Summerschool on **Liquids, Freezing and the Glass Transition**, J.P.Hansen and D.Levesque (editors), North-Holland, Amsterdam (1990).

[23] Proceedings of workshop on **Cellular Automata and Modelling of Complex Physical Systems**, P. Manneville(editor), Springer, Berlin, 1989.

[24] M. E. Colvin, A. J. C. Ladd and B. J. Alder, Phys. Rev. Lett. **61**:381(1988)

[25] L. P. Kadanoff, G. R. McNamara and G. Zanetti, Phys. Rev. **A40**:4527(1989)

[26] R. F. Fox, Phys. Rev. **A27**,3216(1983).

ON THE COUPLING BETWEEN THE INTRINSIC ANGULAR
MOMENTUM OF MOLECULES AND THE FLUID VORTICITY

Hong XU*, Jean-Paul RYCKAERT**

Faculté des Sciences
Université Libre de Bruxelles
* Service de Physique Statistique
Plasmas et Optique non-linéaire, CP231
** Pool de Physique, CP 223

ABSTRACT

A Transient Non Equilibrium Molecular Dynamics (NEMD) experiment is performed on an isotropic diatomic fluid originally at equilibrium: at t=0, the molecular centre of mass velocities are given an additional systematic component corresponding to a purely rotational flow of vorticity γ. This velocity field is then kept constant in time by appropriate (periodic) boundary conditions. The so called 'vortex viscosity' coefficient characterising the coupling between the intrinsic (spin) angular momentum of the molecules and the vorticity of the flow is obtained from the time evolution of the average angular velocity of the molecules which starts from zero (at equilibrium) and reaches its asymptotic value γ.

1. INTRODUCTION

In the hydrodynamic description of polyatomic fluids, various phenomelogical coefficients describing the coupling between the velocity field $u(r)$ and the internal degrees of freedom of the molecules can be introduced [1,2,3]. In their book on Nonequilibrium Thermodynamics, De Groot and Mazur [1] discuss in particular the correlation between the intrinsic angular momentum of the molecules (rotation of molecules around their centre of mass) and the local flow vorticity. A measure of the coupling between these two forms of angular momentum is provided by the so-called 'vortex viscosity' coefficient η_r which enters into the theory as a phenomelogical coefficient defined by the constitutive law

$$\Pi^a = - 2\eta_r ((\nabla u)^a - \omega) \qquad (1a)$$

where Π^a, the antisymetric part of the pressure tensor, is the thermodynamic flux while the term in parentheses, the thermodynamic force, named the 'sprain rate' [4], gives the imbalance between the fluid vorticity $(\nabla \mathbf{u})^a$ and the molecular average spin angular velocity ω (to be treated here as the antisymetric tensor dual to the usual axial vector).

On the basis of the conservation of the total angular momentum and the linear law eq.(1a), De Groot and Mazur showed that when an isotropic fluid at equilibrium is suddenly subject to a purely rotational flow of constant vorticity γ (say around the z axis), the average angular velocity of the molecules ω_z evolves asymptotically towards γ according to

$$\omega_z(t) = \gamma (1 - \exp(-t/\tau)) \qquad (2\,a)$$

with characteristic time

$$\tau = \frac{\rho <I>}{4\,\eta_r} \qquad (3a)$$

where ρ is the molecular number density and $<I>$ is the average moment of inertia of the molecules around their centres of mass in the equilibrium phase.

In this paper, we use an original transient Non Equilibrium Molecular Dynamics (NEMD) method [5] to perform on a diatomic liquid (chlorine) the experiment discussed by De Groot and Mazur. The flow is induced and maintained by an appropriate external field: we discuss the method in some details in the next section.

As it could be expected [4,6] for dense phases of anisotropic molecules, the time evolution of ω_z turns out to be non exponential. Our results thus confirm that the constitutive law (eq.1a) should include memory effects, according to

$$\Pi^a = -2\int_0^t ds\, \eta_r(t\text{-}s)\, [(\nabla \mathbf{u}\,(s)\,)^a - \omega\,(s)] \qquad (1b)$$

as the sprain rate itself varies on a microscopic time scale in the experiment. The 'vortex viscosity' is then identified with $\bar{\eta}_r = \int_0^\infty \eta_r(t)\, dt$, the zero frequency Laplace transform of the memory function. It can be shown by linear response theory [5] that, in the transient experiment described above, eq.2a generalizes to

$$\omega_z(t) = \gamma(1 - C_{ss}(t)) \qquad (2b)$$

where $C_{ss}(t)$ is the normalized time correlation function of a collective variable $\mathbf{S}$, namely the total intrinsic (or spin) angular momentum of the N individual molecules around their own centers of mass. Its explicit formula is

$$\mathbf{S} = \frac{1}{N} \sum_{\alpha=1}^{N} \mathbf{I}_\alpha\, \omega_\alpha \qquad (4)$$

where $\mathbf{I}_\alpha$ and ω_α are respectively the moment of inertia and the angular velocity of molecule α. The intensity of the coupling between the spin angular momentum and the fluid vorticity is thus basically characterised by τ_{ss}, the relaxation time given by the area under $C_{ss}(t)$ that can be related [6,5] to the vortex viscosity by

$$\tau_{ss} = \frac{\rho <I>}{4\,\tilde{\eta}_r} \tag{3b}$$

2. MOLECULAR DYNAMICS SIMULATION OF A ROTATIONAL FLOW

Our molecular dynamics experiment simulating fluids at the molecular level under rotational flow is developed on the basis of the Sllod equations of motion [7,8] which read

$$\dot{\mathbf{r}}_{i\alpha} = \frac{\mathbf{p}_{i\alpha}}{m_i} + (\nabla\,\mathbf{u})^T\,\mathbf{R}_\alpha \tag{5a}$$

$$\dot{\mathbf{p}}_{i\alpha} = \mathbf{F}_{i\alpha} - \frac{m_i}{M}\,(\nabla\,\mathbf{u})^T\,\mathbf{P}_\alpha \tag{5b}$$

for a set of N molecules ($\alpha = 1,N$) of mass M composed of n atoms ($i=1,n$) of mass m_i, coordinate $\mathbf{r}_{i\alpha}$, linear momenta $\mathbf{p}_{i\alpha}$ and force $\mathbf{F}_{i\alpha}$. The perturbation terms involve the coupling between the translational degrees of freedom of the molecules ($\mathbf{R}_a, \mathbf{P}_a$) and the velocity gradient ($\nabla\,\mathbf{u}$) taken uniform in space but possibly varying in time. To represent a pure rotational flow, we have

$$\nabla\mathbf{u} = \begin{matrix} 0 & +\gamma\theta(t) & 0 \\ -\gamma\theta(t) & 0 & 0 \\ 0 & 0 & 0 \end{matrix} \tag{6}$$

where the Heaviside step function $\vartheta(t)$ indicates a uniform vorticity γ for $t>0$. That these equations are effectively inducing a vortex flow around the z axis (characterised by unit vector $\mathbf{1}_z$) can be shown as follows:

a) around $t=0$, the linear momenta $\mathbf{p}_{i\alpha}$ are continuous, but the atomic velocities of molecule α are augmented by a systematic rotational component $\gamma\,\mathbf{1}_z \times \mathbf{R}_\alpha$.

b) for $t>0$, the perturbed eqs of motion are equivalent to

$$\ddot{\mathbf{r}}_{i\alpha} = \mathbf{F}_{i\alpha} - \gamma^2\,\mathbf{R}_\alpha \tag{5c}$$

i.e. the usual Newton's equations of motion to which an additional centripetal force is added to preserve the homogeneity of the system: note that this term is second order in γ so that this term does not affect the linear response.

What is remarkable with Eqs. (5) is that periodic boundary conditions (PBC) can be perfectly accommodated to these homogeneous flows [9]. However, it generally implies that the basic cell (the molecular dynamics 'box' and by extension the infinite 3 dimensional array of replica of the system obtained by translational symmetry operations) changes in time according to the imposed macroscopic flow. We consider here the onset of a purely rotational flow around the z axis in a system at equilibrium which is

293

originally enclosed in a cubic box with sides pointing along the three cartesian axes with usual PBC: the system remains periodic [5] with a cubic MD basic cell which simply rotates at constant angular velocity γ around the z axis.

A NEMD experiment is conducted by performing typically a few hundreds perturbed trajectories starting from uncorrelated equilibrium configurations generated by a traditional MD experiment (the 'mother' trajectory). At small strain rates (within the linear response regime), it is necessary to use the subtraction technique [10] to get a sufficient signal to noise ratio. An additional advantage of working at low gradient is that heating up effects, appearing uniformly in space, are negligible.

3. DIATOMIC LIQUID IN ROTATIONAL FLOW

Our experiment has been performed on liquid chlorine using a model of N=216 rigid LJ dumbell of mass M and length $l/\sigma = 0.63$ and working at a thermodynamic point $\rho\sigma^3 = 0.544$ and $k_b T/\varepsilon = 0.97$, corresponding exactly to a rather old chlorine model used in previous studies [11,12]. We integrated the atomic cartesian equations of motion using the velocity version of the Verlet Algorithm [13,14] with a time step $\Delta t / (\sigma \varepsilon^{-1/2} M^{1/2}) = 0.0015$.

Applying an extremely small vorticity $\gamma \sigma \varepsilon^{-1/2} M^{1/2} = 5. \, 10^{-5}$ around the z axis, the average angular velocity of the molecules along the non-equilibrium trajectories, i.e. $< \omega_z(t) >_{n \, eq}$, was obtained by averaging $S_z(t)$ given by eq.4 over 400 segments, exploiting the noise reduction provided by the subtraction technique and normalizing by a factor $<I> / M\sigma^2 = 0.06615$.

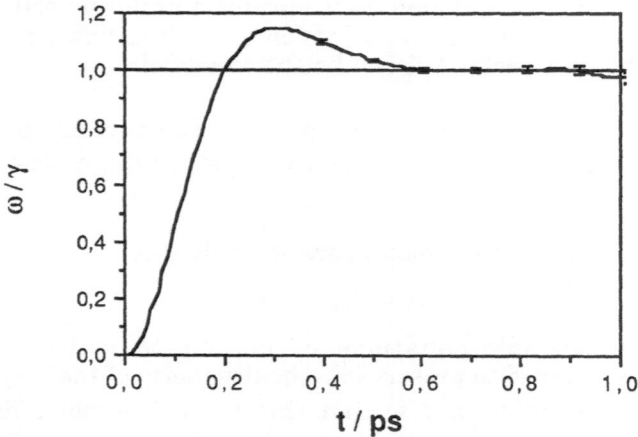

Figure 1. Time evolution of the mean angular velocity of chlorine molecules (normalised by γ) in an isotropic liquid phase subject at t=0 to a rotational flow of vorticity γ.

As shows figure 1, the resulting transient behavior turns out to be highly non exponential. On the basis of eqs. 2b and 3b, we get a relaxation time τ_{SS} σ^{-1} $\varepsilon^{1/2}M^{-1/2}$ = 0.034 leading to $\bar{\eta}_r$ σ^2 $\varepsilon^{-1/2}M^{-1/2}$ = 0.262. Note the rather exceptional small error bars on a NEMD response function obtained by subtraction technique.

At the occasion of this work, we observed that the total spin angular momentum of the system $\mathbf{S}$ (defined by eq.4) and the single molecule spin angular momentum $\mathbf{s}_\alpha = \mathbf{I}_\alpha \, \omega_\alpha$ lead to almost identical normalised time autocorrelation functions, as we show in figure 2: this implies small cross effects $<\mathbf{s}_\alpha(t)\mathbf{s}_\beta(0)>$ for different molecules α and β.

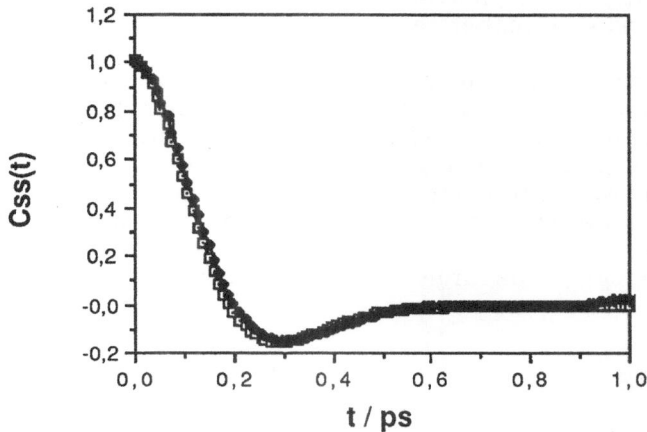

Figure 2. Normalised spin angular momentum time correlation functions. a) collective $C_{SS}(t)$ derived from the (linear) NEMD response eq.2b (black triangles). b) single molecule $C_{ss}(t)$ obtained from equilibrium trajectory (white squares).

This property, shown for chlorine in figure 2 has also been observed in liquid nitrogen in a separate experiment [5]. When these cross terms are negligible (one can wonder whether this is true for all liquid phases of rigid non spherical molecules), it involves that an estimate of $\bar{\eta}_r$ can be derived directly from eq.(4) using the single molecule relaxation time, i.e. a quantity often quoted in past MD results of diatomics in a different context: this observation also leads naturally to an interpretation of the vortex viscosity as a quasi single molecule rotational friction coefficient.

Quite surprisingly, the value that we obtained for the vortex viscosity of chlorine differs strongly (by a factor 10!) with an estimate obtained on the same system by a NEMD stationary experiment where this phenomelogical coefficient was obtained by direct exploitation of the linear law eq.1 using artificial constraints to keep the sprain rate constant in time [12]. This disagreement is in our opinion related to an accidental error in [12] as our value is supported by the following arguments :

1) calculations on liquid Nitrogen that we have recently performed by the same transient technique [5] suggest a reasonable agreement of $\bar{\eta}_r$ with NEMD results obtained using constrained sprain rate techniques or with results based on equilibrium correlation functions [15]

2) our single molecule relaxation function $C_{ss}(t)$ (fig.1) for chlorine is in perfect agreement with older results [11].

4. FINAL REMARKS

In this paper, we have investigated the response of a diatomic fluid to the onset of a purely rotational flow. For this purpose, we have devised a special technique where the system remains homogeneous thanks to rotating periodic boundary conditions. We restricted ourselves to the linear regime of Non Equilibrium response by choosing a rather small vorticity. However, the same method can be used to study the response of polyatomic fluids in other flows with homogeneous velocity gradient [9] (elongational flow for example) and could be used with higher velocity gradient to identify non linear effects.

We observed on a chlorine liquid model that intrinsic and external angular momentum are indeed coupled and that both the local mean angular velocity of molecules and the local fluid vorticity tend to equalise on a very short time scale of the order of the pico-second. Therefore, this coupling can probably be taken into account in hydrodynamics of polyatomic fluids on a longer time scale by simply considering that vorticity and the average molecular angular velocity are always identical locally (considering local averages on a length scale large with respect to the molecular size).

5. REFERENCES

1. De Groot S.R., Mazur, P.,(1962) Nonequilibrium Thermodynamics (North-Holland).
2. Snider R.F., Lewchuk K.S., (1967), J. Chem. Phys. **46**, 3163.
3. Evans D.J., Streett W.B., (1978), Mol. Phys. **36**,161.
4. Evans D.J., (1979), J. Stat. Phys. **20**, 547.
5. Xu H., Ryckaert J.-P., to be submitted.
6. Ailawadi, N.R., Berne, B., Forster, D (1971), Phys. Rev. A **3**,1462.
7. Evans D.J., Morriss G.P.,(1984) Comp. Phys. Rep.**1**, 299.
8. Ladd, A.J.C., (1984) Mol. Phys.**53**, 459.
9. Ryckaert J-P, to be published in Berichte Bunsengesellschaft fuer Physicalische Chemie (Conference proceedings on 'Transport properties of Mobile Phases' 1989 Aachen).
10. Ciccotti, G.C., Jacucci G., Mc Donald I.R., (1979) J. Stat. Phys.**21**, 1.
11. Singer,K ,Taylor, A.,Singer, J.V.L., (1977), Mol. Phys.**33**, 1757, (1979), Mol Phys.**37**,1239.
12. Edberg R., Evans D.J., Morriss G.P., (1987), Mol. Phys. **62**, 1357
13. Andersen H.C., (1983),, J. Comp. Phys. **52**, 24.
14. Marechal G., Ryckaert J.P., Bellemans A.,(1987), Mol. Phys.**61**, 33.
15. Evans D.J., Hanley H.J.M., (1982), Phys. Rev. A, **25**,1771.

PART V

APPLICATIONS (III): SOLIDS

AMORPHIZATION INDUCED BY CHEMICAL DISORDER IN CRYSTALLINE NiZr2 :

A MOLECULAR DYNAMICS STUDY BASED ON A N-BODY POTENTIAL

C. Massobrio[1] and V. Pontikis[2]

[1] ICSMA, LTPM (U.A. 446), Université Paris XI
91400 Orsay, FRANCE

[2] Centre d'Etudes Nucléaires de Saclay, Section de Recherches de
Métallurgie Physique, 91191 Gif sur Yvette Cedex, FRANCE

INTRODUCTION

Solid state crystalline-to-amorphous transformations received recently considerable attention[1] due to the possibility they offer to produce bulk amorphous materials and the challenging questions that arise when trying to understand the underlying mechanisms. Various physical processes such as irradiation[2], interdiffusion[3], annealing[4] or mechanical alloying[5] are able to transform a crystalline material into an amorphous solid, provided the temperature does not exceed the crystallization temperature of the amorphous system at the same composition. A common feature of these processes is the intimate intermixing of the constituents and the resulting chemical disorder. For systems where important size effects exist the chemical disorder leads to amorphization[6] possibly related to an elastic instability as suggested by recent theoretical work[7,8].

This paper presents the results of a molecular dynamics study (MD) of the effects of chemical disorder on the stability of crystalline NiZr$_2$. This alloy is chosen as a model system since it is known to amorphize either by electron beam irradiation[9] or by ball milling powders of Ni and Zr of the appropriate composition[5]. Our study shows that when the crystal is subjected to a chemical disorder, introduced by interchanging Ni and Zr atoms, amorphization occurs provided the amount of disorder exceeds a critical value of the long range order parameter. This nonequilibrium transformation does not involve mass transport over long distances, only local relaxation is observed around the antisite defects which leads to the breakdown of the crystalline lattice. On the contrary, amorphization by interdiffusion is a process occuring at equilibrium, without external forces acting on the system, which cannot be understood without taking into account the possible role of heterophase interfaces. Indeed, the crystal being the minimum free energy state at equilibrium, amorphization by interdiffusion may correspond to a mechanical instability made possible by the presence of interfaces. A first investigation of such effects is also reported here consisting in the study of the stability and time-evolution of a NiZr$_2$ slab, part of which has been disordered. The disordered, amorphous region of this system propagates and tends to invade the entire model.

In the following sections details of the models we used, the computations we performed and the results we obtained are presented. The final section is devoted to a short discussion and some conclusive remarks.

Microscopic Simulations of Complex Flows   297
Edited by M. Mareschal, Plenum Press, New York, 1990

MODEL AND COMPUTATIONS

The structure of the crystal is of the $CuAl_2$ (C16) type, a body centered tetragonal lattice with lattice cell dimensions a=0.6483 nm, c=0.5267 nm at room temperature[10]. Our system consisted of N point particles enclosed in a parallelepipedic box. We studied systems of three different sizes : 3x3x3 (Ni: 108 atoms, Zr: 216 atoms), 4x4x4 (Ni: 256 atoms, Zr: 512 atoms) and 12x4x4 (Ni: 768 atoms, Zr: 1536 atoms) lattice cells. The particles interact via a n-body potential constructed in the framework of the second moment approximation of the tight binding scheme, well adapted to transition metals[11]. Accordingly the total energy of the system is given by :

$$E = \sum_{\alpha} \sum_{i_\alpha=1}^{N_\alpha} \left\{ \sum_{\beta} \sum_{\substack{j_\beta=1 \\ i_\alpha \neq j_\beta}}^{N_\beta} A_{\alpha\beta} \exp\left[-p_{\alpha\beta}\left(\frac{r_{ij}^{\alpha\beta}}{d_{\alpha\beta}} - 1 \right) \right] - \sqrt{ \sum_{\beta} \sum_{\substack{j_\beta=1 \\ i_\alpha \neq j_\beta}}^{N_\beta} \xi_{\alpha\beta}^2 \exp\left[-2q_{\alpha\beta}\left(\frac{r_{ij}^{\alpha\beta}}{d_{\alpha\beta}} - 1 \right) \right] } \right\}$$

where $r_{ij}^{\alpha\beta} = \left| \vec{r}_{i_\alpha} - \vec{r}_{j_\beta} \right|$ and the indexes $i_\alpha(j_\beta)$, run over all the particles. The parameters $p_{\alpha\beta}$, $q_{\alpha\beta}$, $A_{\alpha\beta}$ and $\xi_{\alpha\beta}$ are determined as follows[12] : i) for $\alpha=\beta$, a fit is performed on the cohesive energy, equilibrium condition and bulk modulus at zero temperature of the pure elements Ni and Zr. Only nearest neighbors (NN) contributions at distances $d_{\alpha\alpha}$ are taken into account. ii) similarly for $\alpha\neq\beta$, the corresponding parameters are fitted on the elastic constants[13] and the cohesive energy of the alloy and $d_{\alpha\beta}$ is taken equal to the Ni-Zr NN distance in $NiZr_2$. The value of the cohesive energy is obtained from that of the ideal mixture, using the experimental cohesive energies of the pure metals Ni and Zr, modified by including the experimental formation enthalpy of the $NiZr_2$ crystal[14]. To avoid disturbing artifacts of MD related to short cutoff distances[15], we empirically extended the range of the potential by including in the fitting procedure interactions up to r_c=0.53 nm. Values of the potential parameters, physical quantities we used for the fit and a comparison between the experimental and calculated elastic constants of $NiZr_2$ are reported on table 1.

Table 1. Values of the tight-binding potential parameters for $NiZr_2$. The experimental values of cohesive energy, E_c, bulk modulus, B, and NN distances, δ, extrapolated at T=0 K, are used for the fit of the potential parameters $p_{\alpha\beta}$, $q_{\alpha\beta}$, $A_{\alpha\beta}$ and $\xi_{\alpha\beta}$. The calculated elastic constants are compared with the corresponding experimental values (between parentheses). E_c, $A_{\alpha\beta}$ and $\xi_{\alpha\beta}$ are expressed in eV/atom, elastic constants in units of 10^{12} dyn·cm^{-2} and distances in nm.

	Ni	Zr	NiZr$_2$
E_c	4.44[16]	6.17[17]	6.0*
B	1.88[18]	0.97[18]	1.19[13]
δ	0.249[16]	0.3179[19]	0.2761[10]
C_{11}	2.57(2.61[18])	1.53(1.55[18])	1.61(1.59[13])
C_{12}	1.65(1.51[18])	0.74(0.67[18])	1.05(1.34[13])
C_{13}	–	0.55(0.65[18])	0.95(0.85[13])
C_{33}	–	1.68(1.73[18])	1.58(1.47[13])
C_{44}	0.93(1.32[18])	0.45(0.36[18])	0.54(0.24[13])
C_{66}	–	–	0.61(0.06[13])
p	10.0	9.3	8.36
q	2.7	2.1	2.23
A	0.1368	0.1615	0.2166
ξ	1.756	2.34	2.139

* see text

A good agreement is obtained between predicted and experimental elastic constants. As expected, our potential reproduces the violation of the Cauchy conditions exhibited by the experimental data. Constant temperature and pressure were obtained by the Nosé-Andersen MD technique[20,21]. The equations of motion were integrated using periodic boundary conditions and a fifth order predictor-corrector algorithm with a timestep $\delta t = 10^{-15}$s.

MODEL VALIDATION

The model is further tested by comparing an estimate of its melting temperature with the experimental one $T = T_m^{exp} = 1440$ K. This is obtained by computing the atomic mean square displacement and the potential energy per atom as a function of temperature.

For the largest system we studied, the former (figure 1) reaches the value $<u^2> = 6.26 \cdot 10^{-4}$ nm^2 at $T = T_m^{exp}$ which corresponds to a Lindemann ratio $\delta \approx 8\%$[12,22]. It is therefore likely that the actual melting temperature of the model is not very different from the experimental one. This assumption is strengthened by the potential energy per atom profile, U, displayed on the same figure. An abrupt change of U is observed at $T \approx 1600$ K followed by high diffusion values, $D \approx 10^{-5} cm^2 s^{-1}$, for both constituents thus indicating that melting occured. Indeed, solid state diffusion cannot take place in our system since sinks and sources of point defects, such as free surfaces, are absent and the barrier of formation of Frenkel pairs is too high to produce diffusion on the space and time scales of the simulation. Thus the model should melt at $T_m^{model} \leq 1600$ K some overheating being unavoidable for an infinite extension perfect crystal. The reasonable $<u^2>$ value obtained at $T = T_m^{exp}$ and the fact that melting is observed only ≈ 160 K above T_m^{exp} suggest that the dynamical properties of our model at $T \neq 0$ K are not dramatically different from those of $NiZr_2$.

RESULTS

To simulate the chemical disorder induced by the aforementioned physical processes leading to the solid state amorphization, a given number of Ni and Zr atoms, randomly choosed, are instantaneously exchanged in a equilibrated configuration of the system at T=300 K. The disorder is conveniently quantified by the long-range order parameter, $S = (p-r)/(1-r)$, where p and r are respectively the probability of presence of a A-type atom (A=Ni or Zr) on an ordered lattice site and the molar ratio of A atoms in the system. Values S=0 and S=1 correspond respectively to the full chemical disorder or to the perfectly ordered crystal. Five values of S=0.0, 0.2, 0.4, 0.6 and 0.9 were investigated and the partial and total pair distribution functions of the system, g(r), are computed for each. The immediate effect of this perturbation is a volume increase illustrated on figure 2 for different S-values. Before the system volume reaches a stationary value, a short relaxation occurs which lasted up to 10^4 time steps for the largest possible perturbation S=0. The atomic structures of the perfect and the disordered system differ drastically. The total and partial pair distribution functions, g(r),

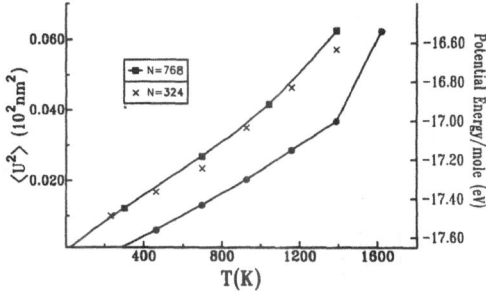

Figure 1 . Variation of of the potential energy per molecule (full dots) and of the atomic mean square displacements (full squares: N=768, crosses: N=324) in crystalline $NiZr_2$ as a function of temperature. Full line is a guide for the eye.

displayed on figure 3 for S=0.0, provide a convenient tool to reveal those differences. Many of the peaks characterizing the crystalline structure disappear whereas others are smeared out thus suggesting that the system becomes amorphous. This conclusion is qualitatively and to a lesser extent quantitatively confirmed when comparing the g(r) of the rapidly quenched liquid at T=300 K (figure 3) (quenching rate q=10^{13} K/s) .

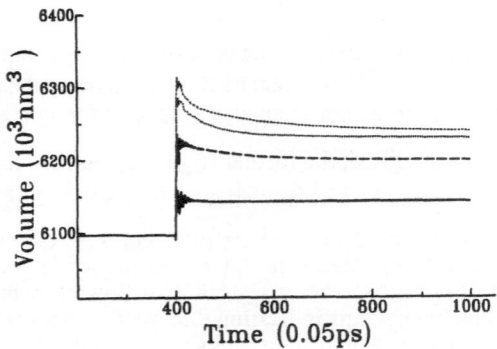

Figure 2 . Average volume of the system at T=300 K before and after the chemical disorder is introduced for different values of the long range order parameter S. Full line: S=0.9, dashed line: S=0.6, dotted line: S=0.2, short dashed line: S=0.0.

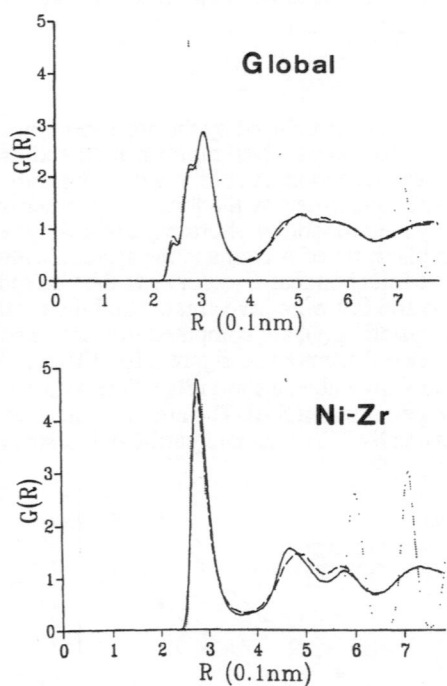

Figure 3 . Partial and global radial pair distribution functions at T=300 K averaged over 40 ps. Full line: after the chemical disorder is introduced into the system, S=0. Dotted line: crystalline $NiZr_2$. Dashed line: quenched liquid.

The total g(r) thereby obtained is identical to that of the disordered solid while small differences persist between the corresponding partial pair distribution functions. These differences may be attributed to the different structural relaxations occuring when quenching the liquid or during the amorphization process of the disordered crystal. The amorphous system we produced is stable against recrystallization on annealing, at least on the time scale of our simulations : no noticeable structural changes are observed over a $4 \cdot 10^4$ timesteps equilibrium trajectory at T=300 K. For values of the long-range order parameter S<0.6 amorphization is always observed. However for S=0.9, despite the small volume increase (figure 3), the crystal is stable on the same time scale. The narrow range of S-values, 0.6<S<0.9, separating the stability and instability regions of the crystalline state indicates first-order-like features for the crystal-to-amorphous state transition in NiZr$_2$. Rapid diffusion of one of the components of the alloy, in our case Ni atoms in a Zr crystal, has been invoked as being a necessary condition to decide on whether or not a solid state amorphization reaction is possible [3]. It is worth noticing that in the present case the amorphization occurs without long-range diffusion and is exclusively due to local relaxations starting when the chemical disorder is introduced into the system.

A comparison of our results with experimental data is possible due to recent work on amorphous NiZr$_2$ prepared by rapidly quenching the liquid[23-25] and to experiments in which ion irradiation induces chemical disorder and the subsequent amorphization of the alloy Zr$_3$Al[6]. In the former, partial pair distribution functions have been determined by neutron scattering. Although a comparison between the shapes of experimental pair distribution functions and those we computed cannot be made, due to the large uncertainties of deconvoluted experimental data, the positions of first and second neighbors peaks compare favourably well with our results as shown on table 2.

Table 2. Partial pair distribution peak positions, expressed in nm, and number of nearest neighbors in the amorphous NiZr$_2$ alloy. Comparison of present computations with experimental data[24,25]. Experimental values are given between parentheses.

	1st maximum	1st minimum	2nd maximum	n° NN
Ni-Ni	0.24 (0.25-0.27)	0.32 (0.33-0.34)	0.45 (0.42)	2.75 (2.3-3.3)
Zr-Zr	0.31 (0.32-0.33)	0.42 (0.40-0.43)	0.52 (0.52)	10.46 (9-11)
Ni-Zr	0.28 (0.27-0.29)	0.37 (0.4)	0.47 (0.45)	3.9 (2.9-4.8)

A partially disordered system has been settled up by introducing full chemical disorder, S=0, in the centre of the model. The disordered region extends over eight successive nickel atomic planes normal to the X-axis, that is along the largest dimension of the simulation box. The disordered, amorphous region is conveniently visualized by the density profile of nickel atoms illustrated by figure 4a. To check on whether or not this disorder will propagate, longer MD trajectories were produced and the structure of the system has been monitored by computing density profiles and pair correlation functions on a local basis. After $4 \cdot 10^5$ time steps annealing the disorder extends over a region twice larger than the spatial extension of the disordered system portion at the beginning of the annealing process (figure 4b). Owing to the progressivity of the process, this behavior cannot be attributed to a relaxation of the interfaces which, in absence of propagation, will occur in a much shorter delay as is usually observed in MD simulations of grain boundaries[26]. This result demonstrates that although the crystal is the most stable thermodynamical state of our model system the presence of a crystal-amorphous solid interface provides to it a mechanical instability path leading to the glassy state. Further investigation of this behavior is presently undertaken and will be reported elsewhere[12].

DISCUSSION AND CONCLUSIVE REMARKS

Rehn et al.[6] observed that when a Zr$_3$Al alloy sample is submitted to ion irradiation, amorphization occurs when the irradiation dose exceeds a critical threshold correponding to a long-range order parameter value S≤0.2. The volume expansion of their system at the value S=0.2 above which amorphization does not occur, x=δV/V≈2%, is close to that we found at the same S-value : x=1.8%. However in the system we studied amorphization is observed up to S=0.6 and x=δV/V ≈1.6%, this difference being probably related to the different stability regions of Zr$_3$Al and NiZr$_2$ crystals. In addition Rehn et al.[6] as well as previous work[27] suggested that the solid-state amorphization may be a first order phase transition monitored by an elastic instability following the chemical disorder. The results of the present study show

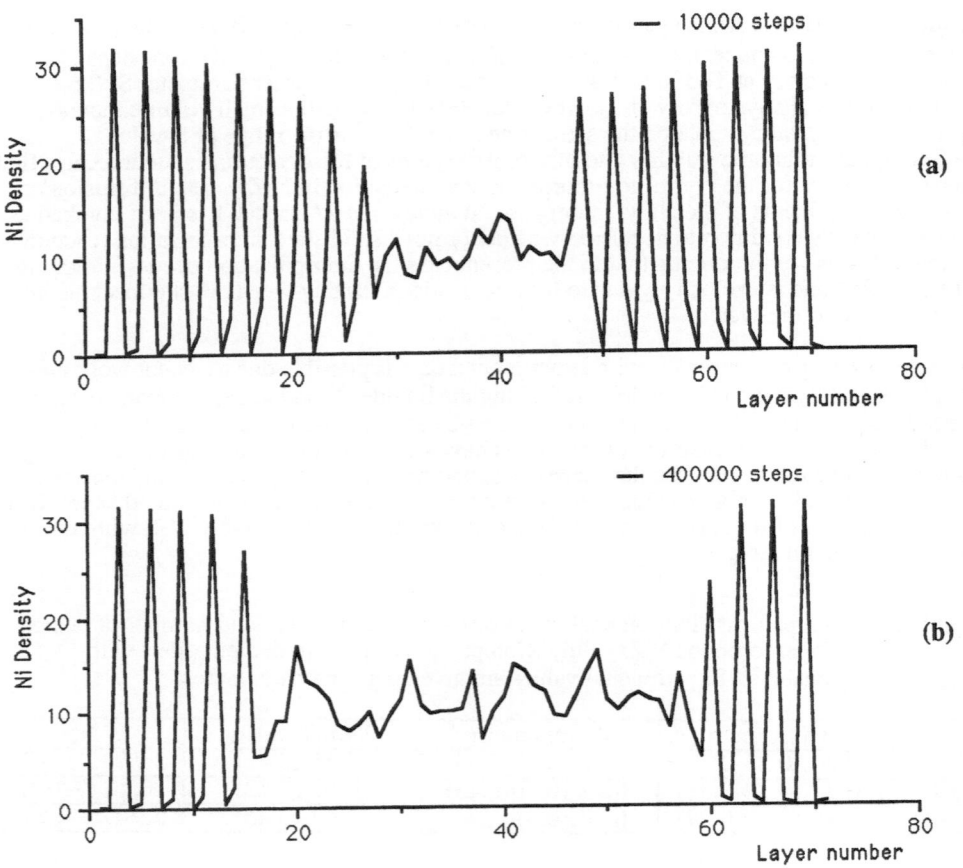

Figure 4 . Nickel density profiles of a partially disordered system at T=1160 K averaged over 1000 time steps. (a) : initial configuration after 10^4 time steps annealing, (b) : propagation of the disorder through an interface reaction after $4 \cdot 10^5$ time steps annealing.

similarly that amorphization may be related to the strains introduced by the antisite defects when their concentration exceeds a critical value.

The importance of interfaces in the amorphous phase formation by solid-state reaction has been often invoked in the litterature. Recent results obtained by Vredenberg et al.[28] give more particularly emphasis to the role of grain boundaries as nucleation sites for the amorphous phase in thin films. Our results suggest that crystal-to-amorphous material interfaces may have another important role : amorphization by interdiffusion can be an interface-driven process. However more work is needed to clarify the details of the amorphization mechanisms.

ACKNOWLEDGEMENTS

We are pleased to thank Y. Adda for his constant interest in this work.

REFERENCES

1. R. B. Schwartz and W. L. Johnson, Eds., Proc. of the Conf. on *Solid State Amorphizing Transformations* (Los Alamos, NM, August 10-13, 1987), Elsevier Sequoia, Lausanne, J. of the Less-Common Met., **140** (1988).
2. K. C. Russell, Prog. Mat. Sci., **28**, 229 (1985).
3. M. Gerl and P. Guilmin, Solid State Phenomena, **3&4**, 215 (1988).

4. A. Blatter and M. von Allmen, Phys. Rev. Lett., **54**, 2103 (1985).
5. E. Gaffet, J. Mat. Sci. and Eng., to be published (1989).
6. L. E. Rehn, P. R. Okamoto, J. Pearson, R. Bhadra and M. Grimsditch, Phys. Rev. Lett., **59**, 2987 (1987).
7. W. L. Johnson, Progress in Mat. Sci., **30**, 81 (1986).
8. R. W. Cahn and W. L. Johnson, J. Mat. Res., **1**, 724 (1986).
9. H. Mori, H. Fujita, M. Tendo and M. Fujita, Scripta Met., **18**, 783 (1984).
10. E. E. Havinga, H. Damsa and P. Hokkeling, J. of the Less-Common Met., **27**, 169 (1972).
11. F. Ducastelle, J. de Physique, **31**, 1055 (1970).
12. C. Massobrio, V. Pontikis and G. Martin, Phys. Rev. Lett., **62**, 1142 (1989); (b) ibid, to be published.
13. F. R. Eshelman and J. F. Smith, J. Appl. Phys., **46**, 5080 (1975).
14. M. P. Henaff, C. Colinet A. Pasturel and K. H. J. Buschow, J. Appl. Phys., **56,** 307 (1984).
15. V. Rosato, M. Guillopé and B. Legrand, Phil. Mag., **A59**, 321 (1989).
16. C. Kittel, *Introduction à la Physique de l'état Solide*, Dunod, Paris (1972).
17. G. B. Skinner, J. W. Edwards and H. L. Johnston, J. Am. Chem. Soc., **73**, 174 (1951).
18. G. Simmons and H. Wang, *Single Crystal Elastic Constants and Calculated Aggregates Properties*, MIT Press, Cambridge (1971).
19. A. R. Kaufmann and T. T. Magel, in *Metallurgy of Zirconium*, B. Lustman and F. Kerze, Eds., Mc Graw Hill (1955) p.377.
20. S. Nosé, J. Chem. Phys., **81**, 511 (1984).
21. H. C. Andersen, J. Chem. Phys., **72**, 2384 (1980).
22. F. A. Lindemann, Z. Physik, **11**, 609 (1910).
23. A. E. Lee, G. Etherington and C. N. J. Wagner, J. of Non-Cryst. Sol., **61&62**, 349 (1984).
24. A. E. Lee, S. Jost, C. N. J. Wagner and L. E. Tanner, J. de Physique, **46**, C8 (1985).
25. T. Mizoguchi, S. Yoda, N. Akutsu, S. Yamada, J. Nishioka, T. Suemasa and N. Watanabe in *5th Int. Conf. on Rapidly Quenched Metals*, Eds. F. Steeb and H. Warlimont (Wurtzbourg, Germany, September 3-7, 1984) North-Holland, p. 483 (1985).
26. M. Guillopé, G. Ciccotti and V. Pontikis, Surf. Sci., **144**, 67 (1984).
27. T. Egami and Y. Waseda, J. of Non-Cryst. Sol., **64**, 113 (1984).
28. A. M. Vredenberg, J. F. M. Westendorp, F. W. Saris, N. M. van der Pers and Th. H. de Keijser, J. Mater. Res., **1**, 774 (1986)

HIGH TEMPERATURE CORE STRUCTURE AND PIPE DIFFUSION

IN AN EDGE DISLOCATION OF COPPER: A MOLECULAR DYNAMICS STUDY

J. Huang[1,2], M. Meyer* [1] and V. Pontikis[2]

[1]LPM, CNRS 1, Pl. A. Briand, 92195 Meudon Cedex, France
[2]SRMP, CEN Saclay, 91191 Gif sur Yvette Cedex, France

INTRODUCTION

The transport of matter in crystalline solids involves point defects diffusing with characteristic times rather large when compared to simulation times. So with the exception of some cases (e.g. superionic conductors) the direct study of bulk diffusion in solids is generally beyond the scope of Molecular Dynamics simulation without constraints. However this limitation does not exist when diffusion occurs in extended defects.

It is well established that the diffusivity is enhanced along extended defects such as grain boundaries and dislocations. The compilation of experimental results obtained in metals[1] shows that the diffusion rates along dislocations are several orders of magnitude higher than in the lattice and reach liquid-like values for the melting point. In order to analyze the experimental results, simplified models are commonly used, in which the dislocation is represented by a narrow cylindrical region of high diffusivity (the pipe). This region corresponds roughly to the core of the dislocation where the disorder is important; it is assumed to have an average diffusivity D_p (the pipe diffusion coefficient) and an effective radius δ. The pipes are embedded in a matrix of lower diffusivity which is that of the perfect lattice. Some models allow the direct determination of D_p but in most cases they only yield the integrated flux $D_p \delta^1$. Thus it is necessary to take an assumed value for δ in order to determine D_p, the usual value chosen for δ is equal to 0.5 nm[1,2]. Fast diffusion along dislocations has been extensively investigated in fcc metals[3]. Beside the high diffusivities, it has been shown that the associated activation energies are smaller than those of lattice diffusion[3]. In these experiments, pipe diffusion has been investigated for several types of dislocations[3]. The general behaviour is the same (e.g. the D_p values are several orders of magnitude higher than those of the lattice diffusion coefficients) but there are some differences related to the nature of the dislocations. The diffusion tends to be slower and the activation energy increases when the dislocations are dissociated; some authors also conclude that δ increases[4,5] with the length of dissociation.

Different mechanisms have been proposed for dislocation diffusion[6,7] (namely vacancy or interstitial mechanisms). The diffusion by interstitials

_____

* To whom correspondence should be addressed.

Microscopic Simulations of Complex Flows
Edited by M. Mareschal, Plenum Press, New York, 1990

is generally discarded, since it is believed, by extrapolating the results of bulk diffusion, that the formation energy would be too high. There is another argument against this mechanism, since the migration along the dislocation could be slower than in the bulk[2]. Indeed it is assumed that the interstitial can be trapped on some sites of high binding energy. For these reasons the vacancy mechanism remains the most probable. Moreover the low experimental values of the activation energies for dislocation diffusion are more likely consistent with this mechanism[2]. This conclusion is supported by the calculations of binding energies E_b performed at zero K on several configurations of vacancies or interstitials interacting with dislocations [8-10]. The maximum values of E_b are obtained when the defect is located near the dislocation, this shows that the formation energies of the point defects should decrease significantly in the core of the dislocation with respect to the bulk. However, it should be noted that these calculations give only an information limited to the variation of the formation energy. Thus, since there is no indication about migration, one can actually not compare the activation energies for diffusion of the different mechanisms. Another limitation of these calculations is due to the fact that they have been performed using empirical potentials which are not adequate to describe metals. A more detailed investigation of the vacancy mechanism has been made by Fidel'man et al.[11] who also performed calculations by energy minimization at zero K as the previous authors did, but used a pseudopotential to model fcc and bcc metals. The formation and migration energies were calculated for vacancies located near a dislocation and in the bulk[11]. Both formation and migration energies are reduced near the dislocation, moreover when the dislocation is dissociated the activation energy increases with respect to the value obtained when it is not dissociated.

There are still questions which are posed regarding the mechanism of pipe diffusion: there is no direct experimental proof of the existence of point defects. If they are involved, which one is it? what is the jump mechanism, does it evolve with temperature? and finally what is the pipe radius? The answers to these questions are not straightforward but one can expect to have some insight into pipe diffusion by Molecular Dynamics (MD) simulations[12]. They provide useful informations since they allow a direct study of the microscopic behaviour at high temperature and not at zero K.

In this paper we report the results of such a study aimed at the characterization of the dislocation core structure and at the determination of the atomistic mechanism of diffusion along the dislocation. Specifically, the study considers the diffusion of vacancies and interstitials along the core of a [110] ($\bar{1}$11) edge dislocation in copper. In a first part, we describe the model used to simulate by MD the properties of this dislocation in a large temperature range. The method of relaxation used to allow for the dissociation of the edge dislocation into two partials is also explained in this section. Then we describe a technique used to study the core structure and to determine the average positions of the partial dislocations. The trajectories of the atoms displaced by the migration of the point defects diffusing along the dislocation at various temperatures (up to the melting point) are given and discussed in the third section. The migration energies for the vacancy and the interstitial are deduced from the analysis of the temperature dependence of respective jump frequencies. To conclude, the activation energies for the vacancy or interstitial mechanisms are evaluated by calculating the respective formation energies of the two point defects.

COMPUTATIONAL DETAILS AND POTENTIAL MODEL

We have used the technique of molecular dynamics at constant volume to study a [110] ($\bar{1}$11) edge dislocation dissociated in its glide plane. The simulated system is a parallelepiped with edges parallel to the [110], [$\bar{1}11$], [$1\bar{1}2$] directions corresponding respectively to X,Y,Z (figure 1). Periodic boundary conditions are applied in the [$1\bar{1}2$] direction parallel to the dislocation line. In the two other directions the dynamical core of the system is surrounded by a rigid lattice, whose thickness is larger than the cutoff radius of the potential used to calculate the interactions in the dynamical part of the system. The dimensions L_x, L_y, L_z of the dynamical core vary with the temperature and are approximately equal to 10.2 nm, 5 nm, 1.8nm. In order to study the properties of a dislocation in copper we have chosen the pseudopotential derived by Dagens[13] using a resonant model. This potential proved a suitable one to reproduce various properties of copper such as phonon dispersion curves[14], elastic moduli[13], formation energies of point defects in bulk material[15] and atomic mean square displacements[16]. Unfortunately, with this pseudopotential, the value of the stacking fault energy Υ tends towards zero when all the atomic pair interactions are summed upon distances large enough to obtain convergence[17]. To overcome this difficulty, we chose an empirical solution by taking a cutoff radius r_c= 2.3a (a lattice parameter) for the potential. With this value of r_c, we obtain a stacking fault energy Υ equal to 73 ergs/cm^2 (experimental value : Υ = 55 ergs/cm^2). The corresponding elastic constants $C'= 1/2(C_{11} - C_{12})$ = 17 GPa and C_{44} = 86 GPa are still in good agreement with the experimental ones, respectively 25.6 and 81.8 GPa.

It is well known that in fcc metals, perfect edge dislocations with a Burgers vector $\vec{b}$=1/2 [110] dissociate in their slip plane ($\bar{1}11$) into two Shockley partials[18] separated by a stacking fault ribbon (figure 1). In order to introduce the two partial dislocations in our model, we first calculate the distance of dissociation R_0 according to the elasticity theory[18] and using the stacking fault energy Υ and the elastic constants reported before. We then remove the atoms belonging to the two half planes located respectively at + and - $R_0/2$ distances from the centre of the system (see figure 1); the remaining atoms are displaced from the perfect lattice positions according to the values given by the elastic theory of dislocations[19]. The atoms located in the dynamical part of the system are relaxed using the quasidynamic technique[20] and a new dissociation distance R_1 is thus obtained.

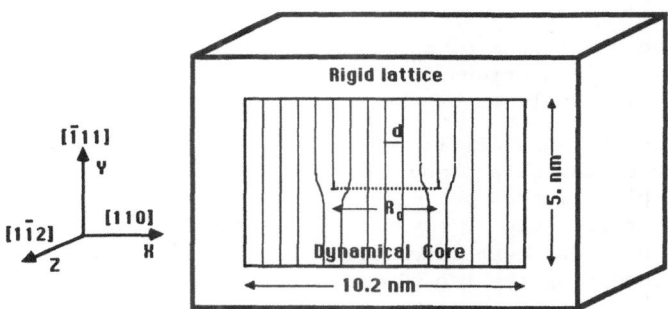

Fig. 1. Schematic representation of the simulated system. d is the
interplanar distance in the [110] direction and R_0 is the dissociation distance between the two partials indicated by ⊥. The
dotted line shows the trace of the stacking fault plane.

The atoms belonging to the rigid lattice are displaced again to take into account the elastic deformation corresponding to R_1. This procedure is repeated in an iterative way until the equilibrium configuration is reached (i. e. when the difference between two successive values of the dissociation distance is inferior to $2|\vec{b}|$). This relaxation procedure is performed for each temperature studied by MD. The starting configuration is prepared using the lattice parameter corresponding to the chosen temperature. This change of parameter results in a density variation which is accounted for in the potential interactions. The potential parameters are modified according to the prescription given by Dagens[16]. The equations of motion of the atoms belonging to the dynamical core are integrated with a time step $\delta t = 2.5 \times 10^{-15}$ s.

STRUCTURE AND LOCALIZATION

The values of the local strain field ϵ_{xx} and ϵ_{yy} calculated in the two directions X and Y respectively parallel to [110] and [$\bar{1}$11] are used to identify the position of the partial dislocations at high temperature. To obtain these values we have calculated the density of particles P(x) and P(y) in the X and Y directions. As we are interested in local values, these distributions are calculated in restricted domains (e. g. close to the partials and/or to the stacking fault). In the X direction for example P(x) is defined as follows:

$$P(x) = <\sum_{i \in \omega} \delta(x-x_i)>$$

where x_i indicate the atomic coordinate in the [110] direction and ω the restricted domain. At zero K, in the perfect lattice, this function is a succession of δ functions corresponding to the (110) planes. The average distance between two successives δ functions correspond to the interplanar distance $d_{0[110]}$ in the [110] direction. When the temperature is increased, the δ functions broaden because of the thermal vibrations of the atoms but it is still possible to measure the interplanar distances $d_{T[110]}$ corresponding to the defect-free lattice. The distances $d_T(x)$ between two neighbouring maxima of P(x) are also evaluated locally as a function of x. The local values of the strain field at a temperature T are given by:

$$\epsilon_{xx}(x) = \frac{d_T(x) - d_{T[110]}}{d_{T[110]}}$$

To investigate the variations of ϵ_{xx} and ϵ_{yy} as functions of x and y, we have calculated the density functions P(x) and P(y) by time averaging over some 10000 time steps. The analysis of the results shows that the strain field ϵ_{xx} in the vicinity of the partials is important in the two ($\bar{1}$11) planes located just above and below the slip plane and then drops down rapidly when the distance to this plane increases. ϵ_{yy} has a maximum value near the slip plane, but this strain field is always smaller than ϵ_{xx}. So one can conclude that most of the disorder associated with the dislocation core is located in a planar region involving the two planes situated just above and below the stacking fault. We will then use the values of ϵ_{xx}^a (above) and ϵ_{xx}^b (below) corresponding to these two planes to characterize the dislocation structure by plotting the relative deformation $\rho(x) = (\epsilon_{xx}^a(x) - \epsilon_{xx}^b(x))$ at various temperatures. Two of these distributions $\rho(x)$ are represented in figure 2, they correspond to temperatures respectively in the low (T= 680 K) and high

(T=1309 K) temperature range with respect to the experimental melting temperature T_m (T_m= 1356 K). As we need an objective criterion to do a quantitative comparison of the data obtained at various temperatures, we have fitted the calculated distributions with analytical curves. A good fit is obtained with the choice of two identical gaussian distributions separated by an adjustable distance R (figure 2). The maxima of the two gaussian curves decrease and their widths increase when the temperature is raised. This shows clearly that the core size is larger at high temperature. The average distance R between the maxima which correspond to the average positions of the partials remains practically unchanged. Thus we can conclude that the dissociation length does not vary with the temperature.

DEFECT FORMATION AND MIGRATION

The temperature range chosen to study the structure of the dislocation is large and extends to temperatures close to the melting point. However diffusion is not observed and point defects are not spontaneously created in the system during the MD runs. So, to study diffusion it is necessary to introduce a point defect in the system either by removing an atom to create a vacancy or by adding an extra atom to obtain an interstitial. To calculate the formation energy E_f of a vacancy or an interstitial, the systems containing either the dislocation or the dislocation and the point defect are allowed to relax at constant volume to their minimum energy configuration. The difference in potential energy of the two systems plus a virial term accounting for the expansion or compression of the lattice after creation of the defect yields the value of E_f[15]. Several sites have been chosen to introduce the extra atom in order to create the interstitial. The minimum energy corresponds to a dissociated interstitial located in the tensile region on a site neighbour to the dislocation line. The orientation of this split interstitial is close to a <110> direction and its formation energy E_f^i is equal to 1.75 eV. For the vacancy the situation is less complicated and the minimum energy corresponds to a site on the dislocation line in the compressive region. The corresponding formation energy E_f^v is equal to 1.05 eV.

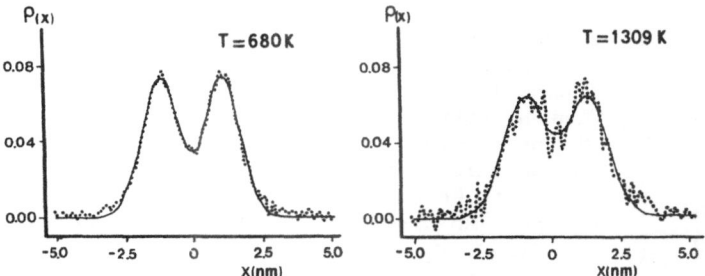

Fig. 2. Relative deformation of the lattice $\rho(x)$ calculated in the vicinity of the stacking fault plane. Simulation performed at T = 680 K, T = 1309 K. The dotted lines correspond to the results of the simulation while the solid lines refer to the two gaussian distributions fitted on the calculated results.

To study the defect migration, the system relaxed at zero K is heated at the required temperature and equilibrated during several thousand time steps. The motion of defects is analyzed in term of jumps by counting the number of atoms moving out of their average position by a given distance. Due to the disorder introduced by the dislocation, it is difficult to define the criterion of the length of jump.

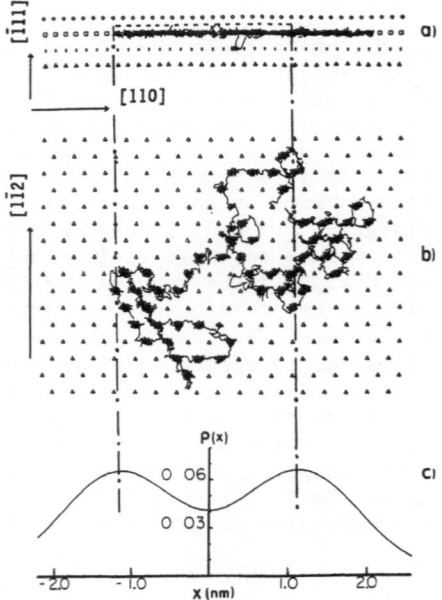

Fig. 3. Trajectories of the atoms displaced by the migration of an interstitial at 1035 K.

a) Projection onto a ($1\bar{1}2$) plane, the dashed line indicates the SF plane.

b) Projection onto a ($\bar{1}11$) plane. The markers ($\Delta, \Diamond, \circ, \Box, x, \Delta$) correspond to the average positions of the non displaced atoms. The area displayed on this drawing corresponds to a restricted part of the simulated system (e.g. the length in the [110] direction is equal to 4.4 nm to be compared to L_x=10.3 nm).

c) Gaussian distribution fitted on $\rho(x)$ see Fig.2. The dashed dotted lines indicate the average positions of the partial dislocations.

This is specially true in the case of dissociated interstitials where the defect jump involves the displacement of three atoms moving quasi simultaneously. So this distance criterion was used to establish a list of atoms which may have been displaced by a defect jump and then a graphical analysis of the trajectories of these atoms was used to count the real jumps.

The migration path of the defects is represented by the trajectories of the displaced atoms which are displayed in figures 3 and 4. These trajectories are projected on two planes, one is parallel to the Stacking Fault (S.F.) and the dislocation lines while the other is perpendicular to them. By looking at these trajectories, one can notice that the migration occurs preferentially in the directions parallel to the S.F. plane. This behaviour is quite clear for the interstitial migration simulated at 1035 K, where the diffusive motion is almost bidimensional as shown in figure 3 a) and b). In the directions [110] and [1̄11] parallel to the S.F. plane (fig. 3b), the migration is nearly isotropic with apparently equivalent jump frequencies in the two directions parallel and perpendicular to the dislocation lines. The analysis of the trajectories also shows that the jump directions correspond to <110> directions. The trajectory is located in the partials' cores and in the region extending between them. The average positions of the partial dislocations are indicated by the relative deformation profile (fig. 3 c). This behaviour is also observed on vacancy trajectories as shown in figure 4.

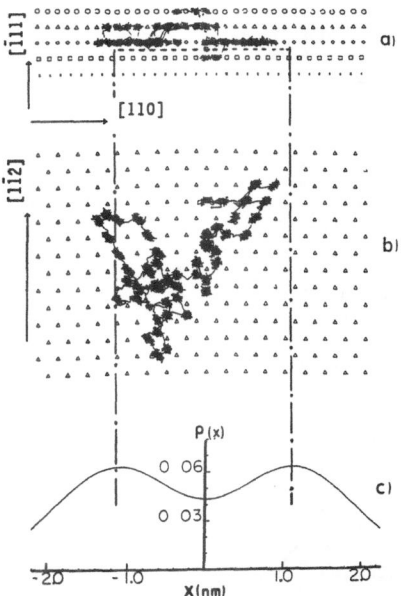

Fig. 4. Trajectory of the atoms displaced by
the migration of a vacancy at 1352K.

Table 1. Jump frequencies

	Vacancy				Interstitial		
T (K)	1197	1310	1352	1423	737	1035	1344
N	56	64	66	78	40	60	102
$\Gamma(10^{11}$Hz)	2.2±0.3	4.3±0.5	5.9±0.7	7.2±0.8	5.3±0.8	8.1±1.05	13.6±1.35

The comparison of the trajectories plotted in figures 3 and 4 shows that the motion of the interstitial occurs in the tensile region under the S.F. plane on fig. 3 a) while the vacancy diffuses preferentially in the compressive region (fig. 4 a). This is consistent with the locations of the minimum energy sites obtained for the two types of point defects.

The results obtained from the analysis of the trajectories are collected in table 1 where the numbers of jumps N and the corresponding frequencies Γ are listed. The temperature dependence of the jump frequencies for the interstitial and vacancy are plotted in figure 5 and the migration energies are estimated by linear regression of LnΓ versus 1/T. This analysis yields values for the migration energies E_m^i and E_m^v equal to 0.13 and 0.77 eV for the interstitial and the vacancy respectively. As expected the interstitial migration energy is considerably lower than the vacancy migration energy, but E_m^i and E_m^v are not significantly different from the bulk values calculated by quasidynamic relaxation[15] or by Molecular Dynamics[22] (see Table 2). This result about the migration of the interstitial differs from those obtained in a pipe diffusion study of MgO where the value of the migration energy of an interstitial is higher in the dislocation core than in the bulk[23].

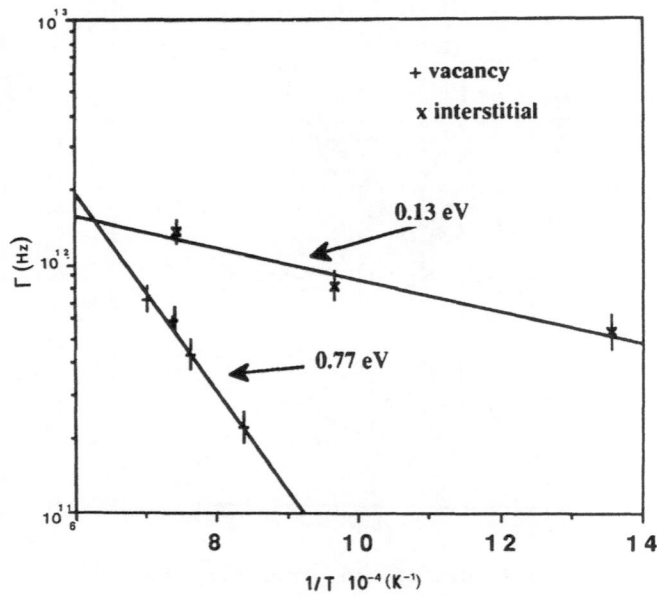

Fig. 5. Jump frequencies Γ plotted as a function of the inverse of the temperature. The continuous line corresponds to the linear regression on calculated data.

Table 2. Formation and migration energies
for vacancy and interstitial

		Dislocation	Bulk	
E_m^i	(eV)	0.13	0.11 [a]	0.14 [b]
E_m^v	(eV)	0.77	0.82 [a]	0.85 [b]
E_f^i	(eV)	1.75	2.61 [a]	
E_f^v	(eV)	1.05	1.42 [a]	

[a] Ref.[15], [b] Ref.[22].

Table 3. Activation energy for pipe and bulk diffusion

E_D	Interstitial	Vacancy	Experimental
Dislocation	1.88	1.82	1.53 [d]
Bulk	2.72-2.75 [b,c]	2.24-2.27 [b,c]	
Ratio [a]	0.7	0.8	0.8 [d]

[a] ratio of activation energies for pipe and bulk diffusion
[b] Ref.[15], [c] Ref.[22], [d] Ref.[3].

If we now compare the activation energies $E_D = (E_f + E_m)$ calculated for a vacancy (1.82 eV) or an interstitial (1.88 eV), we can only conclude that the two mechanisms are equally able to control the diffusion mechanism along a dissociated edge dislocation. Thus on the basis of the energy criterion we cannot conclude as in ref.[2], that the vacancy mechanism is the most probable for pipe diffusion. In order to unambiguously evaluate the roles of the two mechanisms one should also take into account the entropy terms.

The values of activation energy compare rather well with the experimental one derived from the results obtained for Ge pipe diffusion in copper (see Table 3)[3]. It is also interesting to compare the calculated and experimental values of the ratio between the activation energies for pipe and bulk diffusion. To estimate the bulk diffusion energy, we used the results of the calculations of formation and migration energies in perfect lattices[15,22] (Table 2). The ratios are equal to 0.8 for a vacancy and 0.7 for an interstitial, which is comparable with the experimental values ranging between 0.7-0.8 for dissociated dislocations. It is also worth noticing that the extrapolation of the Ge diffusion results in copper yields a value of 0.8. The experimental results showed, as indicated above, that the diffusion is slower in dissociated dislocations and the explanation for that was either a decrease of the Burgers vector or a possible diffusion in the stacking fault ribbon. Our results show clearly that the second hypothesis is the correct one since we observe that all trajectories of interstitial and vacancy include the region which is located between the two partials. This expansion of the migration path in the stacking fault region results in a higher dimension for the region of fast diffusivity. The radius of the pipe usually taken (0.5 nm) is too small to cope with dissociated dislocations when the distance of dissociation is high (e.g. 2.2 nm in copper).

REFERENCES

1. N. A. Gjostein, "Diffusion," Publisher American Society for metals, Metals Park, Ohio, p. 261 (1973).
2. R. W.Balluffi, A. V. Granato, in "Dislocations in solids," Vol. 4. F. R. N. Nabarro ed., Publisher North Holland Amsterdam p.1 (1979).
3 R. W. Balluffi, Physica Status Solidi. 42:11 (1970).
4. M. Wuttig, K.H. Birnbaum, Phys. Rev. 147:495 (1966).
5. C. Baker, M. Wuttig and K. H. Birnbaum, Trans. Japan Inst. Metals. Suppl., 9:268 (1968).
6. E. C. Oren and C.L. Bauer, Acta Met., 15:773 (1967).
7. T. E. Volin, K. H. Lie and R. W. Balluffi, Acta Met., 19:263 (1971).
8. M. Doyama and R. M. Cotterill, Trans. Japan Inst. Metals. Suppl., 9:55 (1968).
9. M. Doyama and R. M. Cotterill, in "Lattice defects and their interactions," R. R. Hasiguti ed., Publisher Gordon Breach New York p 144 (1967).
10. R. C. Perrin , A. Englert and R. Bullough in "Interatomic potentials and simulation of lattice defects," P.C. Gehlen et al. ed., Publisher Plenum press p. 509 (1972).
11. V. R. Fidel'man and V. A. Zhuravlev , Fiz. metal. metalloved, 46: 106 (1978).
12. J. Huang , M. Meyer, and V. Pontikis, Phys. Rev. letters, 63:628 (1989).
13. L. Dagens , J. Phys. F: Met. Phys., 7:1167 (1977).
14. J. C. Upadhyaya and L. Dagens, J. Phys. F: Met. Phys. 8:L21 (1978).
15. N. Q. Lam, L. Dagens and N. V. Doan, J. Phys. F; Met. Phys., 13: 2503 (1983).
16. J. Huang, M. Meyer and V. Pontikis to be published.
17. J. Huang, M. Meyer and V. Pontikis, Scripta Metallurgica, 22: 463 (1988).
18. J. P. Hirth and J. Lothe in "Theory of dislocations," Publishers J. Wiley and sons New York p. 298 ff. (1982).
19. Ref.18 p. 76 ff.
20. C. H. Bennett, in "Diffusion in solids - recent developments," A. S. Nowick and J.J. Burton ed., Publisher Academic Press, New York, p 73 (1975).
21. D. J. H. Cockayne and V. Vitek, Phys. Stat. Sol. (b) 65:751 (1951).
22. S. W. de Leeuw and M. Dixon, Phil. Mag. 52 A:279 (1985).
23 J. Rabier and M. Puls, Phil. Mag. 52 A:461 (1985)

SIMULATION OF MECHANICAL DEFORMATION *VIA*

NONEQUILIBRIUM MOLECULAR DYNAMICS*

W. G. Hoover[1,2], C. G. Hoover[3], I. F. Stowers[4],
A. J. De Groot[5], and B. Moran[6]

1. Department of Applied Science, University of California at Davis/Livermore
2. Department of Physics[#]
3. National Magnetic Fusion Energy Computer Center[#]
4. Precision Engineering Program[#]
5. Engineering Research Division[#]
6. Earth Sciences Division[#]

ABSTRACT

We are developing two- and three-dimensional pair-force and embedded-atom simulations of mechanical deformation processes--indentation, machining, and inelastic ballistic-impact collisions--related to current nanometer machining practice. Here we describe these problems and their implementation using both mainframe and parallel-processor computers.

1. INTRODUCTION

Nanometer technology is an engineering term for the design and fabrication of precision parts with spatial tolerances comparable to microscopic interatomic spacings. Applications include the transmission of electromagnetic waves through large accurately-shaped optical structures as well as the electromagnetic storage of information using nano- (as opposed to micro-) circuitry.

<u>Macroscopic</u> thermodynamics, hydrodynamics, and solid mechanics describe equilibrium and nonequilibrium states of continua. These disciplines make up "continuum mechanics". Their application, through the solution of *partial* differential equations, is the traditional approach to designing and fabricating engineering structures. Conservation of mass, momentum, and energy are fundamental to either the macroscopic

---

[#] All at the Lawrence Livermore National Laboratory, Livermore, California, 94550, United States of America

continuum aproach or to the alternative microscopic atomistic approach. The distinguishing feature of the <u>macroscopic</u> approach is the use of "constitutive equations" which describe the response of particular materials to gradients of the fundamental conserved quantities.

The alternative <u>microscopic</u> approach follows the history of individual atoms by solving the *ordinary* differential equations of motion. The distinguishing feature of the microscopic approach is the use of action-at-a-distance force functions acting among sets of mass points. The macroscopic continuum approach can be thought of as describing averages of microscopic atomistic properties to obtain constitutive properties for *regions* containing many atoms (at least millions). But when the *scale* of a part or feature is *small*, as at a crack tip, dislocation, vacancy, or impurity, the continuum approach fails[1]. This failure shows up in a fundamental way in what is called "size dependence", the failure of material properties to obey simple scaling relationships. If stress were solely a function of strain then the time-and-space-dependence of the macroscopic equations of motion could be scaled. Structures could be built with confidence based solely on the performance of scale models. The fact that large specimens break under smaller stresses than do small specimens of identical shape indicates the presence of a characteristic length (the atomic spacing).

The computer industry is profitting from miniaturization, and the result is a scaleup in the size of simulations it is feasible to make. It does not seem possible to gain much speed by further reduction of the fundamental computer cycle time. But what *can* be done is to increase, greatly, and at little cost, the number of degrees of freedom being treated. The increase makes the difference between a few-body caricature and a truly many-body *simulation*. Many small computers, working in parallel, can now accomplish the same tasks as can the large computers, but 1000 times more cheaply and in the same clock time.

Because the crystal dislocations fundamental to plastic flow have long-range interactions, flow results can be affected by boundaries lying many atomic diameters away. For this reason relatively large (nanometer-scale) simulations are necessary to the atomistic simulations of flow and failure processes in materials science. Such simulations are becoming feasible. Because multiprocessors with millions of processors are on the horizon there is no doubt that billion-particle simulations will eventually be carried out, but still only for times up to about a microsecond. This is exactly the physical scale required for an understanding of shockwave deformation, fracture, and high-strain-rate plasticity. It is because large-scale simulations are becoming a reality that we feel it worthwhile to develop computational tools for addressing these problems.

In this paper we describe our progress and plans.

2. SIMULATION METHODS

Forces and boundary conditions, together with a computer, graphic output, and an integration alogrithm, are the requirements for an atomistic simulation. Hooke's-law and Lennard-Jones forces are instructive and useful models for two-body "pair" forces, and we began with these two. The Hooke's-law crystal obeys exactly the same motion equations as does a two-dimensional bilinear finite-element representation of an elastic continuum[2], for which many analytic and numerical solutions are available. The anharmonic Lennard-Jones potential has a known and relatively-simple phase diagram and has been the subject of hundreds of equilibrium and nonequilibrium simulations ever since the development of the Monte Carlo method at Los Alamos after the Second World War.

In the past ten years there has been increasing emphasis on simulating "real materials". Only a part of this increasing emphasis comes from the shift toward applied and away from basic research. Improvements in computation and experiments are responsible too. Increasing computer power makes it feasibile to use ten-parameter potential functions with angle-dependent forces in combination with disordered structures and sophisticated boundary conditions to model such diverse materials as water, glasses, and proteins under both equilibrium and nonequilibrium conditions. Ever-more-precise experimental tools, such as the field-ion, force-balance, and scanning-tuneling microscopes are providing ever better tests to challenge such simulations.

Pair potentials can describe simple materials, such as the rare gases, relatively well. But pair potentials cannot describe directional chemical bonds, or account for the observed large differences between the elastic moduli C_{12} and C_{44}, or reproduce the differences between the cohesive energy and the vacancy energy. Daw[3] had the idea of introducing a local many-body potential that incorporated these features at low cost, the "embedded-atom" potential.

Pair forces have been studied for nearly a century and the corresponding macroscopic properties are fairly well understood. Because the embedded-atom idea is relatively new there is not yet a correspondingly good understanding of the correlation between potential parameters and macroscopic properties such as yield and fracture toughness. But this will come with experience. We have followed Daw's lead in investigating this low-cost approach to metal simulation.

The basic idea is to calculate **Particle i**'s contribution, $\phi(r_i)$, to the coordinate-dependent potential energy $\Phi(\{r_k\})$ as a function of the density at that particle due to the influence of the neighboring particles at $\{r_j\}$:

$$\rho_i = \Sigma\rho(r_{ij}).$$

Particle **i** can be thought of as being "embedded" in a field provided by its neighbors. So long as this embedding is viewed as a pragmatic and phenomenological low-cost procedure for avoiding prohibitively-expensive quantum simulations, it represents the best method for predicting the properties of metals and their defects.

Boundary conditions are dictated by the corresponding experimental situation. The boundaries can either be free of forces, or they can be subject to prescribed time-dependent displacements or forces. In Vineyard's pioneering work[4] viscoelastic boundaries were used to absorb the wave energy incident on walls. In continuum mechanics corresponding "quiet" boundaries are sometimes used. The main advance in treating boundaries since Vineyard's work has been Nosé's reversible and deterministic method[5] for introducing temperature and stress into the microscopic equations of motion. We have incorporated this method in our work.

The simplest reasonable integration method for conservative systems is the time-reversible centered second-difference algorithm in which the accelerations are replaced by differences:

$$d^2r/dt^2 \equiv [r(t+dt) - 2r(t) + r(t-dt)]/(dt)^2.$$

But because many interesting equations of motion, such as Nosé's, are *first*-order in time, rather than *second*, and also because the same accuracy can be obtained easily with fewer force evaluations, and significantly reduced computer time, we prefer the classic "Fourth-Order" Runge-Kutta approach. With Runge-Kutta[6,7], four easy-to-program estimates of dr/dt and dp/dt are averaged, with a resulting error, over a fixed interval of time, proportional to the *fourth power* of the timestep. For the simple second-difference scheme above the integration error over a fixed time interval is larger, varying as the *square* of dt.

The recent book **Numerical Recipes**[7] claims that the *Bulirsch-Stoer* integration method is likely the best for integrating ordinary differential equations. This method uses a relatively-large timestep (a complete vibrational period for an oscillator, for instance) and evaluates a series of approximate integrals over the interval, using 2, 4, 6, 8, 12, 16, 24, ... function evaluations. These results are then extrapolated, using a Padé (or continued-fraction) approximant, to the limit of an infinite number of evaluations.

On the strength of this recommendation, we investigated the Bulirsch-Stoer integrator. As is usual, such an investigation is hampered by the failure of common library routines to allow either the order or the timestep to be specified. In "packaged software" these parameters are varied internally, at some "overhead" cost, in order to minimize the error. But, because most molecular-dynamics simulations proceed with a roughly constant degree of anharmonicity there is no reason to consider varying the order or timestep to improve accuracy. Diligent comparison of the packaged International Mathematics and Statistics Library (IMSL) Bulirsch-Stoer routine "IVPBS" with the **Numerical Recipes** version showed that the two do indeed, apart from the error and order controls, implement precisely the same method.

A detailed investigation for the one-dimensional harmonic oscillator (See also Reference 6) showed that Bulirsch-Stoer approach, with optimized timestep and number of force evaluations, matched the performance of the popular Gear integrator when the required maximum coordinate error was set at one part per million. This level of error required about one hundred force evaluations per oscillator period. With fewer evaluations the Bulirsh-Stoer performance degrades much more rapidly than the others. Encouraged by the oscillator results we applied the same technique to Lennard-Jones crystals. Here the Bulirsch-Stoer performance is disappointing, even relative to the simple Runge-Kutta integrator, and we therefore abandoned the Bulirsch-Stoer technique.

Today the Lawrence Livermore National Laboratory emphasizes "Supercomputing", computing with mainframe machines costing tens of millions of dollars. But, for molecular dynamics, this approach has the look of a dinosaur. Interesting low-cost alternatives are becoming available. The SPRINT (Systolic Processor with Reconfigurable Interconnection Network of Transputers) computer[8] is about 1000 times cheaper than a state-of-the-art CRAY. It was developed as a doctoral thesis project at the Lawrence

Livermore National Laboratory under the auspices of the University of California's Davis/Livermore Department of Applied Science.

The SPRINT has 64 "Transputer" chips, each a 32-bit microprocessor with about three times the speed of a VAX-11/780. The processors can be connected up in a variety of topologies. In the checkerboard configuration appropriate to two-dimensional molecular dynamics simulations the SPRINT performed with a speedup over a single transputer between 16 and 32, carrying out 1000-atom molecular dynamics simulations at the same clock speed as a CRAY-1 computer. The loss of speed, relative to that of a single transputer, is due to the need for communication of information among the processors. Because these transfers only involve neighboring parts of the problem, the overall efficiency of the multiprocessor should remain roughly unchanged with increasing problem size.

The rapid display of computed results is still a bottleneck, even at Livermore. A thousand-frame ten-thousand particle movie runs for less than a minute, but requires the processing of about 10^8 particle or pixel coordinates. This is too much information for a single tape to hold. The portability and lower cost of videotape make that medium preferable to conventional movies. To make a finished videotape requires several hours of processing effort, with a considerable portion of that time devoted to transporting data from one machine to another. (CRAY to VAX to STELLAR to a tape editor, at Livermore.) But the results are worthwhile. Cooperative motion and correlations that would be difficult to see in still pictures stand out in movies.

Though presentday mathematicians may claim that Poincaré "understood" the complexity of chaotic mechanics through his analysis of the intricacies of homoclinic points, there is no doubt that a short videotape of the Lorenz attractor provides a relatively effortless and accessible understanding every bit as sure and reliable as Poincaré's.

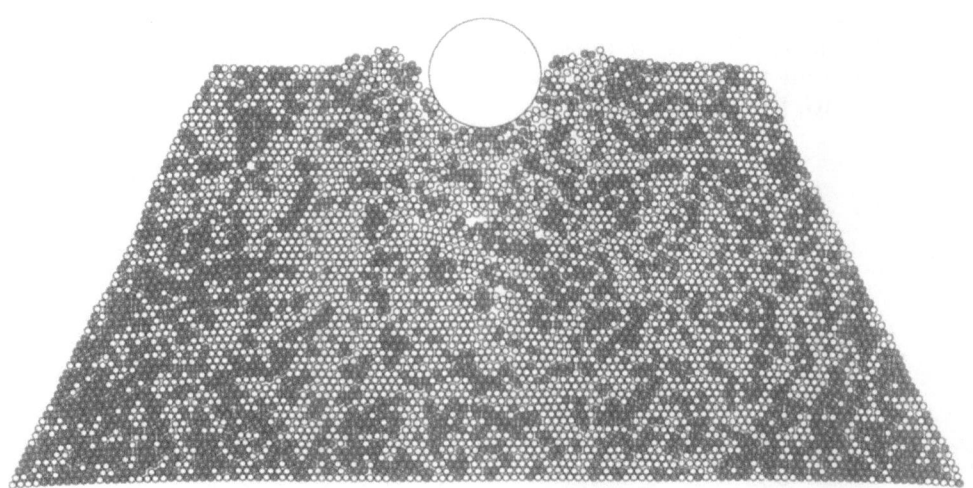

FIGURE 1. Indentation of 5430 Lennard-Jones atoms at about two-thirds the melting temperature ($kT/\varepsilon = 0.30$). The indentor speed increases linearly with time, up to a maximum value of $(\varepsilon/m)^{1/2}$, and then decreases linearly back to zero.

All of our simulations have been two-dimensional, but we are developing, for the SPRINT, the straightforward three-dimensional extension of these models. It appears that the current 64-processor SPRINT should be capable of efficient molecular dynamics simulations with 50,000 atoms. At Los Alamos Brad Holian and Art Voter expect to be able to treat one million particles on the 65,536-processor "Connection Machine".

3. SIMULATION RESULTS

We began our <u>indentation studies</u> with triangular and round indentors. See **Figure** 1. In testing real materials the yield strength is determined by dividing the applied load by the area of permanent deformation. We carried out sample two-dimensional calculations for a variety of indentor sizes and workpiece temperatures. We used both fixed-load and programmed-displacement boundary conditions. The resulting yield stresses (load divided by length) are *size-dependent*, still decreasing with increasing workpiece size for two-dimensional crystals with 5000 atoms. But the presence of many dislocations in the deforming crystal indicates that convergence should be possible in two dimensions. We intend to pursue convergence using a plane-strain (two-dimensional) representation of copper and to compare the resulting stress and strain fields with macroscopic finite-element simulations.

We followed our indentation studies with simulations of the <u>diamond-turning process</u>, in which strips only a few atoms thick are cut from a spinning workpiece through contact with diamond chips embedded in a resilient matrix. Computational specimens only a few atoms thick produced voids and extended defects at their bases, but larger specimens produced realistic chips, particularly when an embedded-atom model was used. The Lennard-Jones potential produced substantially more vapor (or dust?) in the cutting operation. A comparison of the two kinds of simulations is indicated in **Figure** 2. Note that in the evolution of these simulations a cutter radius of curvature was introduced. The embedded-atom potential we use has a functional form $\Phi = \Sigma \rho_i \ln \rho_i$, with $\rho(r)$ given by a parabola vanishing just short of the triangular-lattice second-neighbor separation. We intend to extend this work also to a more-realistic model for copper. In both the indentation and the cutting studies the peak deformation velocities were about one-tenth the sound velocity.

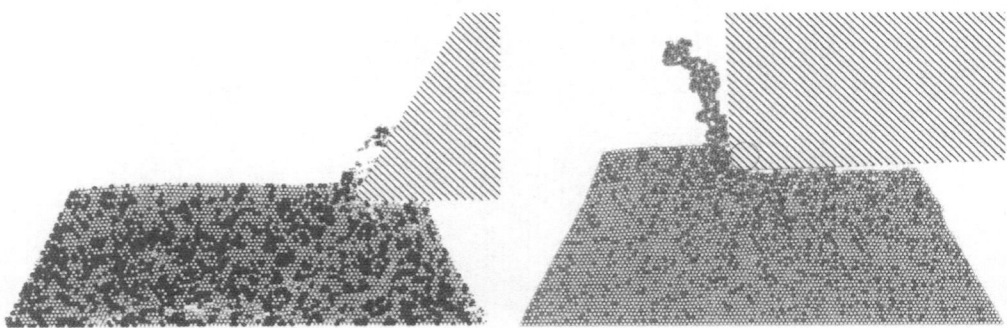

FIGURE 2. Cutting of Lennard-Jones (left) and Embedded-Atom (right) Crystals with a cutting tool moving at approximately one-tenth the sound velocity.

The simplest <u>impact problem</u> is the symmetric collision of identical particles. Ever since Newton's time the "coefficient of restitution" has been used to describe the result of such "inelastic" collisions, that is collisions in which the outgoing particle speed is less than the incoming one[9]. The ratio of these speeds *is* the coefficient. Because an explanation of the coefficient's magnitude is hard to find in the literature (despite considerable analysis by Rayleigh[10] and experiments carried out by Raman[11]) we carried out a short investigation. Initial trials revealed that roughly-circular disks, containing a few hundred to a few thousand atoms, bounce from a *mathematical* hard wall (at which incoming atomistic velocities are reflected) with a restitution coefficient approaching unity at low speeds, in agreement with Rayleigh's analysis. We obtained similar results for macroscopic continua using a finite-element approach.

But the collisions of particles with a mathematical "wall" seemed unrealistic. Why not simply allow two similar particles to collide? We did, and the results were amusing. *There was generally no "restitution" at all.* Instead the two particles would come together and stick, permanently cold-welded together. The dynamics underlying this phenomenon is reminiscent of the phase-space structure of the strange attractors of dynamical systems theory. The colliding two-particle system starts out with most of its kinetic energy directed. Then the collision occurs. Next, the anharmonicities and elastic wave reflections from the curved boundaries dephase, scatter, and thermalize the directed energy. The "Poincaré recurrence time" required for the momenta to again line up, allowing the particles to separate, is impossibly long. It is comparable to the time required for a similar number of particles in a box to simultaneously seek out one half or the other of the box. Thus the coefficient of restitution studies indicated rather strongly the importance of impurities on the surface. Without them restitution won't occur.

These collisional studies are not at all without applied interest. Rayleigh's calculations and Raman's pendulum experiments with "Hertzian-contact" dynamics showed that the stress reached between two colliding ball bearings is several kilobars, comparable to the yield strength of steel. Today it appears that the corresponding experiments, properly analyzed could be a useful source of high-strain-rate constitutive models. We found that our embedded-atom model, again with the potential function taken as $\Sigma \rho_i \ln \rho_i$, could plastically deform, as shown in the **Figure** 3, even with a relatively small kinetic energy, of order a few percent of the melting energy. We view this qualitative dependence of the deformation on the forcelaw as strong empirical evidence for the usefulness of the embedded-atom approach.

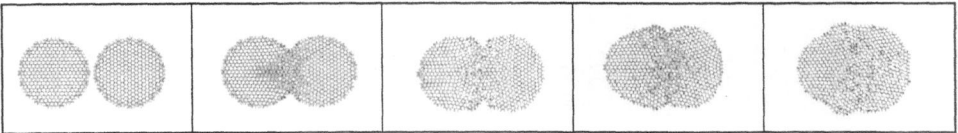

FIGURE 3. Snapshots of the low-speed inelastic cold-welding of two embedded-atom particles.

ACKNOWLEDGMENT

It is a pleasure to acknowledge the help and advice of Brad Holian, Art Voter, and Murray Daw. Dan Nikkel and Jim Belak have likewise contributed to our understanding of these problems.

REFERENCES

* This work was supported by the Department of Energy and performed at the Lawrence Livermore National Laboratory under the auspices of the University of California pursuant to Contract W-7405-Eng-48.

1. D. Scott, W. W. Seifert, and V. C. Westcott, Sci. Am. **230** (May), 88 (1974).; E. Rabinowicz, Sci. Am. **218** (June), 91 (1968).
2. A. Hrennikoff, J. Appl. Mech. **8A**, 169 (1941).
3. M. S. Daw and M. I. Baskes, Phys. Rev. **B29**, 6443 (1984); M. S. Daw and S. M. Foiles, Phys. Rev. Letters **59**, 2756 (1987); S. M. Foiles, M. I. Baskes, and M. S. Daw, Phys. Rev. **B33**, 7983 (1986).
4. G. Vineyard, J. Appl. Phys., cover (August, 1959); J. B. Gibson, A. N. Goland, M. Milgram, and G. H. Vineyard, Phys. Rev. **120**, 1229 (1960).
5. S. Nosé, J. Chem. Phys. **81**, 511 and Mol. Phys. **52**, 255 (1984); W. G. Hoover, Phys. Rev. **A31**, 1695 (1985).
6. G. D. Venneri and W. G. Hoover, J. Comp. Phys. **73**, 468 (1987).
7. W. H. Press, B. P. Flannery, S. A. Teukolsky, and W. T. Vetterling, _Numerical Recipes_ (Cambridge University Press, Cambridge, 1986).
8. A. J. De Groot, E. M. Johansson, J. P. Fitch, C. W. Grant, and S. R. Parker, IEEE Transactions on Nuclear Science, **34**, 873 (1987).
9. G. Barnes, Am. J. Phys. **26**, 5,9 (1958).
10. Lord Rayleigh, Phil. Mag. **11**, 283 (1906).
11. C. V. Raman, Phys. Rev. **12**, 442 (1918).

LATTICE INSTABILITY AND COLD FUSION IN DEUTERATED METALS

Alexander Tenenbaum (a) and Eugenio Tabet(b)

(a) Dipartimento di Fisica, Università "La Sapienza", Roma
(b) Laboratorio di Fisica, Istituto Superiore di Sanità
and INFN-Sezione Sanità, Roma (Italy)

INTRODUCTION

In this paper, we present a model which could help in explaining cold fusion processes on the basis of a lattice collapse in a deuterated metal. We shall show that a thermodynamic instability can, under favourable conditions, trigger a coherent and concentric displacement flow in the metal, which can accumulate in a small region the excess elastic energy originally distributed in an expanded domain. Conditions allowing nuclear fusion processes can thus be created.

DEUTERATED METALS

It is well known that deuterated metals are subject to relevant volume changes, depending on the concentration of deuterium in the host metal lattice. For example, the volume of palladium can expand by as much as 20% for a 1:1 concentration of deuterium[1,2] . This expansion is accompanied by an accumulation of elastic energy, which for the above mentioned case is of the order of 0.16 eV per atom of Pd[3].

The phase diagram of the system Pd–D (Fig. 1) shows that above the critical temperature T_c = 276 $^{\circ}$C the lattice expansion is uniform because the D atoms are uniformly distributed. Below T_c three different regions appear, depending on the concentration c_0 of deuterium. In the coexistence region $\alpha + \beta$ the system is composed by domains of low (c_-) and high (c_+) concentration, giving rise to the mean nominal concentration c_0.

In regions α (β) the lattice parameter is a_- (a_+), while above T_c it is a_0, with $a_- < a_0 < a_+$.

THERMODYNAMIC INSTABILITY

Let us now suppose that the system is rapidly driven from a state at $T > T_c$ and concentration c_0 to a state with $T < T_c$, keeping the concentration constant (see Fig. 1). As the new state lies in the coexistence region, the system will not be in thermodynamic equilibrium, and will evolve towards a state where domains of the α and β type coexist.

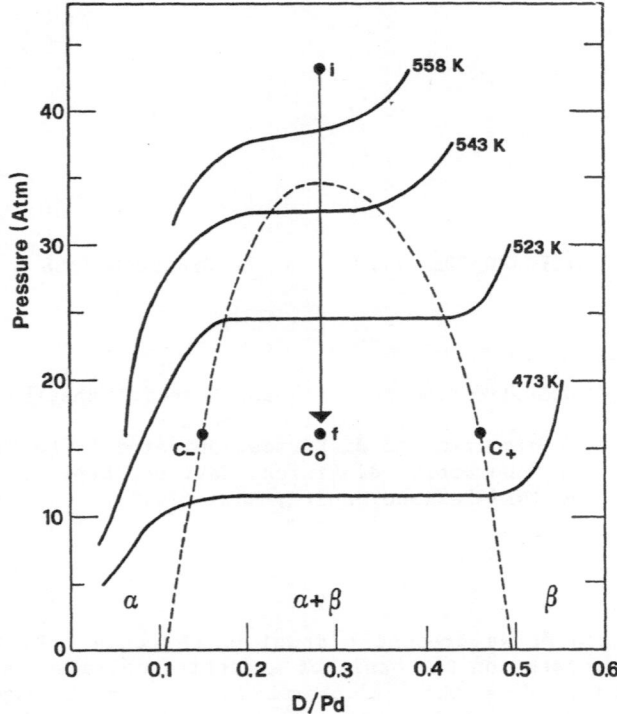

Fig. 1   Phase diagram of deuterated palladium. The arrow indicates the transition from the initial (i) to the final (f) state, discussed in the text.

During this evolution deuterons will migrate in the metal, nucleating domains of low and high deuterium concentration; the duration of this process is determined by the diffusion rate of the D atoms.

In the domains which are progressively depleted of deuterons the lattice will tend to relax from the original lattice parameter a_o to the new value a_-. It may be assumed that, at the beginning of the growth of a low concentration domain, the surrounding lattice, which is still at concentration c_o, will frustrate the tendency to relax of the domain. Therefore, up to a certain size, the domain will be kept in a metastable state with the old lattice parameter a_o.

The maximum size of the metastable region is reached when the excess elastic energy entailed in it equals the energy needed to displace the atoms in the outer layer of the domain to a new equilibrium position, corresponding to the relaxed state.

Assuming for simplicity that the domains grow isotropically, the excess elastic energy in one of the metastable domains of low concentration is given (for a f.c.c. lattice) by

$$E_R = 8 \pi u^2 m \, (R/a_o)^3 \, \varepsilon^2$$

where u is the average velocity of sound, m is the atomic mass of the metal, $\varepsilon = (a_o - a_-)/a_o$ is the relative contraction of the lattice parameter, and R is the radius of the domain. It can be shown[4] that the energy necessary to displace each atom of the outer layer to the relaxed position is $\Delta e = mu^2\varepsilon/2$; when R reaches the value a_0/ε the elastic energy is sufficient to shift to the relaxed position the atoms of the outer layer of the domain.

LATTICE COLLAPSE

This coherent shift produces two effects: on one hand, dislocations are formed at the border between the domain and the surrounding unrelaxed lattice. On the other hand, a displacement wave will move inwards from the border of the domain, giving raise to a plastic flow. At the end of this collapse[5] all the atoms in the domain will have reached their relaxed position, corresponding to the lattice parameter a_-.

However, one must take into account that single atoms in the outer layer can relax due to thermal motion, if the temperature is high enough. If a significant fraction of atoms undergoes this process, the domain will relax smoothly and no coherent collapse will occur. Keeping in mind that Δe is proportional to ε, it is evident that, for a given temperature of the final state, coherent collapse can be achieved only if ε exceeds a lower limit, which turns out to be[4] of the order of $k_B T/mu^2$.

The duration of the collapse can be estimated by dimensional arguments to be of the order of $E_R^{-1/2}\,\varrho^{1/2}\,R^{5/2}$, where ϱ is the density of the metal. The time the displacement wave needs to cross a layer of thickness a_0 results thus to be typically of the order of one ps.

The collapse process is sketched in Fig. 2.

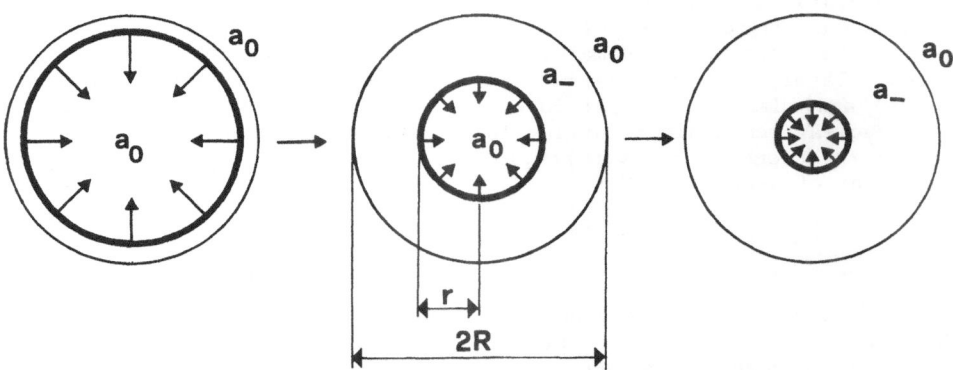

Fig. 2   Sequence of initial, intermediate and final stages during the collapse of a spherical domain.

At a distance r from the center, the imploding wave will have collected an energy $E'_r = \eta 8\pi u^2 m\varepsilon^2 (R^3 - r^3)/a_0^3$ from the region it has already swept. η is an efficiency factor which measures the fraction of the excess elastic energy entailed in the front of the imploding wave.

The increasing energy of the wave thermalizes to increasing local temperatures the deuterons still present in the region crossed by the wave. The diffusion velocity of the deuterons will thus be enhanced as the wave proceeds toward the center, and it can be seen[4] that it can reach values of the order of the velocity of the wave itself. On the other hand, the heated layer is the border between the outer region where the lattice parameter is (already) a_-, and the inner region where the lattice parameter is (still) a_0 (see Fig. 2). It is known that in such a situation the migration of the deuterons takes place preferentially toward the inner region (Gorski effect[6]). A microscopic analysis of this phenomenon[7] shows that for ε values not too small a fraction β higher than $1/2$ of the thermalized deuterons will diffuse inwards, so that the effective number n^* of deuterons in a layer of thickness a_0 at a distance r_i from the center will be:

$$n^*(r_i) = \sum_k n(r_k)\, \beta^{\,k-i}, \quad k = i \ldots q.$$

Here n is the original number of deuterons in the layer and q is the index of the layer in which the diffusion velocity reaches the velocity of the wave.

NUCLEAR REACTIONS

The over-all effect of the collapse process will thus be to produce at the core of the imploding domain a high temperature and high deuteron density state. As a point of reference, for a typical transition like the one depicted in Fig. 1, the excess elastic energy E_R is of the order of 10 KeV; the amount of this energy effectively concentrated in the core will of course depend on the value of η .

The energy of the deuterons inside the core can reach values of the order of some 10^2 eV, so that nuclear fusion reactions can occur with a non negligible probability[8]. From the knowledge of the energy $E_r^!$, one can compute the average velocity of the deuterons in each spherical layer of the contracting domain, and from this the nuclear reaction rate in that layer. Taking into account the activation time of each layer ($\simeq 1$ ps) and summing over all layers in the domain, one gets the number N_r of nuclear reactions occurring in a given domain. The total number of nuclear reactions in a given sample will thus be $N_n = N_r N_d$, where N_d is the number of α-domains [4].

RESULTS

If one considers the case of palladium, taking into account only the channel $D + D = n(2.45$ MeV$) + {}^3$He $(0.82$ MeV$)$ for the nuclear reactions between deuterons, and extrapolating to low energies the experimental nuclear cross section data[8,9] , one gets the neutron yield (N_n) as a function only of the concentration c_0 and of the final temperature T. In Fig. 3 we report the values of N_n for one gram-atom of palladium; the curve is given for an ideal case, in which $\eta = 1$ and the colliding deuterons are endowed with the whole energy available in the surrounding metal lattice. It can be seen that, for the given temperature, the neutron yield depends strongly on c_0 . This thermodynamic parameter may not be well defined in non-equilibrium states; this could help understanding the difficulties often encountered in cold fusion experiments.

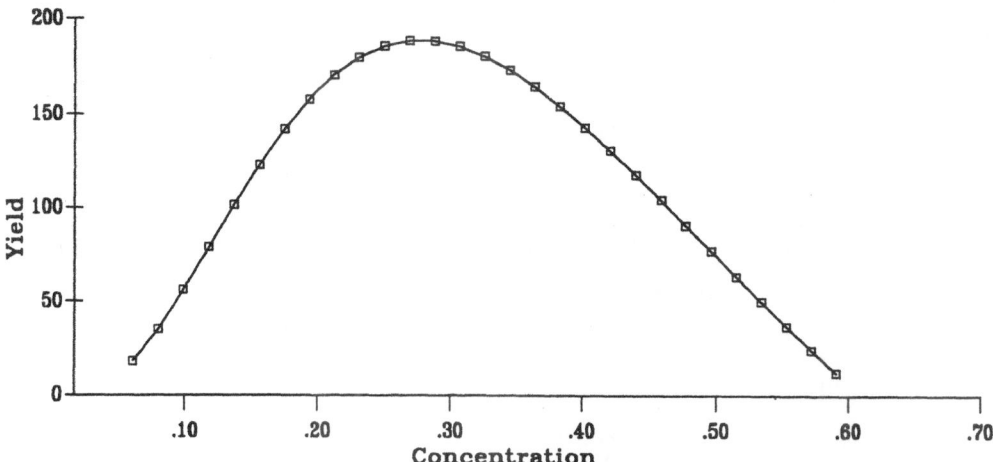

Fig. 3   Neutron yield (in thousands) as a function of concentration c_0 , at T = 273 K for one gram-atom of palladium (when $\eta = 1$).

A comparison of our model with experimental results would require
an evaluation of the time needed for the whole system to reach a new
equilibrium state and of the parameter η : this would allow us to
compute a global reaction rate. The time is related to the diffusion
velocity of the deuterons and to the nucleation time of the α—and
β—domains. On the other hand, the value of η and the detailed mechanism
by which energy is transferred from the metal lattice to the deuterons
can be assessed only by a microscopic computer simulation. Work is in
progress in this direction.

REFERENCES

1.   "Hydrogen in Metals", Eds. G.Alefeld and H.Voelkl, voll. I and II,
        Springer Verlag, Berlin (1978).
2.   See p. 70 in vol. I of ref. 1.
3.   E. Wicke and H.Brodowsky, Hydrogen in Palladium and Palladium
        Alloys, in vol. II of ref. 1.
4.   E.Tabet and A.Tenenbaum, "Displacement flow and deuteron drag in a
        metal: a path towards cold fusion", in: Proceedings of the
        workshop "Understanding of Cold Fusion Phenomena", Varenna,
        September 1989, Il Nuovo Cimento Conference Proceedings (in
        press).
5.   The idea of a lattice collapse in deuterated metals was originally
        suggested in: M. Cassandro, G. Gallavotti and G. Jona-Lasinio,
        Preprint n. 672, Department of Physics, University of Rome, May
        1989.
6.   See, for example, H. Wagner, Elastic Interaction and Phase
        Transition in Coherent Metal-Hydronge Alloys, in vol. I of ref.
        1.
7.   An extended version of the present work, with more details, is in
        preparation.
8.   L.Maiani, G.Parisi and L.Pietronero, "Cold fusion: a first look",
        in: Erice "Workshop on Cold Fusion", Erice, April 1989.
9.   E.Segré, "Nuclei and Particles", The Benjamin/Cummings Publishing
        Company, London (1977).

PART VI

APPLICATIONS (IV): CHEMISTRY

NON EQUILIBRIUM CHEMICAL REACTIONS:
THE MOLECULAR DYNAMICS APPROACH

J. Boissonade

Centre de Recherche Paul Pascal, Univ. Bordeaux I
Château Brivazac, 33600 Pessac, France

1 Introduction

Chemical reactions kept far from equilibrium by a constant feed of fresh reactants can exhibit various fascinating self-organization phenomena such as multiple steady states, periodic or chaotic oscillations, or spatial structures [1-5]. Although most of these phenomena are well understood by the sole use of deterministic nonlinear kinetic laws in the frame of general bifurcation theory , local fluctuations sometimes play a major role, in particular when metastable non-equilibrium states are involved (see for example the paper by F.Baras in this volume). Unfortunately, experimental probes of this local fluctuation behaviour are practically out of reach so that numerical simulation techniques are often preferred.

Although oversimplified in regard to real chemistry, the standard problem of an unstirred reactive medium where the initial reactants are in a large excess and where the final products do not modify the kinetics is the most commonly examined in the theoretical literature. Let us for instance consider a system involving two species A and X, reacting according to the following scheme:

$$A \overset{k_1}{\to} X \tag{1}$$

$$2X \overset{k_2}{\to} Products \tag{2}$$

A is a *major species* which is assumed to be in large excess so that its concentration is *constant and uniform*. The concentration of X, a so-called *intermediate species*, depends on space and time according to the *reaction-diffusion* equation:

$$\frac{\partial X}{\partial t} = k_1 A X - 2k_2 X^2 + D_X \Delta_r X \tag{3}$$

where A and X are respectively the concentrations of A and X, D_X the diffusion coefficient of X and Δ_r the Laplacian operator. The mass action law was used for the kinetics. This model does not present any exotic behaviour but is both simple and general enough to illustrate our discussion. In particular step 2 is autocatalytic, a basic process to generate bistability or concentration oscillations. A more elaborate model will be used in section 4.

The fluctuation behaviour has been extensively studied from a theoretical point of view on the basis of a stochastic description in terms of birth and death leading ultimately to the so-called *"Global Master Equation"* and *"Multivariate Master Equation"*. The former only considers homogeneous fluctuations , the latter takes into account their local character and the fundamental role of diffusion. The basics of these theories can be found in classical books [1,2]. The central result, first established by Nicolis et al.[6], points out an essential difference between short range and long range fluctuations. According to the local equilibrium principle which only requires that the ratio of reactive to non reactive collisions is small, the normal situation in the diluted solutions, short range fluctuations are Poisson-like as they are at equilibrium. Thus the mean square deviation σ of concentration is given by

$$\sigma^2(X) = <X^2> - <X>^2 = <X>\qquad(4)$$

where $<X>$ is the averaged concentration of X On the contrary, long range fluctuations are dominated by the non-equilibrium processes and deviate from the Poisson law so that

$$\sigma^2(X) = \alpha <X> \text{ with } \alpha \neq 1\qquad(5)$$

Thus α is a function of the volume ΔV in which the fluctuations are averaged. In the following we shall only consider stationary asymptotic regimes so that this averaging can be performed on time. When $\Delta V \to 0$, then $\alpha \to 1$; when $\Delta V \to \infty$ then $\alpha \to \alpha_\infty$, where α_∞ is derived from a global master equation and depends on the reaction scheme and the kinetic constants. For Model 1, with $k_1 = k_2$, it comes from general formulas Hanusse [7] that $\alpha_\infty = 0.75$. Analytical solutions of multivariate master equations are limited to specific cases but more general solutions have been obtained by Monte-Carlo techniques [8]. They all rest on the basic assumptions of stochastic theory, i.e. the validity of a description in terms of discrete steps, discrete diffusion and kinetic Markov processes. In order to bypass these prerequisites, test the validity of stochastic theory, and possibly improve the description, several authors have introduced non-equilibrium molecular dynamics techniques. In most earlier works, the authors have come up against the difficulty of keeping *uniformly* constant the major species concentrations since the use of a large excess of particles is excluded due to prohibitive computation costs. In several works [9,10], the concentration was only kept constant in a global way but not uniformly. Explanation for the Poisson-like regime thet Portnow [9] observed for the entire system, in contradiction with stochastic theory and use of a global process, should be found in the excessively small number of particles, this spurious effect being increased by use of specular reflections at walls. Other authors [11,12] avoid these difficulties by introducing different non-equilibrium processes in which the local kinetic constants depend on the surroundings of each particle. Because, in contradiction with the spirit of molecular dynamics, they involve a large amount of stochastic operations and mean field concepts, their results cannot be easily compared to the standard problems defined above. The problem of the standard non-equilibrium conditions has been solved by Boissonade [13-17]. On the same lines, Amellal has recently applied similar techniques to quasi-one dimensionnal systems [18] (such a geometry allows for a significant increase of the spatial range. His results essentially confirm our previous conclusions. After a brief description of the technique , we show that the computed fluctuation amplitudes are in good agreement with the theoretical predictions and report an application to the problem of nucleation and metastability in a chemical bistable system. Eventually, we shall give a brief evaluation of the future of molecular dynamics techniques in the field of far from equilibrium reactions. This paper is a digest of previous works covering several aspects of the field. Technical details will be found in ref. 14 and 15. More results on the fluctuation regime can be found in ref.13 and 15. The nucleation problem is presented in ref. 16 and 17.

2 The molecular dynamics technique

The microscopic description does not try to mimic genuine reactions so that the description of the molecules and reactive processes has been oversimplified to retain only the ingredients essential to the study of the non-equilibrium effects and limit ourselves to a two-dimensionnal space. For simplicity, we shall use a coherent system of arbitrary units

2.1 The medium

The medium is represented by a diluted gas of identical hard disks of mass $m = 1$, radius $R(\sim 0.05 - 0.075)$, and mean kinetic energy $<u^2> = 1$, enclosed in a square box of side $L = 30$ with periodic boundary conditions (total number of particles $N \sim 1850$ to 4500). All collisions, even reactive ones, are assumed to be elastic.

2.2 The chemistry

Particles are labelled according to their species. All reactive steps are bimolecular: they are performed instantaneously at each collision by changing the labels according to the kinetic scheme. Since collisions are elastic, all steps are athermal but in order to tune the reaction rates of each step a symmetric activation energy barrier was assumed so that the reactive step is performed only if the relative kinetic energy is larger than a specified value; the reaction rate follow the mass action law and is related to this energy threshold by the Arrhenius law. A chemical pool of non-reactive ("*Neutral*") particles, labelled N, is used for creation of inert products and as a reservoir in the processing of major species (see further). Model 1 can thus be represented by:

$$A + C \quad \overset{k_1/C}{\to} \quad X + C \tag{6}$$

$$2X \quad \overset{k_2}{\to} \quad N + N \tag{7}$$

where C is a conservative species. All the particles being identical, the diffusion coefficients of all species are equal to a unique value D which can be adjusted by changing the radius R.

2.3 The non-equilibrium conditions

The concentration of major species must be kept constant *everywhere*. Thus when a major species particle, say A, disappears, the nearest neutral particle N is immediately converted into A. Nevertheless, because this operation occurs simultaneously with a specific event (here, step 6), it can add some artificial correlation between A and the created particles (here X), which in turn can introduce a significant bias in the values of the kinetic constants. This correlation results from the absence of a local randomization, a kind of "*chemical thermalization*". Such randomization was found to be properly restored [14,15], eliminating the bias, by slightly changing the preceding rule as follows:

Rule: When a major species particle A is transformed into a particle Z, the nearest particle N is transformed into A, except if this particle is already on a trajectory leading to a collision with the newly created Z particle, before one of them (N or Z) has met a third particle. If the transformation cannot be performed, this rule is applied to the next neighbour N, and so on, until the condition is met.

In these conditions the major species concentrations are kept constant and uniform, except for the spontaneous fluctuations, the relative value of which become small only in the unpractical limit of very large concentrations.

3 The fluctuation regime

The simulation of model 1 has been performed with the following parameters:

  - mass $m = 1$

  - radius $R = 0.075$

  - system size $L = 30$

  - mean energy per particle $< u^2 >$

  - number of particles A: 450, C: 450, total (with X and N): 1800

After the system has relaxed to the stationary state, the mean concentration $< X >$ and the value of $\alpha(l) = \sigma^2(X)/ < X >$ have been computed during 20000 time units with a sampling time $\Delta t = 1$ in square boxes of various size, ranging from $l = 3$ to $l = 30$, this last value corresponding to the entire system. Only these two limiting cases can be compared with the predictions of stochastic theory. For intermediate values the local concentration

Table 1

	β	$<X>$	$\Delta V \to 0$	$\Delta V \to \infty$
Stochastic		450	$\alpha(0) = 0.964$	$\alpha(\infty) = 0.75$
Mol. dyn.	0.032	448	$\alpha(3) = 0.963$	$\alpha(30) = 0.806$
Mol. dyn.	0.10	461.5	$\alpha(3) = 0.950$	$\alpha(30) = 0.773$

fluctuations of the major species A and C should be taken into account in the stochatic theory. Nevertheless these corrections vanish for $l = L$ since the total number of A and C particles is constant and for $l \to 0$ since the local poissonian distribution results from the sole local equilibrium principle. Note that in the latter case, α is not rigourously equal to 1 but to $\alpha(0) = 0.964$ when corrected from small compressibility effects related to the finite size of the particles. This correction vanish at $l = L$ because of the periodic boundary conditions [15]. Two runs with different activation energy level have been performed in order to change the ratio β of reactive collisions to non-reactive ones.

The results, collected in Table 1, are in excellent agreement with the stochastic theory predictions, specially if one consider that the small size of the system can be somewhat counterbalanced by the use of different activation energies. As expected, the short range limit is better approached when β is large and the long range limit when β is small. With $\beta = 0.1$ both limits are approached better than 3%. The molecular dynamics techniques have been applied to different models where the fluctuation regime can be analytically predicted from stochastic theory [13-15,18]. Both theories are in good quantitative agreement which confirms the validity of the basic stochastic assumptions. Molecular dynamics appear to be able to provide information in systems where the fluctuations play a major role. As an illustration, we shall now apply this technique to the transitions in a bistable chemical reaction.

4 M.D. simulation of metastability and nucleation in a bistable system

4.1 The nucleation problem in non-equilibrium chemical reactions

Let us consider the following model, first introduced by Vidal [19], and slightly modified for our purpouse:

$$A + C \overset{k_1}{\to} Y + C \tag{8}$$

$$B + Y \overset{k_3}{\to} X + Y \tag{9}$$

$$2X \overset{k_5}{\to} X + N \tag{10}$$

$$D + X \overset{k_3}{\to} 2X \tag{11}$$

$$X + Y \overset{k_7}{\to} 2N \tag{12}$$

where

- X and Y are the intermediate species

- the kinetic constant are fixed to $k_1 = k_6 = k_7 = 1$, $k_3 = 0.2$, $k_5 = 0.5$

- the major species concentrations are fixed to $B = 1.5$, $C = 2$, $D = 3$, A being used as an adjustable control parameter.

The system has two different branches of stable steady states E_1 and E_2 respectively corresponding to the low and high concentrations of A. However, when $A_1 < A < A_2$ with $A_1 = 1.51$ and $A_2 = 2.51$ both solutions exist and the final state depends on the initial values of the system. The two branches are continuously connected by a branch of unstable states in a characteristic S-shaped curve (these branches are represented in Figure 2). This *"bistability"* phenomenon is the most elementary temporal dissipative structure and is rather common with autocatalytic reactions [3]. There is a striking analogy between this curve and the liquid-gas equilibrium curve in a (p, V) diagram. Several authors have shown [20-25] that in spite of fundamental differences between equilibrium and non-equilibrium systems the transitions between E_1 and E_2 should actually present many common features with a liquid-gas transition [26]. If two regions, respectively in state E_1 and E_2 are set to contact, they cannot coexist except for a critical value $A_1 < A_0 < A_2$ and the *most stable* state—E_1 for $A > A_0$, E_2 for $A < A_0$—propagates and progressively overspread the whole system. Thus, for $A_0 < A < A_1$, state E_2 is not stable, only *metastable* and vice versa. Let us assume that the system is in the metastable state E_2. Small regions of state E_1 permanently form as the result of thermal fluctuations. If large enough, such a region constitutes a nucleus which expands and invades the entire medium, performing the transition. If the nucleus is too small the diffusion processes dominate the chemical processes and the fluctuation is damped. The life time τ_2 of E_2 corresponds to the typical time for a supercritical nucleus to appear spontaneously in the system. This life time increases rapidly with the critical size of the nucleus, i.e. when A approaches A_0—where the relative stability, controled by the sole chemical processes, changes—or when the diffusion rate increases. When τ_2 is large in regard of the typical duration of an experiment t_{exp}, E_2 appears to be *stable* and *unstable* if $\tau_2 \ll t_{exp}$. In the narrow range of A where $\tau_2 \approx t_{exp}$, the transition only begins after an indeterminate but finite amount of time ($\sim \tau_2$), revealing the metastability of state E_2. In practice, the genuine observed bistability range, located around A_0, is the more reduced as the diffusion rate is small. In common chemical bistable systems, such metastability phenomena are extremely difficult to study. Use of numerical simulations is specially advisable. The nucleation process has been evidenced by Monte-Carlo simulations of the master equation [27,28] but molecular dynamics in which fluctuations only results from the dynamics of the internal dynamics of the system are still more appropriate.

4.2   M.D. simulation of the bistable system

We have applied our M.D. technique to the model described aboved . All parameters are identical to those of section 3, except for the total number of particles equal to 4500. The concentration unit is conventionally fixed to 150 particles. Each numerical experiment is performed at a given value of A, the system being initialized on branch E_1 or E_2, generally by continuity with regard to a former result. A map of concentrations unit is builded at every time by dividing the system into 100 square boxes of size 3. To follow the dynamics of the transitions, we have given a binary representation of this map by assigning the identities E_1 or E_2 respectively to boxes where $X > 2$ and $X < 2$. The experiments are carried on until a transition has occured or during at least 900 time units. Two series of runs have been performed respectively with $R = 0.05$ and $R = 0.075$ in order to evidence the role of diffusion (the ratio of diffusion rate to reaction rate behave as R^{-2}). We report only the most significant features; more extensive results and discussions can be found in ref. 17. Figure 1 establishes the existence of a nucleation process : we follow de development of a nucleus in a transition from state E_1 to state E_2. The figure 2, where we report the observed mean values and the stability of the computed stationary states all along branches E_1 and E_2, illustrates the role of diffusion. The observed bistability range (finite diffusion rate) is smaller than the range predicted from macroscopic equations (infinite diffusion rate) which demonstrates the stabilizing role of diffusion. When the the ratio of diffusion rate to reaction rate is decreased approximatively by a factor 2 (changing $R = 0.05$ into $R = 0.075$), this bistability range is approximatively reduced by a factor 4. On figure 3, we evidence the metastability phenomenom in a transition from an unstable state E_1, located close to the bistability limit in order that τ_2 should be comparable to $t_{exp} \sim 900$. The system remains in state E_1 during a long finite time of 650 units before a local supercritical fluctuation occurs and drive the system into state E_2. The metastability of state E_1 is confirmed by the small

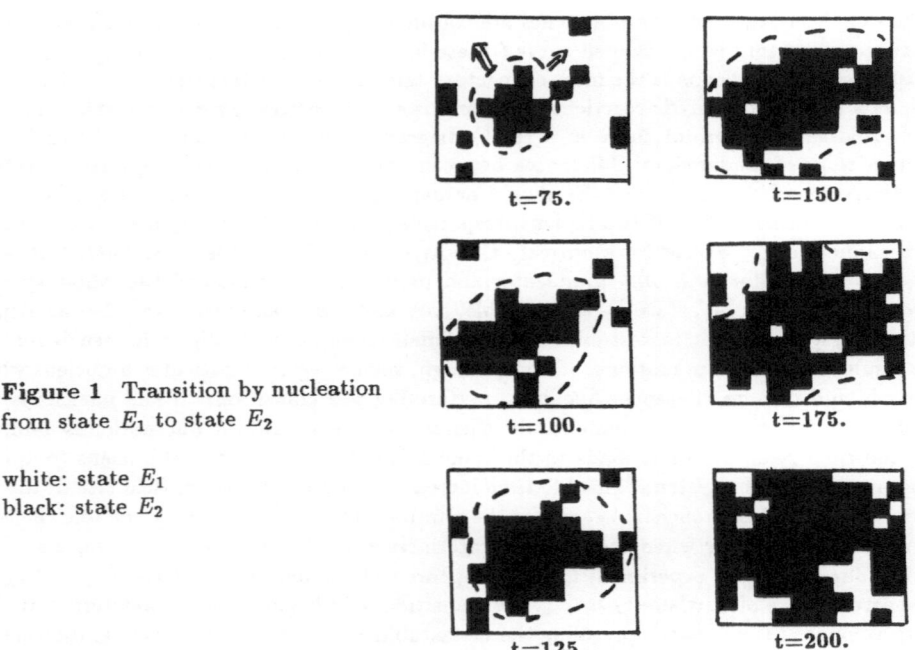

Figure 1 Transition by nucleation from state E_1 to state E_2

white: state E_1
black: state E_2

t=75.

t=150.

t=100.

t=175.

t=125.

t=200.

amplitude of fluctuations before the transition: most local fluctuations are small in regard of the critical size so that they are damped by diffusion and cannot persist during enough time to produce a large amplitude noise.

5   Conclusions and perspectives

In order to study far from equilibrium reactions, we have proposed some time ago a non-equilibrium molecular technique which solves the problem of keeping the major species constant and uniform [13-17]. This technique and similar methods [18] are receiving a new attention. The fluctuation regime observed in the simulations is in excellent agreement with the predictions of stochastic theory and prove their ability to distinguish between long range and short range fluctuations. Eventually we have produced evidence for metastability and nucleation processes that can be expected from theory but cannot presently be studied experimentally. In spite of the rapidly increasing power of large size computers, these simulations remain expensive and their implementation must be limited to well defined standard theoretical problems.In one-dimensional problems and diluted solutions, Monte-Carlo simulations based on the multivariate master equation [27,28] are probably more suitable. In order to increase efficiency, several authors have recently introduced the Bird's method, where the displacement is deterministic but collision processes are stochastic, instead of full molecular dynamics [30]. This approach provides solutions of the Boltzman equation, much the same way the M.C. method does for the master equation. In principle, this method only applies to dilute gases. Shall this drawback be or not a severe limitation is an open question but in most cases the result should be appliable to diluted solutions as well. The field of application of full M.D. seems to be the study of reactions limited by diffusion, strongly exothermic reactions [29], or correlations between diffusion coefficients and reaction related to the presence of cross diffusion coefficients in non diluted systems [30].

Whatever the selected technique is, full M.D. or Bird's method, a different approach should now be used in handling the major species. Keeping major species constant and uniform is a severe even for experimental work. Recent practical experimental studies of

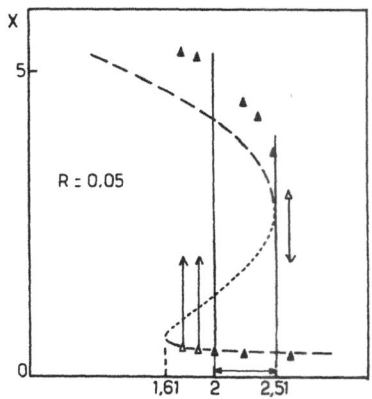
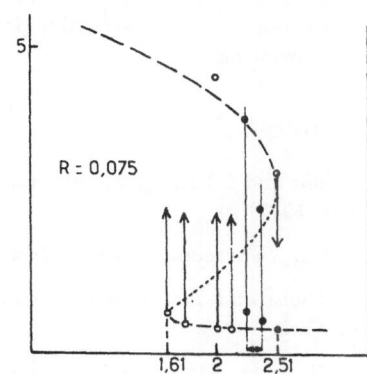

Figure 2 Stability of the stationary states for two different radius R of particles.

Deterministic homogeneous states:
Stable: — — — — —
Unstable: •••••••

M.D. simulation:
Stable: ▲ ●
Unstable: △ ○
Bistability range: ◄───►

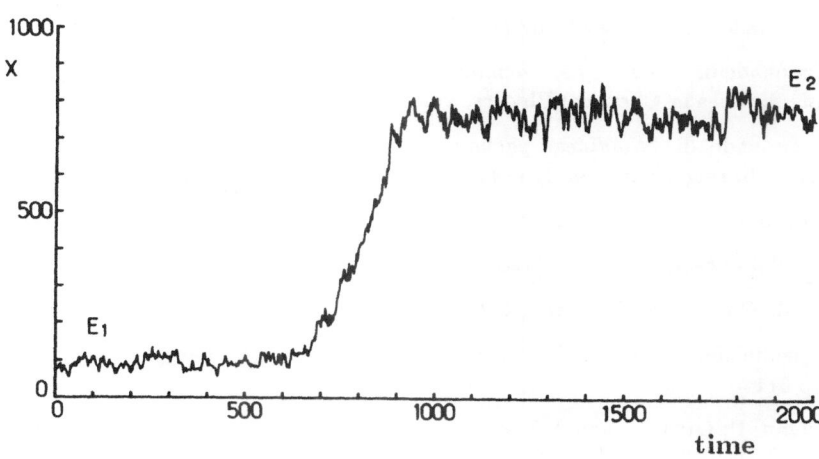

Figure 3 Time series exhibiting a metastability plateau before transition

sustained chemical dissipative structures have been performed in open reactor fed by diffusion from the boundaries (for a short review, see ref. 31). Thus, feeding the simulated medium by fixing different concentrations at opposite boundaries and changing the labels during collisions at walls would certainly be advisable, be more realistic and avoid residual spurious uncontroled effects. The significant increase in the system size, needed in this option, should be rapidly balanced by the increasing power of computers as shown in other fields of N.E.M.D. during this meeting.

References

[1] G. Nicolis and I. Prigogine, *Self-organization in Nonequilibrium Chemical Systems* (Wiley, New York, 1977)

[2] H. Haken, *Synergetics, an introduction* (Springer, 1977)

[3] *Oscillations and Traveling Waves in Chemical systems*, Ed. D.Field and M.Burger, (Wiley, New York, 1985)

[4] A. Babloyantz, *Molecules, Dynamics and Life*, Wiley (1986)

[5] C. Vidal and H. Lemarchand, *La réaction créatrice*, Hermann (1988)

[6] G. Nicolis, P. Allen and A. Van Nypelseer, *Progr. Theor. Phys.*, **52**,1481 (1974)

[7] P. Hanusse, *Phys. Lett.*, **59A**,421 (1977)

[8] P. Hanusse and A. Blanché in *Systems far from Equilibrium*, Ed. L. Garrido, p. 337, Lectures Notes in Physics, Springer (1980); P. Hanusse and A. Blanché, *J. Chem. Phys.*, **74**,6148 (1981)

[9] J. Portnow, *Phys. Lett.*, **51A**,370 (1975)

[10] J. Boissonade and W. Horsthemke, *Phys. Lett.*, **68A**,283 (1978)

[11] P. Ortoleva and S. Yip *J. Chem. Phys.***65**,2045 (1976)

[12] J. S. Turner, *J. Phys. Chem.*, **81**, 2379 (1977)

[13] J. Boissonade, *Phys. Lett.*, **74A**, 285 (1979)

[14] J. Boissonade, *Thesis*, Bordeaux (1980) (in French)

[15] J. Boissonade in *"Modeling of Chemical Reaction Systems"*, K.H. Ebert and P. Deuflhard Eds, Springer Series in Chemical Physics, Vol. 18, p. 138, Springer (1981)

[16] J. Boissonade in *"Nonlinear phenomena in Chemical Dynamics"*, C. Vidal and A. Pacault, Eds., Springer Series in Synergetics, Vol. 12, p. 134, Springer (1981)

[17] J. Boissonade, *Physica*, **113A**, 607 (1982)

[18] A. Amellal, *Thesis*, Bruxelles (1989)

[19] C. Vidal, *C.R. Acad. Sc. Paris*, **C274**, 1713 (1972)

[20] I. Prigogine and G. Nicolis , in *"From theoretical physics to Biology"*, Karger (1971); also in ref. 1, p.311

[21] A. Nitzan, P; Ortoleva and J. Ross, *Faraday Symp. Chem. Soc.*, **9**, 241 (1975)

[22] I. Prigogine, R. Lefever, J.S. Turner and J.W. Turner, *Phys. Lett.*, **51A**, 317 (1975)

[23] C.Y. Mou, *J. Chem. Phys.*, **68**, 1385 (1978)

[24] F. Schlögl and R.S. Berry, *Phys. Rev.*, **21A**, 2078 (1980)

[25] L. Schimansky-Geier and W. Ebeling, *Ann. Physik*, **40**, 10 (1983)

[26] H.B. Callen, *Thermodynamics*, Ch. 9, Wiley (1966)

[27] P. Hanusse, in *"Far from equilibrium"*, Ed. A. Pacault and C. Vidal, Springer series in Synergetics, vol 3, p. 70, Springer (1970)

[28] P. Hanusse and A. Blanché, *J. Chem. Phys.*, **74**, 11 (1981); A. Blanché, *Thesis*, Bordeaux (1981) (in French)

[29] F. Baras, this volume

[30] A. De Witt, *Mémoire de Licence*, Brussels (1989); M. Mareschal, G. Nicolis, *private communications*

[31] J. Boissonade, in *"Dynamics and Stochastic Processes: Theory and Applications"*, Ed. R. Lima, L. Streit and R. Vilela Mendes. Lecture Notes in Physics, Springer, in press.

MICROSCOPIC SIMULATIONS OF EXOTHERMIC CHEMICAL SYSTEMS

F. Baras and G. Nicolis

Faculté des Sciences
Université Libre de Bruxelles
Belgium

1. INTRODUCTION

Molecular dynamics techniques are becoming a valuable tool in bridging the gap between the phenomenological analysis of large scale macroscopic phenomena and their modelling at the microscopic level. As stressed throughout this volume, in recent years a great amount of effort has been devoted to the application of these techniques in the study of hydrodynamic phenomena in non-reactive fluids such as shock-waves, flow past an obstacle or Bénard convection.

So far the microscopic simulation of chemically reacting systems has been primarily devoted to verifying fluctuation theory. The first attempts were reported in 1975 by Portnow[1] and in 1976 by Ortoleva and Yip[2]. More complete work by Boissonade[3] followed in which measurements of the composition fluctuations in nonequilibrium isothermal chemical systems have been reported. This work has been completed by a study of metastability and nucleation in a nonequilibrium chemical system with multiple steady states[4]. Spatial correlations[5] and sustained oscillations[6] generated by chemical dynamics have also been recently investigated.

The first MD simulation of an exothermic chemical system was performed by Chou and Yip[7] for a hard disk system in contact with thermal reservoirs, mimicking the so called Semenov model of combustion[8]. Although the observed temperature profiles and the critical properties were in qualitative agreement with the macroscopic predictions, important quantitative discrepancies were noted by the authors. Our objective here is to show that the deviations from macroscopic law observed by Chou and Yip can be explained by the deformation of the local equilibrium distribution[9].

For the simulation, we use the DSMC algorithm developed by Bird[10] which remains the dominant predictive tool in rarefied gas dynamics[11] and is especially well adapted in dilute gases studies. Its main difference with the traditional MD experiments consists in the stochastic treatment of the collision processes which increases significantly the efficiency of the computation. The extension of this method to include chemical reactions has already been considered by Bird for isothermal reacting flows[12].

The paper is organized as follows. In section two we investigate using kinetic theory, how nonequilibrium effects can appear in chemically reacting gases. Both isothermal reactions and non-isothermal ones are considered. Section three is devoted to the presentation of the phenomenology of exothermic chemical systems. In section four, we briefly discuss how the simulation is carried out and summarize the results for the stationary temperature profile and the critical properties. The main conclusions are drawn in section five.

2. NONEQUILIBRIUM EFFECTS IN A CHEMICALLY REACTING GAS

The macroscopic formulation of chemical kinetics rests on the strict validity of the local equilibrium hypothesis. Since a chemically reacting system is not in a state of complete equilibrium, deviations from local equilibrium can however be expected, even for *isothermal reactions*.

Let us consider for instance a dilute gas undergoing the reaction

$$A + A \xrightarrow{k} B + C \tag{1}$$

where the rate constant k obeys the phenomenological Arrhenius law

$$k \approx \exp(-E/k_B T) \tag{2}$$

E and k_B being the activation energy and the Boltzmann constant respectively and T the temperature of the system. In the case of translational degrees of freedom, a reactive collision arising from encounters between two A molecules occurs if their relative kinetic energy exceeds some threshold related to the activation energy E. Such a reactive collision will give rise to a transfer of translational energy from the incoming A particles to the outgoing ones B and C. If the reaction proceeds slowly enough - i.e. the ratio $E/k_B T$ is large enough - the high velocity tail of the Maxwell-Boltzmann distribution is replenished as fast as it is depleted by reactive events since the rate of reactive collisions is much smaller than the elastic one. There exists however a range of activation energy (or temperature) for which the rate of reaction becomes comparable to the rate of thermal equilibration. Since the rate of reaction is governed by the ratio $E/k_B T$, how small can this ratio be before the equilibrium based rate formula breaks down ?

Kinetic theory gives some answers to these questions for molecules with translational degrees of freedom lacking internal degrees and spherically symmetrical so that the steric factor are absent. Pioneering calculations were made by Prigogine and Xhrouet[13] in 1949 following a procedure originally developed by Chapman and Enskog[14] for the computation of transport coefficients. They solved the Boltzmann equation for a gas undergoing the reaction (1) to obtain the first order correction to the Maxwell velocity distribution $f^{(0)}$

$$f = f^{(0)}\left(1 + \phi\right) \tag{3}$$

and consequently to the reaction rate formula

$$v = v^{(0)}\left(1 + \eta\right) \tag{4}$$

The explicit form of the correction η depends on the specific choice of the reaction cross section $\sigma(g)$ as a function of the relative velocity of the colliding particles, g. For the cases investigated in ref. 13, the correction η is negative which means that nonequilibrium effects produce a slowing down of the reaction speed as expected by the intuitive argument.

In 1959, Present[15] used the same procedure as Prigogine and Xhrouet for a reaction cross section $\sigma(g)$ given by

$$\sigma(g) = 0 \qquad \varepsilon < E$$
$$= 1 - E/\varepsilon \qquad \varepsilon \geq E \tag{5}$$

$\varepsilon = mg^2/4$ being the initial kinetic energy of the relative motion (m is the mass of the A particle). Using this form for the cross section, one recovers the Arrhenius law in the equilibrium limit

$$v^{(0)} = \frac{1}{2} v\, n\, e^{-E/k_BT} \tag{6}$$

where n is the number density and v the collision frequency given by

$$v = 4\, n\, d^2 \left(\frac{\pi k_B T}{m}\right)^{1/2} \tag{7}$$

with d the diameter of A particle. The factor 1/2 takes account to the fact that the reaction involves a pair of identical particles. The correction to the rate equation is given by

$$\eta = -\frac{1}{32}\, e^{-E/k_BT} \left[\left(\frac{E}{k_BT}\right)^4 - 2\left(\frac{E}{k_BT}\right)^3 + \frac{1}{2}\left(\frac{E}{k_BT}\right)^2 + \frac{1}{16}\right] \tag{8}$$

and is small for usual values of the ratio E/k_BT. It would therefore seem that for realistic systems the estimated nonequilibirum deviations are well within the uncertainties associated with collision diameters and steric factors[13,15,16].

The situation may be different in *non-isothermal reactions*. Consider for instance the following gas-phase reaction

$$F + S \xrightarrow{k} B + S + \textit{heat} \tag{9}$$

corresponding to the transformation of "fuel" particles F to "burned" ones B through reactive collisions with "solvent" particles S. The heat released by the reaction will further increase the translational energy of S particles which may well re-collide with F particles before being thermalized by elastic collisions. If the reaction proceeds fast enough, there exists a continuous flow of energy from F to S particles which may induce a non-negligible perturbation of the equilibrium distribution and consequently modify the macroscopic rate formula.

This problem was analyzed in detail by Prigogine and Mahieu[17] in 1950, using a Chapman-Enskog technique similar to the one used in the isothermal case (see ref. 13). For a specific choice of reaction cross section

$$\sigma(g) = \left(1 - e^{-\varepsilon/E}\right) \tag{10}$$

they obtained the first order correction to the Maxwell-Boltzmann distribution (cfr. eq.(3)) and the corresponding macroscopic rate law

$$v = v^{(0)}\left(1 + 1.2\, X_F\, X_S\, \frac{q}{E}\right) \tag{11}$$

where q is the heat of reaction. The deviation from the usual law is positive as expected and scaled by the ratio of the heat of reaction to the activation energy. X_F and X_S are the molar fractions of fuel and solvent particles respectively.

The continuous flow of energy from F to S particles occurring when the exothermic reaction proceeds acts as an "energetic chain" which increases the rate of reaction. This effect may disappear by introducing an inert gas in the medium, since the molar fractions appearing in the corrective term become small in this case.

So far there exists no experimental evidence for these nonequilibrium effects. Indeed combustion processes are often coupled with complicated hydrodynamical effects. Moreover the exact knowledge of the underlying chemical mechanism is difficult to establish. A deviation from macroscopic theory can thus be tentatively attributed to these factors. One of the objectives of the present chapter is to show that molecular dynamics computer experiments allow one to overcome this difficulty.

3. EXOTHERMIC CHEMICAL SYSTEMS: PHENOMENOLOGICAL BACKGROUND

When an exothermic chemical reaction proceeds in a closed vessel in contact with a thermal reservoir, a *thermal instability* may occur. This ignition phenomenon results from the sudden impossibility of balance between the chemical heat production and the heat loss to the walls.

The basic studies[8] to determine the conditions for the onset of thermal explosion are due to Semenov for the homogeneous case and to Frank-Kamenetskii for inhomogeneous systems. The theory is restricted to time-independent situations and rests on the assumptions that the consumption of reactant is ignored and the fluid is motionless. Heat is thus transported only by conduction.

In the framework of this theory, the critical behavior of a self-heating system is discussed in terms of the stationary solutions of the heat conduction equation with an Arrhenius reaction term. Let us consider the simplest geometry of an infinite slab of length $L_x = 2a$

$$\frac{d}{dx} \kappa \frac{d}{dx} T + q v = 0 \tag{12}$$

where κ is the thermal conductivity, q the heat of reaction and v the reaction rate (cfr eq. (6)). The boundary have the form

$$T = T_r \tag{12.a}$$

T_r being the temperature of the reservoir. One can then show that the properties of the stationary solutions of eq. (12) are conditioned by two parameters: δ, measuring the heat generation and ε, proportional to the inverse of the activation energy

$$\delta = \frac{q \, v(T_r) \, a^2}{\varepsilon \, \kappa \, T_r} \tag{13.a}$$

$$\varepsilon = k_B T_r / E \tag{13.b}$$

Most of the works reported in the literature concern the case of constant thermal conductivity and no temperature dependence of the collision frequency in front of the Arrhenius factor. Even in this case, no exact analytic solution of eq. (12) is known although some approximations are obtained[18,19] in the limit of large activation energy ($\varepsilon \ll 1$).

The maximum stationary temperature in the system T_s may be computed numerically for $\varepsilon \neq 0$. The stationary state diagram corresponding to eq. (12) is shown by the solid curve in figure 1. Provided ε is smaller than a critical threshold and for a certain range of values δ, the solution is triple-valued between the ignition and extinction points δ_c and δ^c, the lower and upper branches being stable and the middle one unstable. For a system initially at the reservoir temperature T_r, the temperature will reach a cool stationary state close to T_r for $\delta < \delta_c$ whereas for δ slightly higher than δ_c, the temperature jumps to a much higher value. This upper branch does not describe adequately the explosive state since for this stage the assumption of no consumption of the reactant breaks down.

Although it is usual to neglect the temperature-dependence of thermal conductivity κ and collision frequency ν in thermal explosion theory, there are certain cases where this is not appropriate. This is especially true in gaseous media, where κ and ν have the form

$$\kappa = \kappa_0 \sqrt{T} \tag{14.a}$$

$$\nu = \nu_0 \sqrt{T} \tag{14.b}$$

with κ_0 and ν_0 constant. These expressions are exact for hard spheres gases. The effects of this temperature-dependence on criticality have been studied numerically in different papers[20,21]. Figure 1 depicts the diagram of stationary states (dashed line) of the equation of conduction (12) when the corrections given by expressions (14) are incorporated.

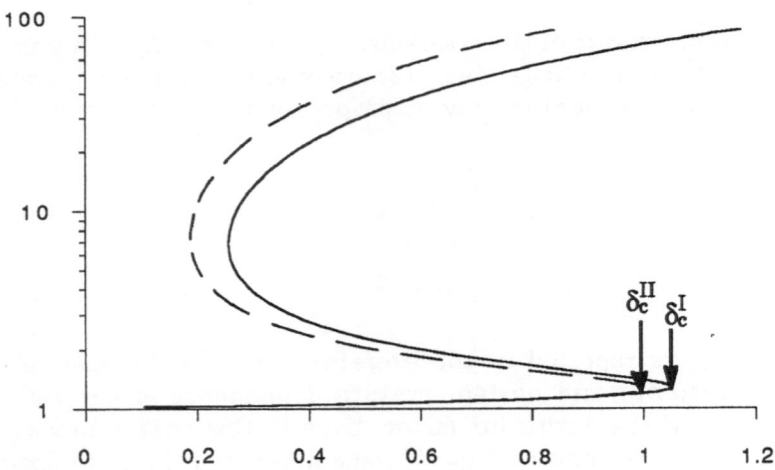

Figure 1. Reduced maximum temperature $\theta_s = k_B T_s/E$ against the Frank-Kamenetskii parameter δ for $\varepsilon = 0.143$. The solid line represents the steady state of eq. (12) in the classical limit while the dashed line denotes the steady solution in gaseous media (see eqs. (14)). δ_c is the critical value for ignition.

4. MICROSCOPIC SIMULATION

4.A DESCRIPTION

For the microscopic simulation, we have used the direct simulation Monte Carlo (DSMC) technique developed by Bird[10]. This method is known to be well adapted for the study of dilute Boltzmann gases. A detailed presentation of the current status and perspectives of the DSMC technique is included elsewhere in this volume.

We consider a 3-dimensional system made of an assembly of $N = 2,000$ (or 4,000) hard spheres of diameter d confined in a rectangular box of 10λ (20λ) of length in the X direction and of a cross section $1 \times 1\lambda^2$ in the Y and Z directions, leading to a number density $n = 200 / \lambda^3$, here λ denotes the mean free path. The walls located at $X = 0$ and $X = L$ act as thermal reservoirs: each time a particle hits one of the walls, it is reinjected with a velocity distribution sampled from a Maxwellian at the reservoir temperature. This corresponds to a purely diffusive reflection on the vessel surface (clean surface, no specular reflection). The energy is scaled by the reservoir thermal energy so that the reservoir temperature T_r is set equal to 1.

The system is divided into 20 (40) "statistical" cells in the X direction over which a space average is performed to give instantaneous values of basic mechanical quantities. A time average leads to the local macroscopic quantities such as temperature in a well established stationary state.

We have extended the DSMC technique to the case of an exothermic reactive hard spheres gas. We implement the chemical reaction as follows:

$$
\begin{aligned}
A + A &\to A + A & &\text{if } K_r < E \\
A + A &\to A + A + heat & &\text{if } K_r \geq E
\end{aligned}
\tag{15}
$$

The collision is *reactive* if the kinetic energy K_r of the relative motion projected along the line of centers of incoming particles is larger than the activation energy E and the collision is *elastic* otherwise. The heat released by the reaction increases the translational velocities of outgoing particles. Momentum and energy are conserved during the collision, even for a reactive one. This microscopic rule corresponds to the reaction cross section $\sigma(g)$ given by eq. (5) which leads to the Arrhenius law in a dilute system at equilibrium.

The microscopic model thus mimics the behavior of a self-heating gas immersed in a thermal bath according the assumptions included in the phenomenological model of combustion presented in section 3. The geometry used corresponds indeed to an infinite slab and reaction (15) satisfies obviously the assumption of no consumption of reacting particles.

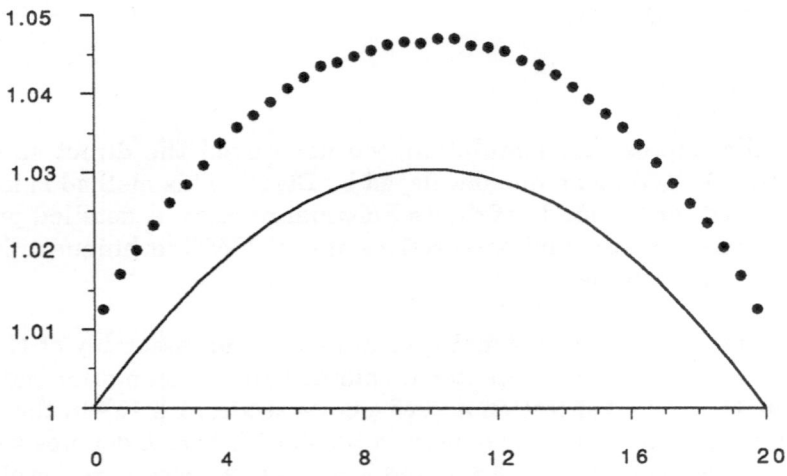

Figure 2. Stationary temperature distribution for the model (15). The dots denote the DSMC data while the full line represents the numerical stationary solution of the macroscopic equation (12) with (12.a). The thermal conductivity coefficient κ is given by eq. (16). $q/k_BT_r = 2$, $L_x = 20\lambda$ and $v^{(0)}$ is given by eq. (6).

In the microscopic simulation, we have measured the rate of reaction of an isothermal system (q = 0), for different values of activation energy. In this limit we recover as expected the usual macroscopic rate law given by eq. (6), since for reaction (15) there is no "coloring" process which might induce nonequilibrium effects.

4.B RESULTS

We start the presentation of the results by the study of the properties of the *cool stationary state* occurring in the subcritical regime. The cool state can be reached by setting the initial value of the system's temperature close to the reservoir one. We have chosen a heat of reaction $q = 2k_BT_r$ for the system of 20λ long. The activation energy E will always be set to $7k_BT_r$. Such parameters allow the existence of a cool state (i.e. $\delta < \delta_c$).

The program was run in a Cyber 855 and a time average over the last 10^5 collisions per particle (cpp) was performed, after the stationary state has been reached (about 10^4 cpp). The statistical error estimated from successive runs of 10^4 cpp does not exceed 0.2%. The measured temperature profile is presented in figure 2 together with the macroscopic profile. The latter is obtained from the numerical solution of eq. (12) in which the rate of reaction is the equilibrium one (see eq. (6)) and the thermal conductivity coefficient corresponding to a hard spheres gas is expressed in terms of the microscopic quantities[22],

$$\kappa = \frac{75}{64d^2} \left(\frac{k_B^3 T}{\pi\, m}\right)^{1/2} \equiv \kappa_0\, T^{1/2} \tag{16}$$

We observe a significant difference between the result of the simulation and the macroscopic prediction and an important slip near the walls. These observations are in qualitative agreement with the results reported by Chou and Yip (see ref. 7).

One way to deal with the slip is to impose as boundary conditions for the macroscopic equation (12) the values of the temperature measured in the extreme cells. The corresponding profile is depicted in figure 3. Despite the improvement, there still subsists a discrepancy with the simulation data. This is particularly clear in figure 4, where the results obtained for a larger value of the heat of reaction $q = 8k_B T_r$ are plotted.

A first explanation that comes to mind is to attribute the above discrepancies to interferences with hydrodynamical effects. In this respect we first notice that although microscopically we are dealing with a 3-D system, the measured data corresponding to macroscopic quantities concern effectively a one dimensional system since $L_x \gg L_y, L_z$. A transport by advection will therefore affect necessarily the pressure profile in the X direction. The latter is found to be constant in all our computer experiments, provided the stationary state is well established[*]. Beside, one of the principal factors at the origin of thermal convection, namely the gravitational field, is absent in our simulations. To be complete, we have also measured the static velocity-temperature correlation function in the X direction and found it to be zero. We therefore conclude that the hydrodynamical effects, if any, could not exceed the statistical errors (0.2%), which is far below the observed discrepancy, especially for $q = 8k_B T_r$ (see figure 4).

We now inquire whether, as suggested in section 2, the observed discrepancy results from a non-equilibrium correction. We have computed the nonequilibrium correction using the procedure developed by Prigogine and Mahieu (see ref. 17) for the reaction cross section given by eq. (5) corresponding to the microscopic prescription implemented in our simulation. The final result for the corrected reaction rate becomes

$$v = v^{(0)} \left\{ 1 + \frac{1}{8} X_A^2\, e^{-E/k_B T} \left(\frac{E}{k_B T}\right)^2 \left[\frac{q}{k_B T} \left(\frac{E}{k_B T} - \frac{3}{2}\right) + \frac{1}{2} \left(\frac{q}{k_B T}\right)^2 \right] \right\} \tag{17}$$

where $v^{(0)}$ represents the usual macroscopic rate law (cfr. eq. (6)) and X_A denotes the molar fraction of the reacting particles A. Since we are dealing with a one component system $X_A = 1$.

---

[*] To be consistent we have taken into account of the measured constant pressure in the numerical integration of equation (12), the number density being inversely proportional to the local temperature.

The correction to $v^{(0)}$ is about 1% for $q = 2k_BT_r$ and increases to 6% for $q = 8k_BT_r$. The macroscopic temperature profiles computed from the corrected rate law are depicted in figures 3 & 4. We observe a very good agreement with the simulation results for these two systems characterized by different heats of reaction.

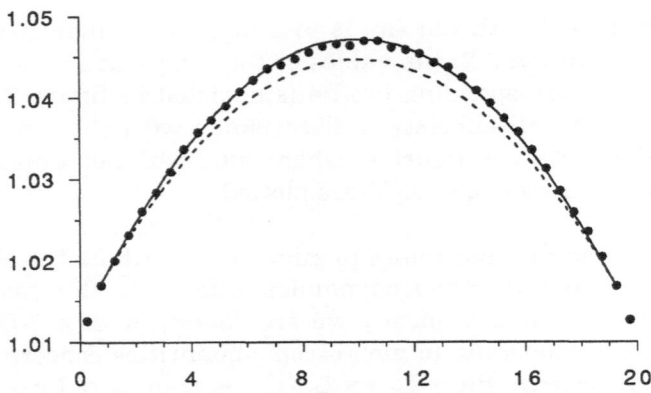

Figure 3. Stationary temperature distribution for the model (15). The dots denote the DSMC data while the full line and the dashed lines represent the numerical solutions of the macroscopic equation with and without correction, eq. (17), respectively. The parameters are the same as in figure 1.

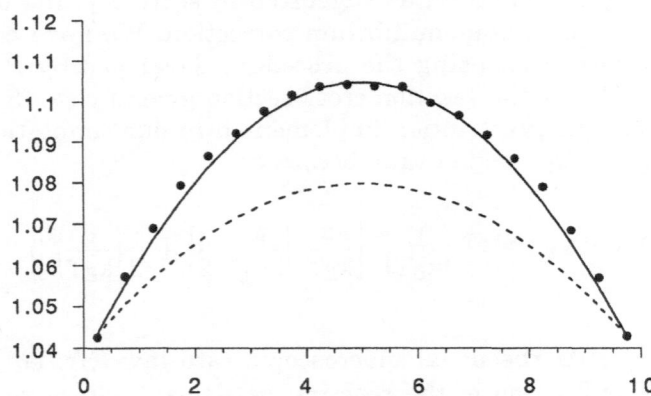

Figure 4. Stationary temperature distribution for the model (15), with $q/k_BT_r = 8$ and $L_x = 10\lambda$. See captions of figures 2. & 3. for details.

Table I. Critical value of the Frank-Kamenetskii parameter (eq. (13.a))
δ_c^{exp}: in the DSMC simulation; δ_c^I: in the classical limit of a self-heating slab (no temperature dependence of κ and ν); δ_c^{II}: for a gas media (κ and ν given by eqs. (14)); δ_c^{NE}: for a reaction rate corrected by nonequilibrium effects (eq. (15)); δ_c with the boundary conditions for eq. (12) given by eqs. (18) & (20) and a nonequilibrium reaction rate (eq. (15)).

E	$6k_BT_r$	$7k_BT_r$	$8k_BT_r$
δ_c^{exp}	.50	.38	.25
δ_c^I	1.09	1.05	1.02
δ_c^{II}	1.04	1.01	1.00
δ_c^{NE}	.62	.50	.39
δ_c^*	.44	.37	.30

Another interesting question refers to the transition between the "cool" and "hot" behavior. The *critical properties* of the simulated system can be compared with the phenomenological predictions presented in section 3.

A series of runs has been performed to locate the critical value of the Frank-Kamenetskii parameter δ_c (eq. (13.a)) at a fixed value of the activation energy E. For each series, all parameters are fixed except the heat of reaction q. The system is initially set at the wall temperature T_r by the usual procedure, and each particle is assigned a velocity sampled from a Maxwellian at T_r. When the reaction proceeds, the maximum local temperature is monitored. From the temporal evolution of the temperature one is able to determine whether the system is subcritical or supercritical for a given value of δ. One can iterate between sub- and supercritical conditions until a fairly precise value δ_c for the transition is obtained.

Table I shows the measured critical values δ_c for different choices of activation energy and the predictions based on conductive theory. We see a large deviation between critical properties obtained from the simulation and the values expected from macroscopic theory. The discrepancy is reduced somewhat if we take into account the nonequilibrium correction of the reaction rate.

One can suggest a plausible explanation for this lack of quantitative agreement. We first observe that the boundary condition $T = T_r$ used in eq. (12) for the estimation of the critical parameter does not allow any temperature slip near the wall as observed in the simulation. The discontinuity of temperature at a wall bounding an unequally heated gas can be expressed by the following type of equation

$$g \left| \frac{dT}{dx} \right|_{wall} = T - T_r \qquad (18)$$

where g is called the temperature jump distance. Elementary kinetic theory[23] gives us an approximate theoretical expression for the jump distance

$$g = (2\pi k_B T)^{1/2} \frac{\kappa}{(\gamma+1) c_v p} \qquad (19)$$

in the case of a purely diffusive wall, κ being the thermal conductivity, γ the ratio of the specific heats (c_p/c_v), T and p the temperature and the pressure near the walls, respectively. Introducing the expressions for γ, c_v and κ characteristic of a hard spheres gas and the undisturbed number density n_o we get a further approximation independent of the state of the system

$$g \approx \frac{1}{2k_B} \sqrt{\frac{\pi}{2k_B}} \frac{\kappa_o}{n_o} \qquad (20)$$

We have checked these approximations by simulating a system submitted to a thermal gradient without chemical reaction. As shown in figure 5 we observe a fairly good agreement between simulation data and the solution of the heat conduction equation with relation (18) as boundary conditions and g given by eq. (19). To use expression (19) for g we need to know the measured values of the pressure p and the temperature T in the system near the walls while expressions (20) is independent of the state of the system.

For our purposes we have to restrict ourselves to the approximate expression (20) since we want to evaluate the critical properties when the temperature near the wall can have a step change. Solving the heat conduction equation (12) with boundary conditions (18) and (20) gives us an estimation of the critical parameter δ_c^* presented in table I for different choices of activation energies. We note that the agreement with the simulation results is con-

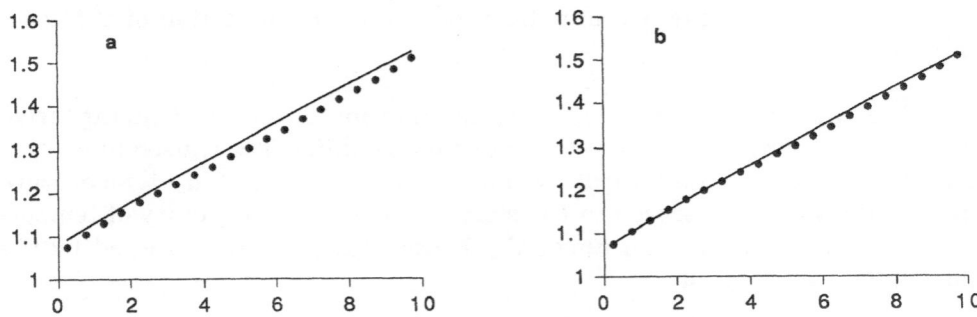

Figure 5. Temperature profile in a system submitted to a thermal gradient $\Delta T=0.6T_r$. The dots denote the DSMC data for a 10λ long system. The full line represents the solution of the heat conduction equation with boundary conditions (18) and g given by relation (19) (curve (a)) or eq. (20) (curve (b)). The thermal conductivity coefficient is given by eq. (16).

siderably improved. Notice that the agreement with microscopic simulation of steady state behavior (figures 2 to 4) remains good when the boundary condition given by eq. (18) is used instead of eq. (12.a) in solving eq. (12).

5. CONCLUSIONS

The extension of the DSMC technique to chemically reactive gases allows us to analyze, by means of microscopic simulations, combustion processes including thermal explosions. Within this framework, we have performed a MD computer experiment of an exothermic chemical system in a reactor closed with respect to mass flow, in thermal contact with a reservoir maintained at a fixed temperature. As predicted by the phenomenological description, we observe two distinct behaviors depending on the value of the heat of reaction: a subcritical regime corresponding to the evolution toward a state close to the reservoir temperature and a supercritical one associated with explosion.

In the subcritical regime, a discrepancy is observed between the results of the simulation and predictions from macroscopic chemical kinetics. The origin of this discrepancy is found to arise form the deformation of the Maxwell distribution by reactive collisions which in turn induces a correction to the usual rate law. In this respect, MD simulations can thus be viewed as an experimental tool to reveal such nonequilibrium effects.

The critical threshold for the onset of explosive behavior revealed in the simulation differs from the phenomenological predictions. The agreement is improved by taking into account the nonequilibrim corrections and by using appropriate boundary conditions. Nevertheless some questions remain open. Corrections to the macroscopic description arising from fluctuations might induce a shift of the critical threshold. Moreover the pressure in the stationary state is slightly different from the initial one, the final state depending on the value of the heat of reaction. The time-independent theory of thermal explosions does not take this fact into account. As noted by Zeldovich[18], when ignition occurs a combustion source or locus appears at the hottest point of the system and a flame propagates from this locus. Microscopic simulations are thus useful to obtain a better understanding of the development of an explosion. The visualization and quantitative measurement of space-dependent properties such as correlation functions would be highly desirable and would allow one to follow the gradual formation of combustion nuclei inducing the explosion phenomenon.

ACKNOWLEDGEMENTS

We wish to thank Dr. M. Malek Mansour for his active participation in this research and for stimulating discussions. The financial support of the Belgian Government through the Actions de Recherches Concertées is gratefully acknowledged.

REFERENCES

1. J. Portnow, Phys. Lett. **51A**, 370 (1975).
2. P. Ortoleva and S. Yip, J. Chem. Phys. **65**, 2045 (1976).
3. J. Boissonade, Phys. Lett. **74A**, 285 (1979).
4. J. Boissonade, Physica **113A**, 607 (1982); ibid, in *Nonlinear Phenomena in Chemical Dynamics*, C. Vidal and A. Pacault Eds., Springer-Verlag, Berlin (1981).
5. G. Nicolis, A. Amellal, G. Dupont and M. Mareschal, J. Mol. Liq. **41**, 5 (1989).
6. F. Baras, J. Pearson and M. Malek Mansour (preprint); see also M. Mareschal, chapter in this volume.
7. D. P. Chou and S. Yip, Combust. Flame **47**, 215 (1982); ibid, Combust. Flame **58**, 239 (1984); ibid in *Chemical Instabilities*, G. Nicolis and F. Baras Eds., Reidel, Dordrecht (1984).
8. D. A. Frank Kamenetskii, *Diffusion and Heat Transfer in Chemical Kinetics*, Plenum Press (1969).
9. F. Baras and M. Malek Mansour, Phys. Rev. Lett., **63**, 2429 (1989).
10. G.A. Bird, *Molecular Gas Dynamics*, Oxford University Press, London (1976).
11. E.P. Muntz, Ann. Rev. Fluid Mech. **21**, 387 (1989).
12. G.A. Bird, in *Rarefied Gas Dynamics*, Eleventh Symposium, R. Campargue Ed., vol. 1, 365, CEA, Paris (1979).
13. I. Prigogine and E. Xhrouet, Physica **15**, 913 (1949); see also C. F. Curtiss, Ph. D. dissertation, University of Wisconsin, unpublished (1948).
14. S. Chapman and T.G. Cowling, *Mathematical Theory of Nonuniform Gases*, Cambridge University Press, London (1939).
15. R. D. Present, J. Chem. Phys. **31**, 747 (1959).
16. J. Ross and P. Mazur, J. Chem. Phys. **35**, 19 (1961).
17. I. Prigogine and M. Mahieu, Physica **16**, 51 (1950).
18. Y.B. Zeldovich, G.I. Barenblatt, V.B. Librovich and G.M. Makhviladze, *The Mathematical Theory of Combustion and Explosions*, Plenum Publ. Corp., New York (1985).
19. N.W. Bazley and G.C. Wake, Combust. Flame **33**, 161 (1978); T. Takeno, Combust. Flame **29**, 209 (1977); W. Gill, A.B. Donaldson and A.R. Shouman, Combust. Flame **36**, 217 (1979).
20. T. Boddington, C.-G. Feng and P. Gray, J. Chem. Soc. Faraday **79**, 1499 (1983).
21. G.C. Wake, Combust. Flame **39**, 215 (1980).
22. P. Résibois and M. De Leener, *Classical Kinetic Theory of Fluids*, Wiley, New York (1977).
23. E.H. Kennard, *Kinetic Theory of Gases*, McGraw-Hill, New York (1938).

SCIENTIFIC COMMITTEE

The scientific committe was composed by: Giovanni Ciccotti (Univeristy of Rome), Brad L. Holian (Los Alamos National Laboratory), Michel Mareschal (University of Brussels) and Grégoire Nicolis (University of Brussels).

PARTICIPANTS

M. P. Allen, H.H. Wills Physics Laboratory, Royal Fort, Tyndall Avenue, Bristol BS8 1TL, United Kingdom
email:mpa@pva.bristol.ac.uk

A. Amellal, Université de Kenitra, Morocco

J. R. Banavar, Department of Physics and Materials Research Laboratory, Pennsylvania State University, University Park, PA 16802 USA
email: jayanth@psuphys1.psu.edu.arpanet

F. Baras, Chimie-Physique, CP 231, ULB, Bld du Triomphe, B1050, Brussels, Belgium
email: flo@bnandp11.bitnet

M. Baus, Chimie-Physique, CP 231, ULB, Bld du Triomphe, B1050, Brussels, Belgium

G. A. Bird, 5, Fiddens Wharf Road, Killara, N.S.W. 2071, Australia

J. Boissonade, Centre de Recherche Paul Pascal, Domaine Universitaire, 33405 Talence Cedex, France
email:ppt005@frors

J. P. Boon, Chimie-Physique, CP 231, ULB, Bld du Triomphe, B1050, Brussels, Belgium
email: r09606@bbrbfu01.bitnet

P. Borckmans, Chimie-Physique, CP 231, ULB, Bld du Triomphe, B1050, Brussels, Belgium

B. Cantaloube, ONERA, 29, Avenue de la Division Leclerc, 92320,Chatillon, France

P. Clavin, URA 1117 CNRS, Université de Provence, Centre St Jérôme s252, Marseille cedex 13, France
email: searby@fromp11.bitnet

G. Ciccotti, Istituto di Fisisca G. Marconi, Universita La Sapienza, Piazzale A. Moro, 2, 00185 Roma, Italy
email: tk5rmx41@icineca2.bitnet

G. Dewel, Chimie-Physique, CP 231, ULB, Bld du Triomphe, B1050, Brussels, Belgium

A. De Wit , Chimie-Physique, CP 231, ULB, Bld du Triomphe, B1050, Brussels, Belgium

J. Dufty, Department of Physics, University of Florida, Gainesville, FL 32611, USA
email: dufty@nervn.bitnet

M. W. Evans, Department of Chemistry, Royal Holloway College, University of London, Egham, Surrey, TW20 0EX, United Kingdom

M. Ferrario, Istituto di Fisisca G. Marconi, Universita La Sapienza, Piazzale A. Moro, 2, 00185 Roma, Italy
email: tk5a@icineca.bitnet

D. Frenkel , FOM- Institute for Atomic and Molecular Physics, P.O. Box 41883, 1009 DB Amsterdam, Nederland
email: frenkel@hasfom51.bitnet or
frenkel@amolf.nl

A. Garcia, Department of Physics, San Jose State University, San Jose, CA 95192-0106, USA
email:tfnaai1@calstate.bitnet

J. Halliday, Faculty of Technology, Pond Street, Sheffield S1 1WB, United Kingdom

L. Hannon, MS#36, IBM Corporation, 1530, Pagemill Road, Palo Alto, CA 94304 Mail Station 36, USA

D. M. Heyes, Department of Chemistry, Royal Holloway College, University of London, Egham, Surrey, TW20 0EX, United Kingdom
email: uhca015@vaxb.rhbnc.ac.uk

F. J. Higuera, E.T.S. Ingenieros Aeronauticos, Pza Cardenal Cisneros 3, E28040, Madrid, Spain
email:c0130003@emdupm11.bitnet

B. L. Holian, Theoretical Division, Los Alamos National Lab., Los Alamos, New Mexico, 87545, USA
email: blh@t12.lanl.gov.arpanet

K. S. Holian, Group C-3, MS-B265, Los Alamos, New Mexico, 87545, USA

C. Hoover , L-794, Lawrence Livermore National Lab., Livermore, CA 94550,
          USA

W. G. Hoover, L-794, Lawrence Livermore National Lab., Livermore, CA
          94550, USA

E. Kestemont, Pool de Physique, CP223, ULB, Bld du Triomphe, B1050,
          Brussels, Belgium
                              email: eddy@bnandp11.bitnet

J. Koplik, Levich Institute, Steinman 202, City College of New York, New
          York, 10031, USA
                              email: koplik@ccnysci.bitnet

A. Ladd, L-207, Lawrence Livermore National Lab., Livermore, CA 94550,
          USA
                         email: ladd@s22.es.llnl.gov.arpanet

W. Loose, Institute fur Theoretische Physik, Technische Universitat Berlin
          (PN 7-1), Hardenbergstr. 36, D-1000, Berlin 12, F.R.G.
                         email: 3101@db0tuz01.bitnet

J. Lutsko, Chimie-Physique, CP 231, ULB, Bld du Triomphe, B1050,
          Brussels, Belgium

M. Malek Mansour, Chimie-Physique, CP 231, ULB, Bld du Triomphe,
          B1050, Brussels, Belgium
                         email:malek@bnandp11.bitnet

M. Mareschal, Chimie-Physique, CP 231, ULB, Bld du Triomphe, B1050,
          Brussels, Belgium
                         email: ulbg036@bbrnsf11.bitnet

M. Meyer, Laboratoire de Physique des Matériaux, CNRS, 1,Place A. Briand,
          Meudon Cedex, 92195, France
                         email: ulrx001@frors31.bitnet

G. Nicolis, Chimie-Physique, CP 231, ULB, Bld du Triomphe, B1050,
          Brussels, Belgium
                         email: r09613@bbrbfu01.bitnet

A. Noullez, Chimie-Physique, CP 231, ULB, Bld du Triomphe, B1050,
          Brussels, Belgium
                         email: r09606@bbrbfu01.bitnet

P. Peeters, Chimie-Physique, CP 231, ULB, Bld du Triomphe, B1050,
          Brussels, Belgium

C. Pierleoni, Pool de Physique, CP 223, ULB, Bld du Triomphe, B1050,
          Brussels, Belgium

V. Pontikis, SRMP, CEN-Saclay, 91191 GIF SUR YVETTE, France
                         email: pontikis@frsrmp11.bitnet

H. A. Posch, Institute for Experimental Physics, University of Vienna,
Boltzmanngasse 5, 1090 Vienna, Austria
email: a8221dab@awiuni11.bitnet

I. Prigogine, Chimie-Physique, CP 231, ULB, Bld du Triomphe, B1050,
Brussels, Belgium

D. C. Rapaport , Physics Department,, Bar-Ilan University, Ramat-Gan,
Israel
email: f67241@barilvm.bitnet

S. Ruffo, Dipartimento di Fisica, Universita di Firenze, Largo E. Fermi, 2,
50125 Firenze, Italy
email: ruffo@vaxfi.infn.it

J.P. Ryckaert, Pool de Physique, CP 223, ULB, Bld du Triomphe, B1050,
Brussels, Belgium
email: ulbg029@bbrnsf11.bitnet

A. Tenenbaum, Dipartimento di Fisica, Universita La Sapienza, Piazzale A.
Moro,2, 00185 ROMA, Italy
email: tenenbaum@vaxrom.infn.it

J. Wallenborn, Chimie-Physique, CP 231, ULB, Bld du Triomphe, B1050,
Brussels, Belgium

H. Xu, Pool de Physique, CP 223, ULB, Bld du Triomphe, B1050, Brussels,
Belgium

G. Zanetti, Department of Applied and Computational Mathematics,
Princeton University, Princeton, New Jersey , USA
email: zag@acm.princeton.edu.arpanet

INDEX